Snowflake Data Warehouse Engineering

Architecture, Modeling, ELT Pipelines, and Operations

Martin Hander

Apress®

Snowflake Data Warehouse Engineering: Architecture, Modeling, ELT Pipelines, and Operations

Martin Hander
Gablingen, Germany

ISBN-13 (pbk): 979-8-8688-2627-6 ISBN-13 (electronic): 979-8-8688-2628-3
https://doi.org/10.1007/979-8-8688-2628-3

Managing Director, Apress Media LLC: Welmoed Spahr
Acquisitions Editor: Shaul Elson
Editorial Project Manager: Gryffin Winkler

Cover designed by eStudioCalamar

Cover Photo by Ray Suarez

Distributed to the book trade worldwide by Springer Science+Business Media New York, 1 New York Plaza, New York, NY 10004. Phone 1-800-SPRINGER, fax (201) 348-4505, e-mail orders-ny@springer-sbm.com, or visit www.springeronline.com. Apress Media, LLC is a Delaware LLC and the sole member (owner) is Springer Science + Business Media Finance Inc (SSBM Finance Inc). SSBM Finance Inc is a **Delaware** corporation.

For information on translations, please e-mail booktranslations@springernature.com; for reprint, paperback, or audio rights, please e-mail bookpermissions@springernature.com.

Apress titles may be purchased in bulk for academic, corporate, or promotional use. eBook versions and licenses are also available for most titles. For more information, reference our Print and eBook Bulk Sales web page at http://www.apress.com/bulk-sales.

Any source code or other supplementary material referenced by the author in this book is available to readers on GitHub. For more detailed information, please visit https://www.apress.com/gp/services/source-code.

If disposing of this product, please recycle the paper

To my lovely wife, whose love, patience, and steady belief in me made every page possible.

Table of Contents

About the Author

Martin Hander, Ph.D. (LIGS University, USA), is a technology professional with a strong focus on data warehousing and cloud analytics. Over his career, he has worked extensively with relational databases and modern cloud platforms, designing and operating data solutions that support reporting and advanced analytics. In combining his academic background with years of practical project work, he bridges theory and implementation. This blend of experience makes him a trusted guide for readers who want to understand Snowflake and modern data warehouse engineering in a clear, hands-on, and practice-oriented way.

About the Technical Reviewer

Flavia Saldanha is Vice President, Data Architect, and Modern Data Strategist at Fifth Third Bank, where she leads enterprise-scale data modernization initiatives focused on regulatory reporting, AI-ready data products, and large-scale banking transformation. With more than 16 years of experience across financial services and technology, she specializes in helping organizations evolve from fragmented legacy data environments to governed, trusted data platforms that enable advanced analytics, responsible AI, and stronger financial integrity.

An **IEEE Senior Member and experienced technical reviewer**, Flavia regularly contributes her expertise to the global technology community by reviewing scholarly publications and technical manuscripts in data engineering, data architecture, and AI-driven systems. She has served as a peer reviewer for IEEE publications and as a technical reviewer for books published by leading technology publishers including Pearson, bringing deep analytical rigor and industry perspective to complex topics such as modern data warehousing, cloud data platforms, and enterprise data governance.

Beyond her industry contributions, she is the **Founding Chapter Lead of Women in Data Cincinnati**, where she has built and led a thriving professional community empowering more than 150 data and AI practitioners to advance their careers and leadership in the field.

Acknowledgments

This book exists because many people helped me learn, test ideas, and keep going when the finish line felt far away. I'm grateful to colleagues and peers who offered thoughtful feedback, caught mistakes, and pushed me to make the explanations clearer.

I also want to thank the broader Snowflake community for the steady stream of real-world patterns, war stories, and best practices that shaped how I think about modern data platforms.

Most of all, thank you to my family and friends for their support, patience, and belief in this project. And thank you to you, the reader; may these pages save you time, spark ideas, and help you ship great work.

Preface

This book began on a call. A dashboard slowed to a crawl during an executive review; a warehouse wouldn't resume; a "harmless" schema change rippled into an overnight incident. Each fix was small and boring—add a date envelope, swap a clone, cap an ad hoc lane—but the pattern was clear: reliable analytics come from habits, not heroics. Snowflake gives you the right primitives—immutable storage, elastic compute, governance at read—if you shape your work so that those strengths matter. This book collects the patterns that make a Snowflake platform fast, safe, and predictable and turns them into code you can run.

You won't find vendor gloss here. The chapters assume real constraints: mixed file shapes, half-migrated pipelines, late-arriving data, tight budgets, auditors, partner shares, and teams that need answers yesterday. The guidance is practical on purpose. We use Time Travel because backfills happen. We prefer views as interfaces because contracts prevent breakage. We tune for pruning and projection before we touch warehouse sizes because doing less work beats buying more.

The audience is broad but specific: analytics engineers who ship models and marts, platform owners who set policy and cost guardrails, data scientists who need reproducible features and governed access, and leaders who must turn "modern stack" into durable outcomes. If you are migrating from Redshift, BigQuery, Synapse, or Teradata, the migration appendix maps concepts without asking you to rebuild the world first. If you run dbt, the structures and tests will feel familiar. If you operate with SLOs, you will see how freshness and latency become simple SQL.

Use the book like a runbook, not a novel. Early chapters build shared foundations—architecture, accounts and environments, data modeling—so later chapters can be short and decisive about streaming, orchestration, performance, FinOps, governance, sharing, and GenAI-ready patterns. Code blocks are designed to be copied, adapted, and versioned. Diagrams fit on a single page so they can be printed, marked up, and taped near a monitor. Tables summarize trade-offs when there isn't one right answer.

A few conventions keep things steady. We write SQL in a style that surfaces shape and intent: envelopes first, explicit types, small CTE "paragraphs," and no `SELECT *` in shared surfaces. We treat views as product interfaces; when a change is breaking, we version and overlap rather than surprise consumers. We prefer small elastic warehouses for bursty BI and right-sized lanes for ELT. We make cost and freshness visible as views so they can be alerted on like uptime. And whenever the book suggests a policy, it comes with the grants you need to attach it safely.

Nothing here requires a greenfield. Start where you are. Add envelope columns to your hot tables. Tag objects so that cost has an owner. Put a resource monitor on the noisiest warehouse. Publish a freshness heartbeat next to your marts. Clone before you change. Each small move will compound; together, they turn a platform into a calm machine that survives end-of-quarter and Monday mornings.

Thank you to the engineers and analysts who shared incidents, dashboards, and scars; to reviewers who asked for fewer buzzwords and more runnable code; and to the teams proving every day that "boring" is a compliment. May this book shorten your outages, reduce your surprises, and make good numbers feel ordinary.

Disclaimer

Although the publisher and the author have made every effort to ensure that the information in this book was correct at press time and while this publication is designed to provide accurate information regarding the subject matter covered, the publisher and the author assume no responsibility for errors, inaccuracies, omissions, or any other inconsistencies herein and hereby disclaim any liability to any party for any loss, damage, or disruption caused by errors or omissions, whether such errors or omissions result from negligence, accident, or any other cause.

Introduction

Data is only useful when it's trusted, accessible, and fast enough to keep up with the business questions being asked of it. Over the last decade, organizations have collected more data from more systems than ever before—SaaS platforms, product analytics, event streams, IoT, partner feeds, and internal applications—while expectations for freshness and reliability have climbed sharply. The result is familiar: dashboards that disagree, pipelines that break quietly, "temporary" tables that become permanent, and teams spending more time firefighting than building.

Snowflake has changed what a modern data warehouse can be. With elastic compute, separation of storage and compute, and a platform designed for cloud scale, Snowflake removes many of the historical constraints that shaped warehouse design. But it also introduces new engineering decisions: how to structure accounts and environments, how to organize databases and schemas, how to model for performance and cost, how to design resilient ELT pipelines, and how to operate the platform safely over time. In other words, Snowflake doesn't eliminate data warehouse engineering—it makes it more important.

This book, *Snowflake Data Warehouse Engineering: Architecture, Modeling, ELT Pipelines, and Operations*, is written for practitioners who want to build Snowflake systems that work in the real world: systems that are maintainable, observable, secure, and cost-aware. It focuses on the intersection of architecture and day-to-day engineering and how design choices show up later in performance, reliability, and team velocity.

What This Book Covers

Rather than treating Snowflake as a set of features, we approach it as an engineering platform. You'll learn how to

- **Design Snowflake architectures** that match your organization's needs, single vs. multi-account strategies, environment isolation, data sharing, and domain ownership

- **Build data models that scale**, from staging to curated layers, with practical guidance on dimensional modeling, Data Vault, and patterns that fit ELT and semi-structured data
- **Create reliable ELT pipelines** using Snowflake-native capabilities and modern orchestration approaches, incremental loading, change data capture concepts, idempotency, late-arriving data, and backfills without chaos
- **Operate Snowflake with confidence**, including workload management, performance tuning, cost governance, monitoring, incident response, and day-2 practices that prevent surprises
- **Apply security and governance** in a way that's enforceable and auditable, roles, least privilege, data classification, and patterns that keep access manageable as your footprint grows

Throughout the book, the emphasis is on decisions and trade-offs: what to do, why it works, when it doesn't, and how to recover when reality doesn't match the diagram.

Who This Book Is For

This book is for

- **Data engineers and analytics engineers** building pipelines and models in Snowflake
- **Data architects and platform engineers** responsible for account structure, governance, and scale
- **Team leads and managers** who need repeatable patterns for reliability, cost control, and operational maturity

You'll get the most value if you already understand basic SQL and common warehouse concepts (tables, views, partitions/micro-partitions at a high level, batch vs. streaming, and general ELT ideas). Snowflake experience helps, but you don't need to be an expert; this book is designed to take you from "I can run queries" to "I can engineer a platform."

How to Use This Book

You can read this book cover to cover, but it's also meant to be a reference you'll return to while designing and evolving your warehouse. Architecture chapters help you set a foundation. Modeling chapters support long-term analytics and interoperability. Pipeline chapters focus on correctness and resilience. Operations chapters keep the system healthy when usage grows and stakes get higher.

As you work through the material, you'll notice a recurring theme: **good Snowflake engineering is about balancing four forces**:

1. **Performance** (fast queries and predictable workloads)
2. **Cost** (efficient compute usage and controlled growth)
3. **Reliability** (pipelines that don't silently fail)
4. **Usability** (a system teams can understand and extend)

Optimizing one at the expense of the others is easy; designing for all four is the craft.

A Note on Examples and Best Practices

Snowflake evolves quickly, and so do the tools around it. The patterns in this book emphasize durable principles, clear separation of concerns, measurable contracts, and operational discipline while showing concrete implementations you can adapt. When there are multiple valid approaches (and there often are), we'll be explicit about the trade-offs, not prescriptive for the sake of it.

Let's Begin

Whether you're migrating from a legacy warehouse, building a new data platform from scratch, or trying to tame an existing Snowflake estate, the goal is the same: a warehouse that delivers trusted data with minimal drama. If you finish this book with fewer late-night pipeline incidents, clearer models, and a Snowflake environment that scales without surprises, it has done its job.

Let's start by building the foundation: architecture choices that make everything else easier.

PART I

Foundations

CHAPTER 1

Why Snowflake?

Snowflake reframes the data warehouse as a fully managed, multi-cloud service where storage, compute, and control are deliberately separated. Instead of wrestling with capacity planning, patching, and appliance quirks, you shape workloads, choose how they should scale, and rely on the platform's optimizer and metadata to do the heavy lifting. This chapter explains what "modern" means in practice, introduces the core building blocks you'll use every day, and clarifies the commercial and regional choices that affect governance and resilience.

The infrastructure and networking examples in this book use **AWS** for concreteness. Snowflake provides equivalent patterns on Azure and Google Cloud, but the SQL and cloud-specific terminology shown here assume an AWS-based deployment unless stated otherwise.

1.1 Modern DWH vs. Legacy

Legacy warehouses—on-prem appliances, clusters of VMs, or hand-rolled Hadoop—were designed for a world of predictable batch loads and a small set of BI users. Capacity was fixed, resizing took weeks, and different teams competed for the same cycles. As data and use cases multiplied, operations became the bottleneck: indexes to manage, statistics to refresh, nodes to patch, and carefully staggered jobs to prevent stampedes.

A modern warehouse flips that equation. In Snowflake, compute is elastic and isolated. You spin up a dedicated engine for ELT, another for dashboards, and another for data science, each paying only while it runs. When demand spikes, you scale out for concurrency or scale up for heavier queries—without touching the data itself. Micro-partition metadata and an aggressive optimizer prune away irrelevant data, so performance flows from how you write and filter rather than from hand-tuned indexes. Safety features like Time Travel and zero-copy cloning make it trivial to branch, backfill, and recover, which shortens release cycles and lowers risk.

M. Hander, *Snowflake Data Warehouse Engineering*, https://doi.org/10.1007/979-8-8688-2628-3_1

For engineers, the craft shifts from hardware tuning to workload design. You right-size warehouses based on real usage patterns, organize pipelines with streams and tasks, and use role-based access and tagging to keep costs and governance in check. Misconceptions linger—bigger warehouses are not always faster, and clustering keys are not mandatory—but the general rule is simple: start small, measure, and let pruning and caching do their work before spending more compute.

1.2 Core Concepts: Virtual Warehouses, Micro-partitions, and Services layer

Snowflake's mental model is deliberately simple: virtual warehouses execute work, micro-partitions store data, and a services layer orchestrates everything with metadata, security, and optimization.

Virtual warehouses are independent compute clusters that read from the same storage. They isolate workloads and make scaling routine: short, bursty dashboards benefit from multi-cluster scale-out; heavy transforms respond better to scaling up. Warehouses auto-resume when a query arrives and auto-suspend when idle, so you don't pay for silence. Tuning begins with usage data—query shapes, queueing, cache reuse—rather than with guesswork.

```
-- Dedicated ELT warehouse (cost-guarded)
CREATE WAREHOUSE wh_elt
  WITH WAREHOUSE_SIZE = 'LARGE'
  AUTO_SUSPEND = 120
  AUTO_RESUME = TRUE
  INITIALLY_SUSPENDED = TRUE
  STATEMENT_QUEUED_TIMEOUT_IN_SECONDS = 60;

-- BI warehouse with bursty concurrency (works in Snowflake Enterprise
Edition or higher)
CREATE WAREHOUSE wh_bi
  WITH WAREHOUSE_SIZE = 'MEDIUM'
  AUTO_SUSPEND = 60
  AUTO_RESUME = TRUE
```

```
  MIN_CLUSTER_COUNT = 1
  MAX_CLUSTER_COUNT = 4
  SCALING_POLICY = 'STANDARD';
```

Micro-partitions are immutable, columnar files, each with rich statistics—min/max values, cardinality hints, and other descriptors. The optimizer consults those stats to skip entire chunks of data. You rarely need indexes. How you ingest data shapes those partitions, so grouping by natural filters—such as load date or entity—often yields excellent pruning out of the box. Clustering keys are a targeted tool for cases where write order fights your most selective predicates.

```
-- Frequent queries: WHERE event_dt BETWEEN ... AND ... AND user_id = ?
-- Native partitioning by load date is suboptimal for user_id filters
ALTER TABLE fact_events
  CLUSTER BY (event_dt, user_id);
```

You can verify whether clustering is paying off instead of guessing:

```
SELECT system$clustering_information('FACT_EVENTS');
```

If you see shallow depth and strong pruning, you may not need to cluster at all.

The services layer is Snowflake's brain. It authenticates users, enforces RBAC, plans queries, manages transactions, and keeps metadata authoritative and fast. Caches accelerate repeat work: an account-wide result cache returns identical queries instantly; metadata and partition stats speed planning; and each running warehouse keeps a local data cache that benefits subsequent queries on the same engine. Zero-copy primitives—Time Travel and cloning—turn risky operations into routine ones: you can restore a table to a point in time, create a full database branch for a release candidate, or run a hotfix without duplicating terabytes.

Change-aware building blocks make incremental data engineering straightforward. Streams track inserts, updates, and deletes; tasks execute SQL or Snowpark on a schedule or dependency graph. Together, they give you a native way to build DAGs that live where the data lives.

1.3 Editions, Regions, and Clouds

Every Snowflake account reflects a few deliberate choices. You select a feature bundle (edition), a cloud provider, and a region. Those decisions shape data residency, compliance posture, and disaster-recovery options.

Editions ladder up by security and governance. Standard covers core warehousing and a shorter Time Travel window. Enterprise extends retention and adds advanced features that matter at scale. Business Critical raises the security bar—think customer-managed keys and stricter encryption—and is the typical destination for regulated workloads. Some organizations go further with dedicated control planes or government-oriented variants. Names and exact bundling evolve, so it's wise to confirm in your console before you architect around a specific feature.

Region selection starts with the laws and latencies that govern your data. Co-locating Snowflake with data producers avoids egress charges and reduces load times; placing it near consumers keeps dashboards snappy. When you need resilience or proximity in more than one place, database and account replication let you copy data and metadata to other regions or even other clouds. Failover and failback workflows can promote a replica during an incident and then return to normal once the primary is healthy. Data sharing is inherently regional, but cross-region or cross-cloud patterns are possible through replication and listings.

A pragmatic multi-cloud approach is to keep ingestion close to source systems while replicating read-optimized copies near major consumer hubs. Security should travel with your data, so standardizing role hierarchies and grant patterns from the outset pays dividends when you replicate or clone entire environments.

1.4 Handy Commands to Ground the Concepts

```
-- Inspect query performance & pruning
SELECT *
FROM TABLE(INFORMATION_SCHEMA.QUERY_HISTORY_BY_SESSION(RESULT_LIMIT
=> 50));

-- Warehouse utilization
SELECT *
```

```
FROM TABLE(INFORMATION_SCHEMA.WAREHOUSE_LOAD_HISTORY(
  WAREHOUSE_NAME => 'WH_BI',
  START_TIME => DATEADD('hour', -24, CURRENT_TIMESTAMP()))
);

-- Time Travel restore (point-in-time)
CREATE TABLE dim_customer_restore CLONE dim_customer
  AT (TIMESTAMP => TO_TIMESTAMP_TZ('2025-09-01 12:00:00 +00:00'));

-- Zero-copy clone for a release branch
CREATE SCHEMA prod_release_2025_09_26 CLONE prod;
```

Snowflake's promise is not just speed; it's composability. You can isolate work without copying data, rewind when things go wrong, and scale with intent rather than panic. The rest of this book builds on those ideas to show how to model, ingest, transform, secure, and operate at pace—without fighting your warehouse.

CHAPTER 2

Architecture Deep Dive

Snowflake's architecture is best understood as a set of boundaries that let you scale what you need—without disturbing what you don't. Storage is durable and shared. Compute is disposable and isolated. Between them sits a services layer that knows where every byte lives and how to reach it efficiently. When this separation is clear, day-to-day engineering choices—like how you size warehouses or lay out tables—become straightforward and predictable.

2.1 Storage vs. Compute Separation

Think of storage as a calmly expanding lake and compute as a fleet of boats you can launch, resize, or retire at will. Data lands once into encrypted, columnar micro-partitions in cloud object storage. Those partitions don't belong to any specific warehouse. Whether you run one warehouse or a dozen, each reads from the same underlying data without copying it.

This detachment has practical consequences. You can pause the ELT warehouse without blocking analysts, because their BI warehouse is independent. You can scale up a heavy transform for an hour and then drop back down, leaving storage unchanged and shared caches largely intact. Upgrades, patching, and disk balancing—classic sources of operational stress in coupled systems—are abstracted away. Your levers are workload design and scheduling, not hardware gymnastics.

Workflows become easier to reason about. You can clone a database instantly for a release branch, run experimental backfills, and discard the clone without ever duplicating the full dataset. Time Travel lets you restore an object to a prior point without offline windows or restoration from backups. Because compute is a consumable resource and storage is content-addressed, experiments are cheap, and reversals are routine.

M. Hander, *Snowflake Data Warehouse Engineering*, https://doi.org/10.1007/979-8-8688-2628-3_2

2.2 Metadata and Services Layer

If storage is where data rests and warehouses are how work happens, the services layer is the navigator. It authenticates users, enforces role-based access, plans queries, and tracks the fine-grained statistics that make pruning possible. Every micro-partition carries rich metadata—min/max values per column, approximate cardinalities, and other descriptors—so the optimizer can decide what **not** to read long before a byte is fetched.

Transactions are handled here as well. Multi-statement operations observe ACID semantics without the engineer having to shepherd locks or maintain indexes. The services layer coordinates streams (to expose change data) and tasks (to schedule work and define dependencies), which means incremental pipelines live alongside the data they transform. Replication and failover are orchestrated from the same plane: databases and even entire accounts can be mirrored to other regions or clouds with metadata preserved, so a promoted replica behaves like the original.

Because the control plane understands lineage and access, governance becomes part of the fabric rather than an afterthought. Tags, masking policies, and row access rules are evaluated centrally and consistently. When you clone or replicate objects, those policies travel with them, which keeps environments aligned as they multiply.

A few SQL touchpoints make the layer tangible during daily work:

```
-- Explore recent queries and their shapes
SELECT query_id,
       query_type,
       user_name,
       warehouse_name,
       database_name,
       schema_name,
       execution_status,
       start_time,
       end_time,
       total_elapsed_time,
       compilation_time,
       execution_time,
       rows_produced,
       bytes_scanned,
```

```
       partitions_scanned,
       partitions_total,
       query_text
FROM TABLE(INFORMATION_SCHEMA.QUERY_HISTORY(
  END_TIME_RANGE_START => DATEADD('hour', -6, CURRENT_TIMESTAMP()),
  RESULT_LIMIT => 200));

-- Inspect change data available to a stream (incremental pipelines)
SELECT METADATA$ACTION, *
FROM my_db.raw.orders_stream;
```

You rarely need to micromanage the services layer, yet its presence is felt in smooth promotions, reliable rollbacks, and plans that do the smart thing more often than not.

2.3 Caching, Pruning, and Clustering Under the Hood

Snowflake moves quickly not because it reads faster disks but because it often avoids reading at all. Three mechanisms work together to make that happen: caches to skip redundant work, metadata-driven pruning to skip irrelevant partitions, and clustering to keep frequently filtered values co-located.

Caching happens at multiple levels. If two queries are textually identical and the underlying data hasn't changed, the result cache will serve the answer immediately. Even when the text differs, the optimizer's metadata and compiled plans reduce overhead. While a warehouse is running, its local cache holds recently accessed micro-partition data: repeated scans within a session benefit, especially in iterative development or dashboard refresh cycles. None of this requires configuration beyond sensible auto-suspend/resume settings—your main job is to avoid needless warehouse churn that would discard warm caches prematurely.

Pruning is the primary accelerator. Because each micro-partition advertises the ranges of values it contains, a filter like `WHERE order_dt BETWEEN '2025-09-01' AND '2025-09-07'` allows the optimizer to exclude partitions whose date ranges don't overlap. The effect compounds with additional predicates. On large fact tables, it's common for a well-shaped query to touch only a tiny fraction of the stored data. This is why warehouse size is a secondary tuning knob: if pruning is doing its job, a modest cluster can outpace a large one that scans too broadly.

Clustering is a deliberate nudge for cases where natural write order fights your most selective filters. Ingest pipelines often land rows grouped by load date, which is perfect for time-bounded analytics but not always for entity-centric queries like "show me all events for user X over the last 90 days." When the mismatch persists and pruning can't exclude enough partitions, a clustering key encourages Snowflake to reorganize data so that correlated values live closer together. You don't need to do this preemptively; measure first and intervene only where it matters.

Two small utilities keep the work honest:

```
-- See whether a table's layout helps your filters
SELECT system$clustering_information('FACT_EVENTS');

-- Add a clustering key only if selective predicates suffer
ALTER TABLE fact_events
  CLUSTER BY (event_dt, user_id);
```

The first call reports how effectively partitions align with the key you care about; the second sets a policy Snowflake can maintain in the background as data evolves. If depth is shallow and pruning percentages are high, you can skip clustering and save compute.

Put together, these mechanisms let you scale by intent. You shape tables so the optimizer can be ruthless about what it ignores. You size warehouses to match the kind of work they do, not the entire size of your data. You make peace with caches that evaporate when a warehouse sleeps, because auto-resume brings them back quickly, and the real wins come from not touching cold storage in the first place.

2.4 Closing Thoughts

Snowflake's design is not a bag of tricks; it's a contract. Storage remains stable and shared. Compute comes and goes without ceremony. The services layer keeps a precise map so that most queries can be answered by reading far less than is stored. Once you internalize that contract, architectural decisions—warehouse per workload, cautious use of clustering, reliance on cloning and Time Travel for safety—become natural extensions of how the system wants to run.

Figure 2-1 provides a high-level view of Snowflake's architecture, showing the separation of cloud services, compute, and storage, along with the supporting roles of observability and caching.

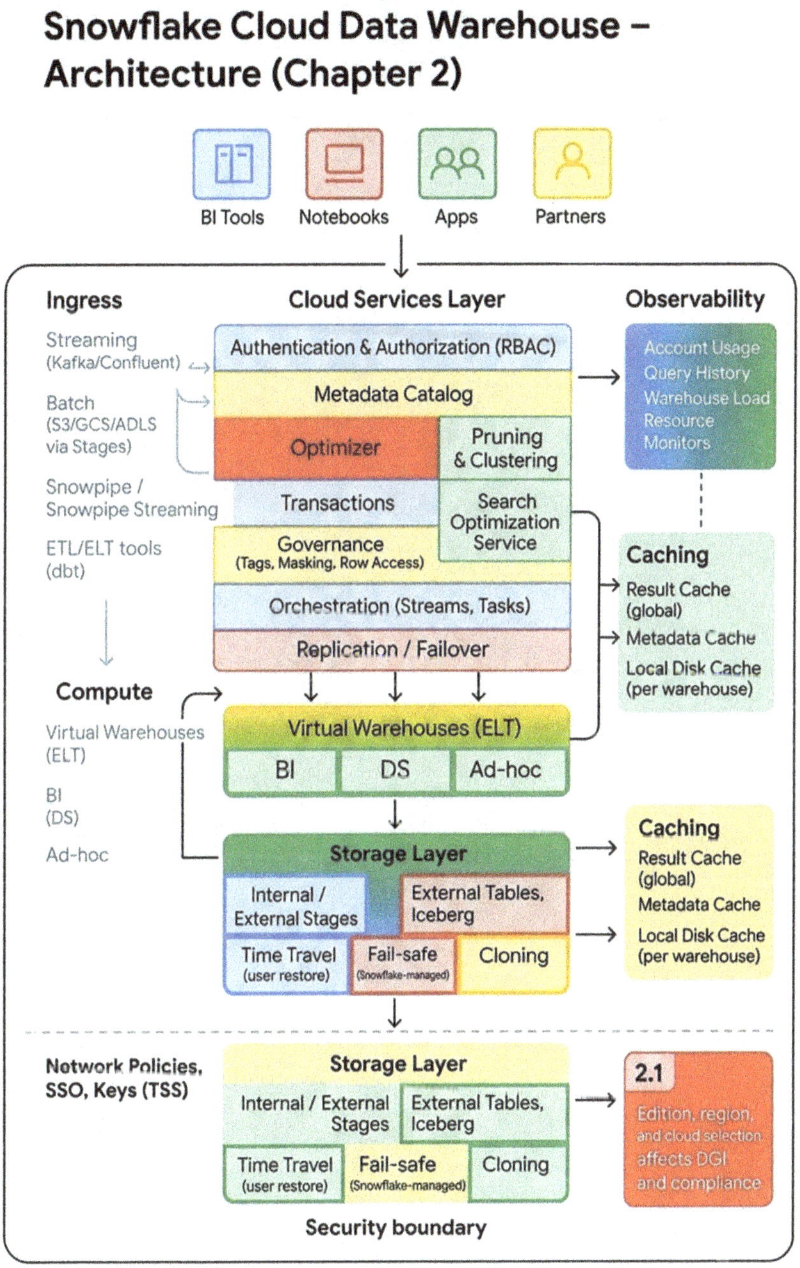

Figure 2-1. *Overview of the architecture of Snowflake cloud data warehouse*

CHAPTER 3

Accounts, Orgs, and Environments

Snowflake scales not just by horsepower but by **boundaries**. At the top sits the *organization*, which owns one or more *accounts*. Each account is a security and billing boundary with its own roles, objects, and warehouses. You'll design these boundaries to match how your company funds work, governs access, and moves code from experiment to production. When the structure is right, promotions feel like switching tracks, not repaving the road.

3.1 Organizations and Multi-account Setups

An **organization** is the administrative shell around your Snowflake presence. It's where global concerns live: consolidated billing, region inventory, account-level lifecycle, and policies that should be consistent everywhere. Within that shell, **accounts** are your operable units. A finance team may have one account with strict governance; a data science platform might own a separate, more permissive account that still inherits the company's standards. The line you draw is less about size and more about **blast radius**—who can affect whom, which workloads should never contend, and how costs should surface on an invoice.

Multi-account designs fall into two broad styles. Some companies go **portfolio-style**: one account per business unit or product, with a shared "platform" account that curates conformed data products through secure data sharing or listings. Others go **lifecycle-style**: a row of accounts that mirror the software lifecycle—sandbox, development, staging, production—so risky work cannot leak into critical environments. Neither is universally better; what matters is that your role model, data sharing, and replication strategy fit the shape you choose.

M. Hander, *Snowflake Data Warehouse Engineering*, https://doi.org/10.1007/979-8-8688-2628-3_3

At the org level, a small set of identities needs broad but narrow-purpose power. The `ORGADMIN` persona creates and manages accounts; `ACCOUNTADMIN` governs within a single account. Day to day, you avoid using those superuser roles by delegating work to domain roles—analytics, engineering, governance—so people hold just enough privilege to do their jobs.

Figure 3-1 and the following code listing make the delegation model concrete. Administrative authority remains concentrated at the organization and account levels, but routine work is pushed downward into domain roles. Those domain roles do not usually hold raw object privileges themselves. Instead, they inherit smaller access roles that map to a clear operational purpose, such as reading analytics data, running ELT pipelines, or administering governance policies. This structure keeps the privilege model modular: access can be changed at the role level without redesigning the whole hierarchy, and users can be reassigned by changing role memberships rather than rewriting grants object by object.

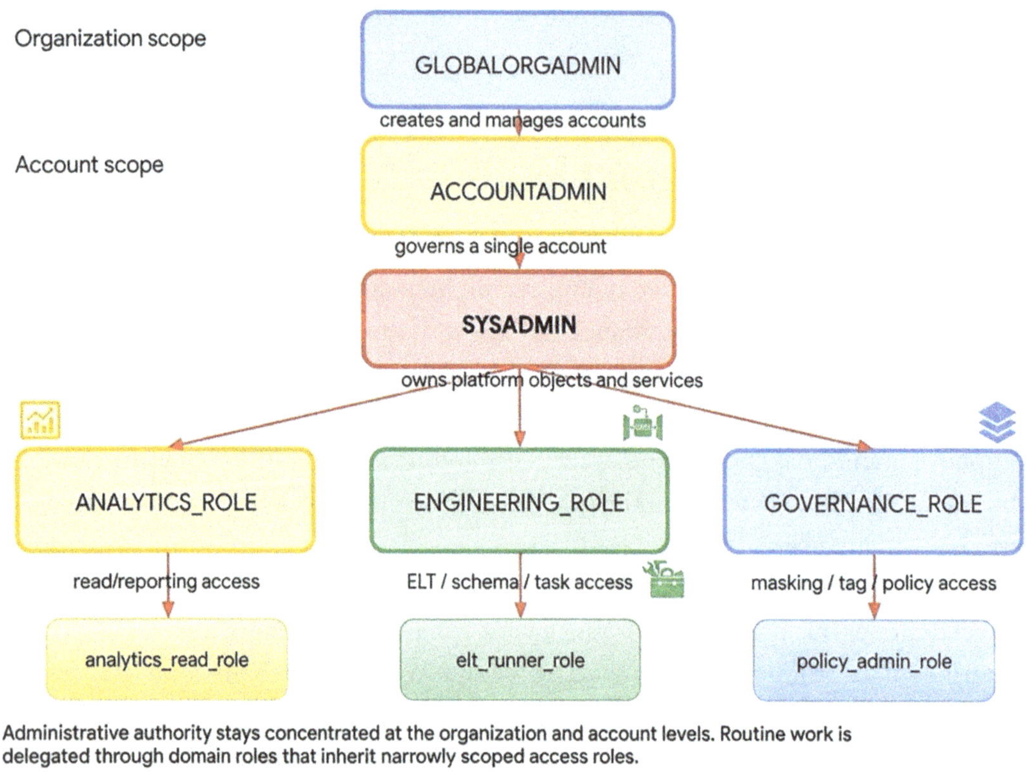

***Figure 3-1.** Example Snowflake role hierarchy*

```
-- Example GRANT chain for domain roles
-- Create domain roles
CREATE ROLE analytics_role;
CREATE ROLE engineering_role;
CREATE ROLE governance_role;
-- Create narrower access roles
CREATE ROLE analytics_read_role;
CREATE ROLE elt_runner_role;
CREATE ROLE policy_admin_role;
-- Grant object privileges to access roles
GRANT USAGE ON WAREHOUSE bi_wh TO ROLE analytics_read_role;
GRANT USAGE ON DATABASE analytics TO ROLE analytics_read_role;
GRANT USAGE ON ALL SCHEMAS IN DATABASE analytics TO ROLE analytics_
read_role;
GRANT SELECT ON ALL TABLES IN DATABASE analytics TO ROLE analytics_
read_role;
GRANT USAGE, OPERATE ON WAREHOUSE elt_wh TO ROLE elt_runner_role;
GRANT USAGE ON DATABASE raw TO ROLE elt_runner_role;
GRANT USAGE ON ALL SCHEMAS IN DATABASE raw TO ROLE elt_runner_role;
GRANT CREATE TABLE, CREATE VIEW, CREATE STAGE ON ALL SCHEMAS IN DATABASE
raw TO ROLE elt_runner_role;
GRANT APPLY MASKING POLICY ON ACCOUNT TO ROLE policy_admin_role;
GRANT APPLY TAG ON ACCOUNT TO ROLE policy_admin_role;
-- Roll access roles into domain roles
GRANT ROLE analytics_read_role TO ROLE analytics_role;
GRANT ROLE elt_runner_role TO ROLE engineering_role;
GRANT ROLE policy_admin_role TO ROLE governance_role;
-- Optionally roll domain roles into a higher administrative path
GRANT ROLE analytics_role TO ROLE sysadmin;
GRANT ROLE engineering_role TO ROLE sysadmin;
GRANT ROLE governance_role TO ROLE sysadmin;
-- Assign domain roles to users
GRANT ROLE analytics_role TO USER analyst_1;
GRANT ROLE engineering_role TO USER data_eng_1;
GRANT ROLE governance_role TO USER data_steward_1;
```

Two quick touchpoints make the hierarchy concrete:

```
-- From ORGADMIN: inventory accounts
SHOW ORGANIZATION ACCOUNTS;
-- Inside an account: keep superuser work minimal
CREATE ROLE data_engineer;
GRANT ROLE data_engineer TO USER alice;
```

Replication makes multi-account practical. You can mirror databases—and, when needed, entire accounts—across regions or clouds and promote a replica during incidents. Because clones and replicas preserve grants and tags, your security posture travels with the data instead of being rebuilt every time.

3.2 Dev/Test/Prod Patterns

"Dev/Test/Prod" is less about three names and more about **predictable promotion**. Snowflake's zero-copy primitives let you move fast without cutting corners. A common rhythm starts with development against **clones** of production schemas: engineers experiment, backfill, and benchmark on objects that look and feel like the real thing, but live on their own storage branches. When work stabilizes, a **staging** account exercises the full pipeline—ingest, transform, publish—against production-like volumes and governance. A final promotion to **production** is a deployment of metadata, not a forklift of data.

Because compute is isolated, you can protect performance with dedicated warehouses for each stage of the pipeline and keep them small by default. The observability that matters most isn't fancy dashboards; it's a habit of checking `QUERY_HISTORY`, `LOAD_HISTORY`, and resource monitors to ensure that real usage, not guesswork, drives sizing.

Versioning belongs in Snowflake as much as in Git. Treat **schemas** as release branches and **databases** as product boundaries. Tag objects with the feature or release they support, so cleanup is obvious. Data contracts—marts, semantic views, masking policies—should live behind stable interfaces, while underlying tables evolve beneath them. When you need to roll back, **Time Travel** restores a table to a safe point without drama.

A short, realistic promotion flow looks like this:

```
-- Create backup before promotion
CREATE TABLE fact_orders_backup CLONE fact_orders;

-- Create an isolated release branch of a production database
CREATE DATABASE prod_candidate_2025_10_01 CLONE prod;

-- Run transforms and tests in isolation (dbt/Snowpark/SQL)

-- Fast rollback if a step misbehaves
ALTER TABLE fact_orders
  SWAP WITH fact_orders_backup;  -- metadata-only swap

-- Once validated, publish new marts into prod
CREATE SCHEMA prod.mart_sales_2025_10_01 CLONE prod_candidate_2025_10_01.
mart_sales;
```

Keep lower environments honest by **refreshing** them from production. Either schedule database replication into dev/test or rebuild from fresh clones on a cadence. Synthetic or masked data can substitute where regulations require it; row access and masking policies help you run the same code across environments while protecting sensitive fields.

Figure 3-2 gives an overview of the Dev/Test/Prod pattern.

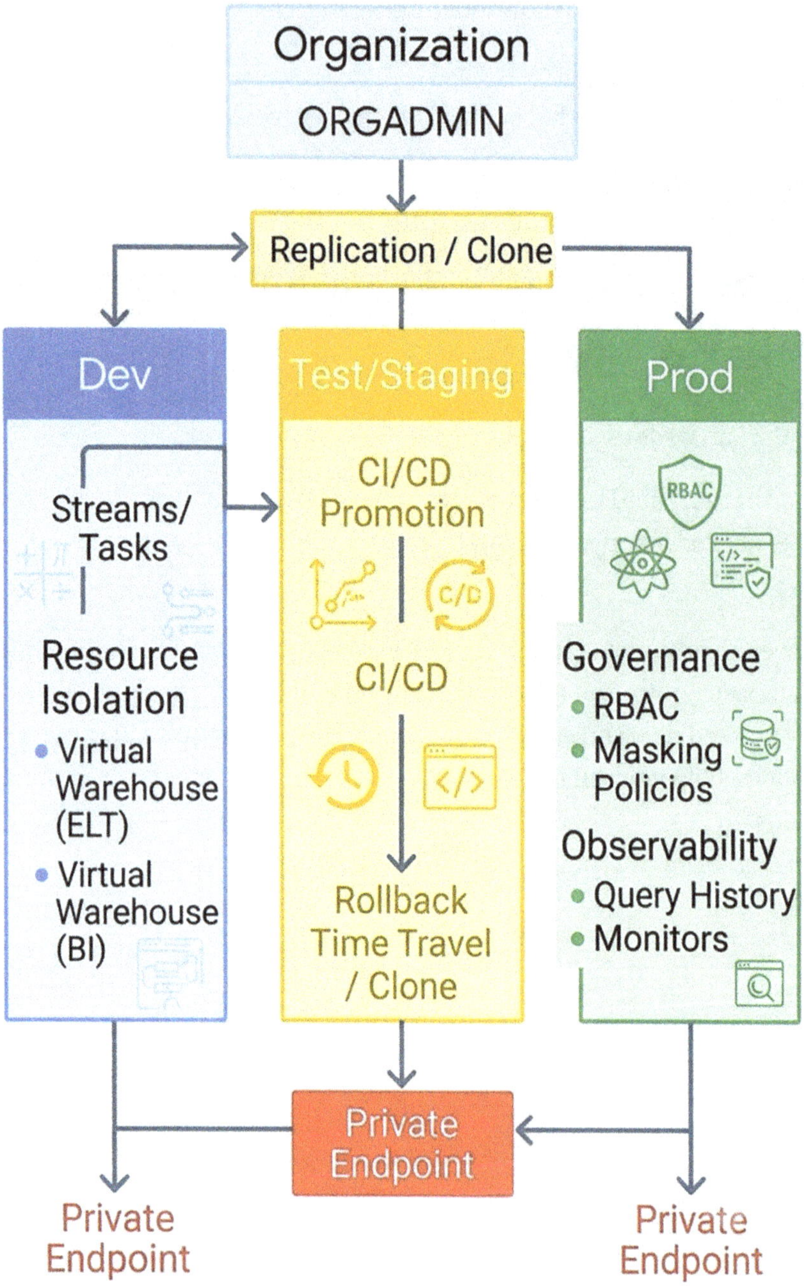

Figure 3-2. *Dev/Test/Prod pattern*

3.3 Networking Options (PrivateLink/Private Link Equivalents, VPC/VNet)

Networking is the other half of isolation. Most teams begin with Snowflake's public endpoint, protected by SSO, MFA, and network policies, and then move to private connectivity when compliance, traffic-control, or internal-routing requirements become stricter. The model is consistent across clouds: on AWS, you use **PrivateLink**, on Azure **Private Link**, and on Google Cloud **Private Service Connect**. In each case, Snowflake provides private-connectivity support and a private account URL so clients can reach the service without traversing the public internet. On AWS, Snowflake documents this as a **Business Critical (or higher)** feature.

Private connectivity does not replace identity or policy. Authentication still happens through Snowflake and your identity provider, while access can be constrained further with **network policies** at the account or user level. Snowflake supports listing and managing these policies with SQL, and the `NETWORK_POLICY` account parameter can be assigned with `ALTER ACCOUNT`. Snowflake also notes that user-level network policies override the account-level policy when both are set.

A useful operational pattern is to validate private-connectivity readiness first, then inspect the policy surface that governs who may connect. For example, the following commands let you confirm whether the account is enabled for private connectivity, retrieve the private-connectivity endpoints and firewall allow-list information, and list the network policies defined in the account. Snowflake provides dedicated system functions for private connectivity, including `SYSTEM$GET_PRIVATELINK` and `SYSTEM$ALLOWLIST_PRIVATELINK`.

```
-- Verify whether the account is enabled for private connectivity
SELECT SYSTEM$GET_PRIVATELINK();
-- Return hostnames and ports to allow through the firewall
SELECT SYSTEM$ALLOWLIST_PRIVATELINK();
-- Show the private connectivity configuration, including private
account URLs
SELECT SYSTEM$GET_PRIVATELINK_CONFIG();
-- List all network policies defined in the account
SHOW NETWORK POLICIES;
```

Once the account is reachable over a private path, the next step is to make the policy model explicit. In practice, that means creating a network policy that reflects your corporate ingress rules and then attaching it at the account level. Snowflake recommends using **network rules together with network policies** rather than maintaining raw IP lists directly inside the policy definition, because network rules are easier to update and reuse.

```
-- Example: create a network policy
CREATE NETWORK POLICY corporate_private_policy
  ALLOWED_IP_LIST = ('203.0.113.0/24')
  BLOCKED_IP_LIST = ('0.0.0.0/0');
-- Inspect the policy after creation
DESCRIBE NETWORK POLICY corporate_private_policy;
-- Assign the policy to the account
ALTER ACCOUNT SET NETWORK_POLICY = corporate_private_policy;
```

If you are using the newer rule-based model, separate the allowed ranges into a network rule and reference that rule from the policy. This keeps the policy readable and makes later changes safer, especially when multiple policies reuse the same address ranges. Snowflake's documentation explicitly recommends the ALLOWED_NETWORK_RULE_LIST and BLOCKED_NETWORK_RULE_LIST approach over embedding large IP lists directly.

```
-- Preferred pattern: define reusable network rules first
CREATE NETWORK RULE corp_office_ipv4_rule
  MODE = INGRESS
  TYPE = IPV4
  VALUE_LIST = ('203.0.113.0/24');
CREATE NETWORK POLICY corporate_private_policy
  ALLOWED_NETWORK_RULE_LIST = ('corp_office_ipv4_rule');
-- Activate the policy for the account
ALTER ACCOUNT SET NETWORK_POLICY = corporate_private_policy;
```

These checks are small, but they make the chapter more concrete. SYSTEM$GET_PRIVATELINK() answers the basic question of whether the account is enabled for private connectivity; SYSTEM$ALLOWLIST_PRIVATELINK() and SYSTEM$GET_PRIVATELINK_

`CONFIG()` expose the hostnames, ports, and private account URLs needed for DNS and firewall configuration; and `SHOW NETWORK POLICIES` plus `ALTER ACCOUNT SET NETWORK_POLICY = ...` make the enforcement side visible in SQL rather than leaving it as abstract platform administration.

Two pragmatic guardrails help avoid surprises. First, standardize private endpoint naming and DNS conventions by region and purpose, because a correct private endpoint with inconsistent DNS is still an outage. Second, test failover the same way you test backups: replicate data to a secondary region, validate the secondary region's private endpoint and DNS path, cut clients over, and then cut back. In private networking, the mistake is rarely the feature itself; it is the untested dependency chain around routing, name resolution, and policy attachment.

3.4 Closing Thoughts

A good Snowflake layout feels quiet. Organizations hold the keys; accounts hold the work; environments advance change without risking production. Networking keeps traffic on trusted roads. The platform's cloning and replication mean your boundaries are light enough to move quickly and firm enough to stay safe. With those fundamentals set, the rest of your engineering—modeling, pipelines, governance—has space to be boring in the best possible way.

PART II

Data Modeling and Ingestion

CHAPTER 4

Data Modeling for Snowflake

Data modeling on Snowflake is less about fighting the engine and more about **shaping data so the optimizer can prune**. Micro-partitions, zero-copy cloning, and semi-structured columns change the old trade-offs: you favor tidy dimensional interfaces, keep raw data close to its original form, and publish governed views or marts that evolve without breaking consumers.

4.1 Star, Snowflake, and Data Vault in Practice

Dimensional models still shine on Snowflake, but you don't need to fear joins the way you might have on legacy MPP. Snowflake's columnar storage, result caching, and automatic pruning make **star schemas** fast and predictable. **Snowflaking** (normalizing out subdimensions) becomes a choice about governance and reuse, not raw performance. **Data Vault** fits well for ingestion-at-scale and auditability, with marts materialized downstream as stars.

M. Hander, *Snowflake Data Warehouse Engineering*, https://doi.org/10.1007/979-8-8688-2628-3_4

4.1.1 When Each Style Earns Its Keep

Modeling style	Where it excels	Trade-offs that matter on Snowflake	Practical pattern
Star (facts + conformed dims)	BI/metrics, stable interfaces, self-service	Wide facts are fine; joins are cheap when predicates prune well	Keep dimensions conformed; surface marts as views or materialized views; avoid premature snowflaking
Snowflake (normalized dims)	Shared reference data, strict governance, slowly growing taxonomies	Extra joins are usually negligible; metadata clarity improves	Normalize only when you have real reuse or ownership boundaries (e.g., `dim_product` → `dim_brand`, `dim_category`)
Data Vault 2.0	High-velocity ingestion, CDC, lineage and audit trail	Raw vault can be verbose; marts are where performance lives	Land to Raw Vault (Hubs, Links, Sats), then build Business Vault and publish star marts

A pragmatic layout many teams adopt:

- **Landing/Raw**: Exact copies from source (files or CDC), possibly in VARIANT
- **Core**: Data Vault (or a lightly normalized core) for history and traceability
- **Marts**: Star schemas per domain (Sales, Marketing, Finance), exposed via views

This keeps ingestion resilient, history complete, and analytics simple.

4.2 Incremental Patterns and Survivorship Rules

Incremental modeling is the default on Snowflake. Streams and Tasks, Time Travel, and cheap clones let you move only what changed and recover when something goes sideways.

Core idea: track deltas, merge idempotently, and encode **survivorship** (which record "wins" when multiple sources disagree).

4.2.1 Typical Incremental Merge (Upsert) into a Dimension

```
MERGE INTO dim_customer d
USING staging_customer s
  ON d.customer_id = s.customer_id
WHEN MATCHED AND s._op IN ('U','D') THEN
  UPDATE SET
    d.email          = COALESCE(s.email, d.email),
    d.status         = s.status,
    d.updated_at     = CURRENT_TIMESTAMP()
WHEN NOT MATCHED AND s._op IN ('I','U') THEN
  INSERT (customer_id, email, status, created_at, updated_at)
  VALUES (s.customer_id, s.email, s.status, CURRENT_TIMESTAMP(),
  CURRENT_TIMESTAMP());
(modified sql):
MERGE INTO dim_customer d
USING staging_customer s
ON d.customer_id = s.customer_id

WHEN MATCHED AND s._op = 'D' THEN
UPDATE SET
d.status = 'deleted',
d.updated_at = CURRENT_TIMESTAMP()

WHEN MATCHED AND s._op = 'U' THEN
UPDATE SET
d.email = COALESCE(s.email, d.email),
d.status = s.status,
d.updated_at = CURRENT_TIMESTAMP()

WHEN NOT MATCHED AND s._op IN ('I', 'U') THEN
INSERT (customer_id, email, status, created_at, updated_at)
VALUES (
s.customer_id,
s.email,
s.status,
```

```
CURRENT_TIMESTAMP(),
CURRENT_TIMESTAMP()
);
```

For history, many teams prefer **Type 2** dimensions (effective dating) but keep the interface stable via views.

```
-- SCD Type 2 (simplified)
MERGE INTO dim_customer_scd d
USING inc_customer s
  ON d.customer_id = s.customer_id
 AND d.current_flag = TRUE
WHEN MATCHED AND (d.email <> s.email OR d.status <> s.status) THEN
  UPDATE SET current_flag = FALSE, valid_to = CURRENT_DATE()
WHEN NOT MATCHED THEN
  INSERT (customer_id, email, status, valid_from, valid_to, current_flag)
  VALUES (s.customer_id, s.email, s.status, CURRENT_DATE(), NULL, TRUE);
```

4.2.2 Survivorship, Stated Explicitly

Put the business rule where engineers can read it—ideally in code and a table.

Attribute	Sources in priority	Survivorship rule	Notes
`email`	CRM > Support > Web	Highest-priority non-null wins	If CRM null, fall back to Support/Web
`status`	Billing	Source of truth; others ignored	Billing events drive `active/suspended`
`phone`	Web > CRM	Most recent update date wins	Use `_last_update_ts` to arbitrate

Implement with a **staging view** that computes winners, then merge the staged winners. Keep the rule pure and idempotent so re-runs don't drift.

For very large facts, prefer **append-only** with late-arriving corrections handled by **compaction jobs** (periodic MERGE of recent partitions), rather than rewriting the entire table.

Figure 4-1 gives an overview of the incremental patterns and survivorship rules.

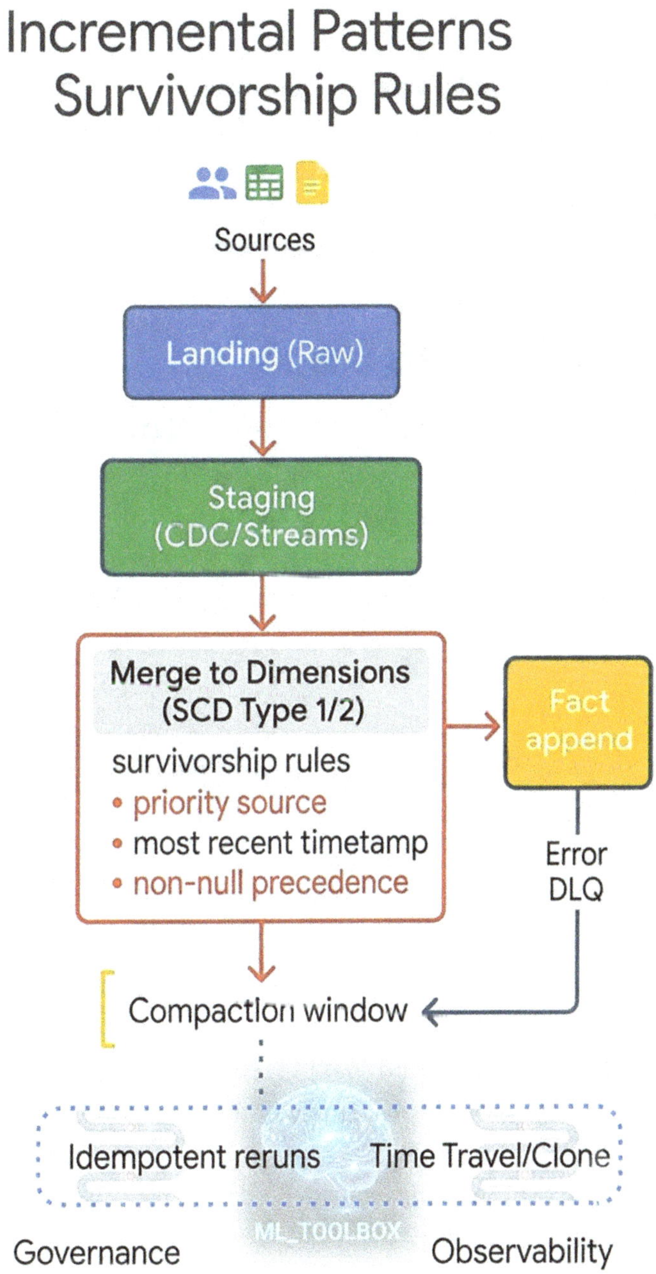

Figure 4-1. *Overview of incremental patterns and survivorship rules*

4.3 Clustering Keys and Micro-partition-Aware Design

Snowflake's first accelerator is **micro-partition pruning**. You don't design partitions directly; you influence them with **how you load** and, when needed, with a **clustering key**. Start by landing data in a way that groups rows by the filters you'll use most (dates, tenant, region). Measure pruning. Only introduce clustering when selective predicates still force too many partitions.

4.3.1 Reading the Table's Shape

```
-- How well does the table align to our common filters?
SELECT system$clustering_information('FACT_EVENTS') AS info;
-- Interpret clustering quality for the table
WITH clustering AS (
    SELECT     PARSE_JSON(SYSTEM$CLUSTERING_INFORMATION('FACT_
    EVENTS')) AS info
)
SELECT
    info:"cluster_by_keys"::STRING AS cluster_by_keys,
    info:"average_depth"::FLOAT AS average_depth,
    info:"average_overlaps"::FLOAT AS average_overlaps,
    info:"total_partition_count"::NUMBER AS total_partition_count,
    info:"total_constant_partition_count"::NUMBER AS    total_constant_
    partition_count,
    ROUND(
        100 * info:"total_constant_partition_count"::FLOAT
              / NULLIF(info:"total_partition_count"::FLOAT, 0),
        2
    ) AS constant_partition_pct,
    info:"notes"::STRING AS notes
FROM clustering;
```

Focus on average_depth and average_overlaps first: lower values usually indicate better clustering, while a higher share of constant partitions suggests Snowflake can prune more micro-partitions during query execution.

If you see high pruning percentages and shallow depth, avoid clustering—let automatic partitioning do its work. If not, add a key that mirrors common predicates:

```
ALTER TABLE fact_events
  CLUSTER BY (event_dt, user_id);
```

Snowflake reorganizes clustered tables incrementally through Automatic Clustering, but that convenience has a cost profile. Reclustering runs only when Snowflake determines the table would benefit, yet each maintenance cycle can still consume credits. For that reason, clustering should be treated as an operational setting, not a one-time design choice. In practice, teams monitor clustering quality, suspend reclustering during periods of heavy ingest or budget sensitivity, and resume it when query performance matters more than short-term maintenance cost. Snowflake supports this directly with `ALTER TABLE ... SUSPEND RECLUSTER` and `ALTER TABLE ... RESUME RECLUSTER`.

```
-- Inspect clustering quality before deciding whether maintenance is worth
the cost
WITH clustering AS (
    SELECT PARSE_JSON(SYSTEM$CLUSTERING_INFORMATION('FACT_EVENTS')) AS info
)
SELECT
    info:"cluster_by_keys"::STRING AS cluster_by_keys,
    info:"average_depth"::FLOAT AS average_depth,
    info:"average_overlaps"::FLOAT  AS average_overlaps,
    info:"total_partition_count"::NUMBER AS total_partition_count,
    info:"total_constant_partition_count"::NUMBER  AS total_constant_
    partition_count,
    info:"notes"::STRING AS notes
FROM clustering;
-- Suspend automatic clustering to control maintenance cost
ALTER TABLE fact_events SUSPEND RECLUSTER;
-- Resume automatic clustering when query performance becomes the
priority again
ALTER TABLE fact_events RESUME RECLUSTER;
-- Review recent automatic clustering activity and credit consumption
SELECT *
```

```
FROM TABLE( INFORMATION_SCHEMA.AUTOMATIC_CLUSTERING_HISTORY(
     TABLE_NAME => 'FACT_EVENTS',
     START_TIME => DATEADD('day', -7, CURRENT_TIMESTAMP())
    )
)
ORDER BY START_TIME DESC;
```

4.3.2 Practical Micro-partition Hints

Situation	Micro-partition-aware approach	Why it helps
Time-series facts	Load by **narrow date windows** (hour/day files)	Natural pruning on `WHERE dt BETWEEN …`
Multi-tenant SaaS	Include **tenant_id** early in file paths or clustering key	Filters by tenant skip most partitions
Semi-structured blobs (VARIANT)	Keep a stable **envelope** (e.g., `event_dt`, `tenant_id`) in scalar columns	Pruning works before JSON is parsed
High-cardinality entity filters	Consider **(date, entity_id)** clustering if queries are "entity over time"	Co-locates hot entity ranges

A small fact, counterintuitively, may run slower on a huge warehouse if pruning is poor—the engine scans more partitions faster instead of fewer partitions smartly. Fix the layout before throwing compute at the problem.

4.3.3 Putting It Together: A Modeling Flow That Fits Snowflake

1. **Land raw** with minimal transformation; preserve source keys and change indicators.
2. **Shape for pruning** in core: batch by time/tenant; add clustering only where selective filters suffer.

3. **Publish marts** as clean stars with conformed dimensions; hide complexity behind stable views.
4. **Incremental by default**: idempotent merges, compaction windows, and survivorship encoded centrally.
5. **Measure, then adjust**: rely on `QUERY_HISTORY`, `TABLE_STORAGE_METRICS`, and `system$clustering_information` to guide changes.

The outcome is a warehouse that feels fast because it reads less, evolves safely because history is preserved, and stays understandable because dimensional interfaces remain simple while the raw truth stays close at hand.

4.4 Closing Thoughts

Good modeling on Snowflake is less about adapting to rigid physical constraints and more about working with the platform's natural strengths. The central idea throughout this chapter is simple: model data so that change is easy to absorb, history is easy to preserve, and the optimizer can avoid reading what does not matter. When those goals are met, performance becomes more predictable and operational trade-offs become easier to manage.

That is why the most effective Snowflake models tend to separate concerns clearly. Raw layers preserve fidelity to the source. Core layers capture history, relationships, and business rules in a form that can be trusted over time. Marts present stable, dimensional interfaces that analysts and applications can use without inheriting the complexity of ingestion or CDC logic. This separation is not bureaucracy; it is what allows pipelines to evolve without forcing downstream consumers to relearn the model every quarter.

Incremental processing is equally central to Snowflake-native design. Rather than reprocessing entire datasets, you move deltas, merge deterministically, and make survivorship rules explicit in code. That discipline pays off twice: once in efficiency, because you avoid unnecessary work, and again in correctness, because re-runs behave predictably. The best pipelines are not only fast on the happy path; they are also understandable after a late-night rollback, a replay, or a source-system correction.

Clustering deserves the same mindset. It is useful when it helps pruning, but it is not a badge of sophistication. A well-modeled table that aligns naturally to common predicates will often outperform a heavily tuned table whose layout fights the workload. Snowflake rewards restraint here: measure first, add clustering only where it clearly helps, and treat automatic reclustering as an operational decision with a real cost profile.

In practice, the chapter's patterns point to one broader lesson: Snowflake works best when modeling stays intentional but lightweight. Keep raw truth close. Publish interfaces that are easy to consume. Encode history and survivorship where engineers can see them. Let measurement guide optimization. When you do that, the warehouse becomes faster not because it is overengineered, but because it is shaped to let Snowflake ignore most of what it stores.

If Chapter 2 explained how Snowflake is built and Chapter 3 explained how it is governed, this chapter shows how data should be arranged so that those platform strengths become visible in daily work. The goal is not just elegant schemas. It is a system that remains explainable as volumes grow, sources multiply, and business rules inevitably become more complicated.

CHAPTER 5

File Ingestion and Staging

In Snowflake, files do not "arrive" straight into tables. They land in **stages**, where file layout and format are explicit, and only then are they loaded with `COPY INTO`. This two-step design turns ingestion into a repeatable contract: put files here, in this shape; the warehouse will pick them up, validate, and either load or park them for inspection.

5.1 Internal vs. External Stages (S3/ADLS/GCS)

Stages are named pointers to file locations plus format metadata. **Internal stages** live inside Snowflake-managed storage; **external stages** point to your cloud buckets via a storage integration. Use internal stages for simple pipelines and secure ad hoc loads; use external stages when files already sit in S3/ADLS/GCS or when you want cloud-native lifecycle controls.

Aspect	Internal stage (`@my_db.%my_table`, `@my_stage`)	External stage (`@ext_stage` → S3/ADLS/GCS)
Storage location	Snowflake-managed	Your bucket/container
Credentials	None (implicit)	**Storage integration** (IAM role/SAS/service account), rotated outside Snowflake
Network path	Snowflake's private path	Your VPC/VNet egress within same region recommended
Cost surface	Storage billed by Snowflake	Storage billed by your cloud
Typical uses	Ad hoc loads, small/medium pipelines, secure "drop-box"	Enterprise pipelines, existing data lakes, cross-system reuse
Extras	Simple lifecycle (PUT/REMOVE)	Native object lifecycle policies, eventing for Snowpipe

M. Hander, *Snowflake Data Warehouse Engineering*, https://doi.org/10.1007/979-8-8688-2628-3_5

Creating each is straightforward:

```
-- Internal named stage with default file format and directory
table enabled
CREATE STAGE stg_sales
  DIRECTORY = (ENABLE = TRUE)
  FILE_FORMAT = (TYPE = PARQUET);

-- Refresh the directory metadata, then inspect the files on the stage
ALTER STAGE stg_sales REFRESH;

SELECT *
FROM DIRECTORY(@stg_sales);

-- External stage to S3 (via storage integration)
CREATE STORAGE INTEGRATION si_raw
  TYPE = EXTERNAL_STAGE
  STORAGE_PROVIDER = S3
  ENABLED = TRUE
  STORAGE_AWS_ROLE_ARN = 'arn:aws:iam::123456789012:role/sf-stage-role'
  STORAGE_ALLOWED_LOCATIONS = ('s3://my-raw/sales/');

CREATE STAGE ext_sales
  URL = 's3://my-raw/sales/'
  STORAGE_INTEGRATION = si_raw
  FILE_FORMAT = (TYPE = PARQUET);
```

For internal stages, enabling a directory table makes staged files queryable as metadata, which is useful for exploratory workflows, validation, and lightweight file inspection before loading data into tables.

Tip Keep file format declarations **on the stage**, so COPY commands stay minimal, and upstream tools can change formats by updating a single place.

5.2 COPY INTO Options, Validations, and Error Handling

`COPY INTO <table>` is the hinge between staged files and relational tables. It maps file structure to columns, proves the data will load, and records what happened so retries are safe. The best runs feel boring: you dry-run first, load deterministically, then inspect history and rejects the same way every time.

5.2.1 A Steady Loading Rhythm

1. **Dry-run on the stage.** Use `VALIDATION_MODE` to surface schema or parsing issues without touching the table.

```
COPY INTO raw.orders
FROM @ext_sales
FILE_FORMAT = (TYPE = JSON, STRIP_OUTER_ARRAY=TRUE)
VALIDATION_MODE = 'RETURN_ERRORS';
```

2. **Load with explicit mapping and error policy.** Favor name-based matching so mild column drift doesn't break you; choose whether bad rows abort or get quarantined.

```
COPY INTO raw.orders
FROM @ext_sales
MATCH_BY_COLUMN_NAME = CASE_INSENSITIVE
ON_ERROR = 'ABORT_STATEMENT'
PURGE = TRUE;
```

3. **Inspect what happened.** Load history and the `VALIDATE` table function explain successes, skips, and rejects.

```
-- Review recent load activity for the target table
SELECT *
FROM TABLE(INFORMATION_SCHEMA.LOAD_HISTORY(
  TABLE_NAME => 'ORDERS',
  START_TIME => DATEADD('hour', -6, CURRENT_TIMESTAMP())
));
```

```
-- Run a load that allows row-level errors to be skipped
COPY INTO raw.orders
FROM @ext_sales
FILE_FORMAT = (TYPE = PARQUET)
ON_ERROR = 'CONTINUE';

-- Validate the last COPY job and return all load errors
SELECT *
FROM TABLE(VALIDATE(raw.orders, JOB_ID => '_last'));
```

5.2.2 What the Key Switches Actually Do

Theme	Option/behavior	Why it matters
Column mapping	`MATCH_BY_COLUMN_NAME = CASE_INSENSITIVE`	Survives harmless header/order drift; avoids positional surprises.
Idempotency	Per-file tracking via name + checksum	Safe re-runs; unchanged files are skipped. Stable file names or manifests help.
Validation	`VALIDATION_MODE = 'RETURN_ERRORS'` or `'RETURN_ALL_ERRORS'`	Proves parse/schema rules before committing.
Error policy	`ON_ERROR = 'ABORT_STATEMENT'` vs. `'CONTINUE'` vs. `'SKIP_FILE'`	Pick strictness by pipeline. Strict for curated feeds; tolerant + quarantine for noisy ones.
Housekeeping	`PURGE = TRUE/FALSE`	Delete files after a successful load when the stage is only a queue.
Semistructured	Declare format on the stage (e.g., `JSON`, PARQUET) and keep an envelope column	Lets you evolve extraction without re-writing ingestion.
CSV hygiene	Explicit delimiters, quoting, nulls, date formats	Prevents silent coercions and unexpected NULLs.

CSV example (clean, explicit):

```
CREATE FILE FORMAT ff_csv_clean AS
  TYPE = CSV
  FIELD_DELIMITER = ','
  SKIP_HEADER = 1
  FIELD_OPTIONALLY_ENCLOSED_BY = '"'
  NULL_IF = ('\\N','NULL','')
  EMPTY_FIELD_AS_NULL = TRUE
  DATE_FORMAT = 'YYYY-MM-DD';

COPY INTO raw.customers
FROM @stg_customers
FILE_FORMAT = ff_csv_clean
MATCH_BY_COLUMN_NAME = CASE_INSENSITIVE
ON_ERROR = 'ABORT_STATEMENT';
```

5.2.3 Handling Rejects Without Drama

Strict pipelines should fail fast; exploratory or high-volume ones often need to pass good rows while parking bad ones. The "good-through, bad-aside" pattern works cleanly in Snowflake:

```
-- Allow load to proceed; record rejects for later triage
COPY INTO raw.orders
FROM @ext_sales
ON_ERROR = 'CONTINUE'
PURGE = FALSE;

-- Materialize rejects (with file/row context) into a DLQ table
CREATE OR REPLACE TABLE ops.orders_rejects AS
SELECT *
FROM TABLE(VALIDATE(
  TABLE_NAME => 'RAW.ORDERS',
  JOB_ID     => LAST_QUERY_ID()
));
```

A small catalog table keeps decisions visible:

Reject code	Typical cause	Action
NUMERIC_VALUE_OUT_OF_RANGE	Overflow (e.g., price too large)	Fix upstream; widen target type if legitimate.
DATE_PARSING_ERROR	Bad date literals	Update file format; add staging cast rules.
INVALID_JSON	Truncated/ill-formed payload	Re-emit file; consider STRIP_OUTER_ARRAY/ALLOW_DUPLICATE flags.

5.2.4 Manifests, Partitions, and "Same File Twice"

Snowflake remembers file names and hashes. If upstream replays the same object key, the loader skips it. When you need deterministic batches independent of "what's in the bucket," use **manifests**:

```
COPY INTO raw.orders
FROM @ext_sales
FILES = ('2025-09-25/orders_001.parquet','2025-09-25/orders_002.parquet')
ON_ERROR = 'ABORT_STATEMENT';
```

Or point COPY at a stable **partition** path so late files don't slip in unexpectedly:

```
COPY INTO raw.orders
FROM @ext_sales/date_dt=2025-09-25/
FILE_FORMAT = (TYPE = PARQUET);
```

5.2.5 Semi-structured Loads That Age Well

A durable approach is to land the whole record in VARIANT alongside typed columns you care about today. You keep provenance and can backfill new attributes later without re-ingesting.

```
CREATE TABLE raw.events (
  tenant_id  STRING,
  event_dt   TIMESTAMP_NTZ,
  payload    VARIANT
```

```
);

COPY INTO raw.events
FROM @ext_events
FILE_FORMAT = (TYPE = JSON, STRIP_OUTER_ARRAY = TRUE)
MATCH_BY_COLUMN_NAME = CASE_INSENSITIVE
ON_ERROR = 'ABORT_STATEMENT';
```

Downstream, extract with views or `COPY INTO` CTAS into typed tables; if extraction rules change, re-run on a **Time Travel** snapshot of `raw.events` rather than recalling files.

5.2.6 Troubleshooting in One Place

Most mysteries resolve by correlating three views:

- `LOAD_HISTORY` for which files were considered and which ones loaded or skipped
- `VALIDATE(..., JOB_ID => ...)` for the exact rejected rows
- `QUERY_HISTORY` for timings, row counts, and warehouse behavior during the load window

If loads slow down, it's rarely a `COPY` switch—it's usually file layout (too many tiny files, inconsistent compression) or external system lag. Fix the shape of the batch, then let `COPY` do the predictable thing.

5.3 Compression, File Sizing, and Parallelism

Snowflake reads in parallel at the **file** level. Performance comes from giving the engine **many reasonably sized files**, compressed wisely, rather than a handful of giants or a spray of tiny fragments.

Concern	Recommended posture	Why it matters
Compression	Prefer each format's native, broadly interoperable compression. For **Parquet**, use **Snappy** as the default practical choice in Snowflake pipelines. For **ORC**, use **ZLIB** or **ZSTD** where your producers support it. For **CSV/JSON**, prefer **GZIP** or **ZSTD**.	Reduces I/O and egress without costly CPU; Snowflake decompresses efficiently.
File size (compressed)	Aim for ~**100–250 MB** per file for Parquet/ORC; **50–150 MB** for CSV/JSON.	Large enough to amortize overhead; small enough to fan out across threads.
File count per batch	Target **hundreds to a few thousand** files per load window. Avoid tens of thousands of tiny files.	Enables wide parallelism without metadata thrash.
Row group/ block size	Parquet row group ≈ **128 MB** (uncompressed) is a solid default.	Preserves predicate pushdown and pruning.
Partitioning	Organize by **date/tenant/source**, not by arbitrary UUIDs.	Aligns with common filters and promotes pruning.
Concurrency	Let Snowflake scale with the warehouse; avoid over-sharding beyond what the warehouse can read concurrently.	Prevents diminishing returns and driver bottlenecks.

In practice, ZSTD is often a strong choice for text-based formats such as CSV and JSON because it can improve compression ratios without making decompression prohibitively expensive, but for **Parquet in Snowflake examples, Snappy remains the safest default because it is the compression option Snowflake documents directly for Parquet file-format handling and unloading behavior.**

```
-- Parquet: use Snappy in Snowflake-oriented examples
FILE_FORMAT = (TYPE = PARQUET, COMPRESSION = 'SNAPPY');

-- CSV: ZSTD is a good modern choice when supported by your producers/
consumers
FILE_FORMAT = (TYPE = CSV, COMPRESSION = 'ZSTD');
```

Use each format's native compression wisely rather than forcing a single rule across all file types. In Snowflake pipelines, Parquet with Snappy is the most portable default and aligns with Snowflake's documented Parquet behavior. For CSV and JSON, GZIP and especially ZSTD are often better choices when you want stronger compression and lower storage or transfer cost. The goal is not maximal compression at any price, but a balance of compatibility, decompression speed, and load efficiency.

Example COPYs that respect these choices are boring—in the best way:

```
-- Parquet, sized ~200 MB per object, date-partitioned in S3
COPY INTO core.fact_orders
FROM @ext_sales/date_dt=2025-09-25/
FILE_FORMAT = (TYPE = PARQUET)
ON_ERROR = 'ABORT_STATEMENT'
PURGE = FALSE;  -- keep data lake as source of truth
```

If you inherit a "tiny files" firehose, batch them into **manifests** or compact upstream. As a last resort, stage into an internal area, **repack** to balanced Parquet, then load:

```
-- (Conceptual) Repack workflow
-- 1) Land tiny CSVs to internal stage
-- 2) Create a transient table, load with CONTINUE, then CTAS → Parquet
CREATE OR REPLACE TABLE work.orders_clean AS
SELECT * FROM raw.orders  -- after validation & casting
;

-- 3) Unload to balanced Parquet for future loads (or downstream sharing)
COPY INTO @stg_sales_balanced
FROM work.orders_clean
FILE_FORMAT = (TYPE = PARQUET, COMPRESSION = 'SNAPPY')
HEADER = FALSE
OVERWRITE = TRUE
MAX_FILE_SIZE = 209715200;  -- ~200 MB
```

MAX_FILE_SIZE is specified in bytes; here, 209715200 targets output files of approximately 200 MB, which aligns well with Snowflake's general guidance to work with files in the roughly 100–250 MB range.

5.3.1 A Quick Tuning Loop You Can Live With

1. Check `LOAD_HISTORY` for rejected files and durations.
2. Inspect a sample of staged objects for size distribution.
3. Adjust upstream chunking (batch windows, `MAX_FILE_SIZE`, Parquet row groups).
4. Re-run with `VALIDATION_MODE` first when formats change.

When in doubt: **fewer, larger, well-formed files beat many tiny ones**; **Parquet beats raw CSV** for analytics; and **stable partitioning keys** (date/tenant) make both loading and querying faster.

5.3.2 Closing Notes

Stages give you a clean seam between upstream systems and warehouse tables. `COPY INTO` makes loading deterministic and recoverable. Balanced compression and file sizing unlock Snowflake's parallelism without heroics. Keep these three in harmony, and ingestion becomes the quiet, reliable foundation the rest of your platform depends on.

CHAPTER 6

Semi-structured Data

Snowflake treats semi-structured data as a first-class citizen. You can land JSON/Parquet/Avro/ORC as-is, keep full fidelity in a VARIANT column, and project only the attributes you need—now or later. The engine stores these values columnarly within micro-partitions and uses path metadata to prune aggressively, so "schema-on-read" doesn't have to mean slow.

6.1 VARIANT, OBJECT, and ARRAY

Snowflake's native trio—VARIANT, OBJECT, and ARRAY—mirrors JSON's model. In practice, you'll store full payloads as VARIANT, dereference with path syntax, and sometimes rebuild smaller OBJECTs/ARRAYs as curated envelopes.

Type	What it represents	Handy constructors and checks	Example access
VARIANT	Any semi-structured value (object, array, scalar)	TO_VARIANT, PARSE_JSON, IS_OBJECT/IS_ARRAY/IS_NULL_VALUE	payload:order_id::STRING
OBJECT	Key/value map (JSON object)	OBJECT_CONSTRUCT, OBJECT_INSERT, OBJECT_KEYS	payload:user:name::STRING
ARRAY	Ordered list	ARRAY_CONSTRUCT, ARRAY_SIZE, ARRAY_SLICE	payload:items[0]:sku::STRING

M. Hander, *Snowflake Data Warehouse Engineering*, https://doi.org/10.1007/979-8-8688-2628-3_6

Key points:

- **Path syntax** uses `:` and `[]`. Quote keys that contain spaces or mixed case: `payload:"User ID"::STRING`.
- **Casting** (`::TYPE` or `TRY_CAST`) matters; use `TRY_*` casts to avoid failing on occasional bad values.
- **FLATTEN** is your join for arrays: it turns each element into its own row.

```
-- Storing and inspecting
CREATE TABLE raw.events (tenant STRING, arrived_at TIMESTAMP_NTZ, payload
VARIANT);

INSERT INTO raw.events VALUES
  ('acme', CURRENT_TIMESTAMP(), PARSE_JSON('{"order_id":"o-1","items":
  [{"sku":"A","qty":2}]}'));

SELECT
  payload:order_id::STRING              AS order_id,
  payload:items[0]:sku::STRING          AS first_sku,
  IS_ARRAY(payload:items)               AS has_items
FROM raw.events;
```

Exploding arrays with `FLATTEN`:

```
SELECT
  e.tenant,
  e.payload:order_id::STRING AS order_id,
  f.value:sku::STRING        AS sku,
  f.value:qty::NUMBER        AS qty
FROM raw.events e,
LATERAL FLATTEN(INPUT => e.payload:items) f;
```

You can also **(re)construct** compact JSON to publish downstream:

```
SELECT OBJECT_CONSTRUCT(
         'order_id', payload:order_id,
         'item_count', ARRAY_SIZE(payload:items)
       ) AS order_summary
FROM raw.events;
```

6.2 Querying JSON/Parquet/Avro/ORC

Snowflake can project fields directly from staged files, query data where it lives through **external tables**, or land records into warehouse tables as VARIANT or typed columns. Pick the entry that matches your goal—explore, federate, or persist—and keep a few envelope columns (ingest_dt, tenant_id, event_type) outside the payload so pruning starts before parsing.

6.2.1 What Each Format Feels Like to Query

Format	Typical entry point	Projection style	Where it excels	Gotchas to watch
JSON	Stage → SELECT $1:... or Table with VARIANT	Path access (:) + casts; arrays via FLATTEN	Logs/events, flexible schemas	Arrays of mixed types; string dates—prefer TRY_TO_TIMESTAMP
Parquet	External table or COPY to typed table	Typed columns, pushdown on row groups	Analytics lake files, columnar ops	Keep row groups consistent; use stable types across partitions
Avro	External table → VALUE (variant) or staged read	Map unions to the "widest" target + TRY_CAST	CDC streams with evolving schemas	Union types (["null","string","int"]) need explicit handling
ORC	External table or COPY	Similar to Parquet; strong typed projojection	Legacy Hadoop exports	Less common in modern pipelines; mind decimal precision/scale

6.2.2 Querying Straight from a Stage (Exploration)

```
-- JSON: one row per JSON object; cast on read
SELECT
  $1:order_id::STRING AS order_id,
  TRY_TO_NUMBER($1:total) AS total
FROM @ext_events (FILE_FORMAT => (TYPE=JSON, STRIP_OUTER_ARRAY=TRUE));
```

```
-- Parquet: columns are already typed in the file
SELECT
  $1:order_id::STRING AS order_id,   -- or project as typed with
  column defs
  $1:total::FLOAT     AS total
FROM @ext_parquet (FILE_FORMAT => (TYPE=PARQUET));
```

Explode arrays with FLATTEN:

```
SELECT
  rec.$1:order_id::STRING AS order_id,
  item.value:sku::STRING  AS sku,
  item.value:qty::NUMBER  AS qty
FROM @ext_events rec (FILE_FORMAT => (TYPE=JSON, STRIP_OUTER_ARRAY=TRUE)),
LATERAL FLATTEN(INPUT => rec.$1:items) AS item;
```

6.2.3 External Tables (Query-in-Place over Your Lake)

External tables are most effective when the lake layout is reflected in the table definition. If your files are partitioned by path, define partition columns explicitly so Snowflake can prune files before reading them. Otherwise, a predicate on a value extracted from the file contents may still require scanning all matching files and filtering rows afterward. Snowflake explicitly recommends partitioning external tables and defining partition columns with PARTITION BY.

```
-- External table over a date-partitioned lake
-- Example path pattern:
-- s3://my-lake/orders/event_dt=2025-09-01/part-0001.parquet

CREATE EXTERNAL TABLE lake.orders_parquet (
  order_date DATE AS TO_DATE(SPLIT_PART(METADATA$FILENAME, '/', 3),
  'YYYY-MM-DD'),
  order_id STRING AS (VALUE:order_id::STRING),
  total NUMBER AS (VALUE:total::NUMBER)
)
PARTITION BY (order_date)
WITH LOCATION = @lake_sales/orders/
```

```
AUTO_REFRESH = TRUE
FILE_FORMAT = (TYPE = PARQUET);

-- Query with partition pruning
SELECT
  order_id,
  total,
  METADATA$FILE_ROW_NUMBER AS row_no
FROM lake.orders_parquet
WHERE order_date >= CURRENT_DATE() - 7;
```

Without a partition column, a filter such as `WHERE value:event_dt::DATE >= CURRENT_DATE() - 7` may still force Snowflake to inspect far more files than necessary; with `PARTITION BY (order_date)`, Snowflake can use external-table metadata to narrow the scan before reading file contents.

Avro with unions benefits from a projection view that normalizes types:

```
CREATE OR REPLACE VIEW lake.orders_avro_norm AS
SELECT
  COALESCE(
    TRY_TO_VARCHAR(value:order_id),
    TO_VARCHAR(value:order_id_int)
  ) AS order_id,
  TRY_TO_TIMESTAMP_NTZ(value:created_at) AS created_at,
  TRY_TO_DECIMAL(value:total, 18, 2)      AS total
FROM lake.orders_avro;  -- external table over AVRO objects
```

6.2.4 Landing for Keeps: VARIANT vs. Typed Columns

A durable pattern is **bronze VARIANT + silver typed view/CTAS**:

```
-- Bronze: lossless ingest
CREATE OR REPLACE TABLE bronze.events (
  tenant_id STRING,
  ingest_dt DATE,
  payload   VARIANT
);
```

```
COPY INTO bronze.events
FROM @ext_events
FILE_FORMAT = (TYPE=JSON, STRIP_OUTER_ARRAY=TRUE)
MATCH_BY_COLUMN_NAME = CASE_INSENSITIVE;

-- Silver: curated, typed interface
CREATE OR REPLACE VIEW silver.orders AS
SELECT
  tenant_id,
  TRY_TO_VARCHAR(payload:order_id)          AS order_id,
  TRY_TO_NUMBER(payload:total)              AS total,
  TRY_TO_TIMESTAMP_NTZ(payload:created_at)  AS created_at
FROM bronze.events
WHERE payload:type::STRING = 'order_created';
```

If you prefer **typed tables at rest** from Parquet/ORC:

```
CREATE OR REPLACE TABLE core.fact_orders
  USING TEMPLATE (
    SELECT ARRAY_AGG(OBJECT_CONSTRUCT(*))
    FROM TABLE(INFER_SCHEMA(
      LOCATION => '@ext_sales_parquet/', FILE_FORMAT => 'PARQUET'
    ))
  );

COPY INTO core.fact_orders
FROM @ext_sales_parquet
FILE_FORMAT = (TYPE=PARQUET)
ON_ERROR = 'ABORT_STATEMENT';
```

`INFER_SCHEMA` helps derive consistent column definitions before a first load.

6.2.5 Performance Notes (Why Queries Feel Fast or Slow)

When you see full scans against large JSON datasets, it's usually because the query applies filters only inside the `VARIANT` payload. Add a few scalar "envelope" columns such as `event_dt` and `tenant_id`, and filter on those first so Snowflake can prune micro-partitions before touching the JSON.

If external-table reads feel slow, the underlying lake probably contains many tiny files or row groups of very uneven size. Compact the data upstream so files are roughly 200 MB and Parquet row groups are about 128 MB; this gives the engine enough parallel work without excessive metadata overhead.

When AVRO fields come back as null or appear garbled, the root cause is often union types that differ from file to file. Normalize those fields in a projection view using TRY_CAST (or related TRY_ functions), and work with the producers to standardize the schema going forward.

If dashboard refreshes pile up and queries start queuing, the problem is typically too few files per refresh cycle or a warehouse that is undersized for the burst. Increase file-level parallelism and enable a multi-cluster policy on the BI warehouse so concurrency spikes are absorbed without contention.

6.2.6 Troubleshooting in Place

```
-- See file set and partitions touched by an external-table query
EXPLAIN USING text
SELECT ...
FROM lake.orders_parquet
WHERE value:event_dt::DATE = CURRENT_DATE();

-- Validate staged records before a load changes table state
COPY INTO bronze.events
FROM @ext_events
FILE_FORMAT = (TYPE=JSON, STRIP_OUTER_ARRAY=TRUE)
VALIDATION_MODE = 'RETURN_ERRORS';
```

The pattern to remember: query quickly where the data sits, but persist clean interfaces for recurring analytics. Keep pruning cheap with envelope columns, normalize types at the edge (especially for Avro unions), and let columnar formats and Snowflake's metadata do most of the work.

6.3 Schema-on-Read and Auto-ingest Patterns

Semi-structured pipelines age well when ingestion is **lossless** and transformation is **declarative**. Land the whole record, extract what you need through views or downstream CTAS, and let ingestion run continuously with Snowpipe or streaming.

6.3.1 A Minimal, Durable Flow

1. **Auto-ingest to a raw table** (or external table) with the full payload in VARIANT.
2. **Publish stable views** that cast and validate fields; consumers query the views.
3. **Backfill safely** by re-running the view logic on Time-Travel snapshots when rules change.

Concern	Recommended stance	Why it helps
Contract with sources	Keep source keys, event time, and a provenance envelope alongside VARIANT.	Traceability and reprocessing without asking producers to replay.
Evolution	Add columns in views, not by rewriting raw. Use TRY_CAST/defaults.	New attributes don't break readers; backfills are deterministic.
Quality	Assert invariants in views (WHERE filters, QUALIFY) and surface rejects in a side table.	Keeps the raw landing pristine while making issues visible.
Cost/perf	Filter early by scalar envelope columns (date/tenant) before digging into VARIANT.	Maximizes micro-partition pruning.

Auto-ingest with Snowpipe (S3 example):

```
CREATE STORAGE INTEGRATION si_events
  TYPE = EXTERNAL_STAGE
  STORAGE_PROVIDER = S3
  ENABLED = TRUE
```

```
  STORAGE_AWS_ROLE_ARN = 'arn:aws:iam::123456789012:role/
  sf-  snowpipe-role'
  STORAGE_ALLOWED_LOCATIONS = ('s3://lake/events/');

CREATE STAGE ext_events
  URL = 's3://lake/events/'
  STORAGE_INTEGRATION = si_events
  FILE_FORMAT = (TYPE = JSON, STRIP_OUTER_ARRAY = TRUE);

CREATE PIPE pipe_events
  AUTO_INGEST = TRUE
  AS
  COPY INTO bronze.events (tenant_id, ingest_dt, payload)
  FROM (
    SELECT
      $1:tenant_id::STRING,
      METADATA$START_SCAN_TIME::DATE,
      $1::VARIANT
   FROM @ext_events
)
ON_ERROR = 'CONTINUE';
```

In production, the pipe should load not only the raw payload but also the envelope columns that make the bronze table queryable and auditable. When possible, extract stable business identifiers such as tenant_id from the payload or file path, and derive ingest timestamps from `METADATA$START_SCAN_TIME` rather than `CURRENT_DATE` inside the pipe definition.

Streaming for very low latency (conceptual):

- Use **Snowpipe Streaming** from a service that reads your bus (e.g., Kafka) and writes rows directly into a table with a `VARIANT` payload and envelope columns.
- The downstream views and `FLATTEN` queries are identical; only the ingress path changes.

Finally, make **pruning cheap** by keeping a few scalar columns outside the JSON—`ingest_dt`, `event_dt`, `tenant_id`, `event_type`. Most filters can hit those first, allowing the optimizer to skip partitions before it ever touches `payload`.

6.3.2 Closing Thought

Treat semi-structured data as an asset, not a nuisance. Keep it intact where it lands, publish tidy, typed interfaces above it, and let Snowflake's path metadata, pruning, and arrays toolkit do the heavy lifting. You'll spend less time arguing about "the schema" and more time shipping reliable analytics that can evolve with your upstreams.

CHAPTER 7

Streaming, CDC, and Near-Real-Time

Near-real-time on Snowflake is less about a single feature and more about how you **land deltas quickly**, **track change safely**, and **publish stable views** without noisy rewrites. Two ingestion lanes do the heavy lifting—Snowpipe (files) and Snowpipe Streaming (rows)—while CDC tools supply the change feed. Inside Snowflake, *Streams* expose those changes as first-class tables, and *Tasks* move them into dimensions and facts with idempotent merges. When the pieces are aligned, dashboards feel fresh, backfills are routine, and recovery is just a re-run.

7.1 Snowpipe and Snowpipe Streaming

Snowflake offers two lanes for "always-on" ingestion. **Snowpipe** watches cloud storage and loads new files moments after they land. **Snowpipe Streaming** bypasses files and writes rows directly from a client or connector. Both preserve idempotency and integrate with Streams/Tasks for downstream processing; they differ in how data arrives, the typical latency you can expect, and how much infrastructure you manage outside Snowflake.

M. Hander, *Snowflake Data Warehouse Engineering*, https://doi.org/10.1007/979-8-8688-2628-3_7

7.1.1 How the Two Lanes Compare

Aspect	Snowpipe (file-triggered)	Snowpipe Streaming (row-ingest)
Arrival path	Files land in S3/ADLS/GCS; cloud events trigger a Pipe to COPY from a stage	Client pushes rows via streaming API/ connector into a target table
Typical latency	Minutes, often 1–5, governed by file emission and event delivery	Seconds to sub-second, governed by client flush/batch size
Ordering and exactly-once	File-level dedupe (name + checksum); within-file order preserved	Offset-based idempotency on the client; per-row upserts if you enforce keys
Operational surface	Manage buckets, events, and file layout; Snowflake runs the loader	Manage a long-lived client/connector; no buckets or events
Best fit	Micro-batches, tool-based CDC (Fivetran/Matillion), upstreams that already write files	Clickstream/IoT, low-latency CDC, services that can host a connector
Cost profile	Storage + egress (your cloud) and Snowpipe per-ingest charges	Warehouse (or serverless) compute for ingestion; fewer storage ops

7.1.2 A Small, Durable Snowpipe Setup (S3 Example)

Create a storage integration and stage that describe where files will arrive and how they are formatted. Then define a pipe that performs the COPY.

```
CREATE STORAGE INTEGRATION si_events
  TYPE = EXTERNAL_STAGE
  STORAGE_PROVIDER = S3
  ENABLED = TRUE
  STORAGE_AWS_ROLE_ARN =   'arn:aws:iam::123456789012:role/snowflake-pipe'
  STORAGE_ALLOWED_LOCATIONS = ('s3://lake/events/');

CREATE STAGE ext_events
  URL = 's3://lake/events/'
  STORAGE_INTEGRATION = si_events
  FILE_FORMAT = (TYPE = JSON, STRIP_OUTER_ARRAY = TRUE);
```

```
CREATE TABLE bronze.events (
  tenant_id STRING,
  ingest_ts TIMESTAMP_NTZ DEFAULT CURRENT_TIMESTAMP(),
  payload VARIANT);

CREATE PIPE pipe_events
  AUTO_INGEST = TRUE
  AS
  COPY INTO bronze.events (tenant_id, payload)
  FROM (
    SELECT
      $1:tenant_id::STRING,
      $1::VARIANT
    FROM @ext_events
)
ON_ERROR = 'CONTINUE';
```

For staged JSON files, $1 represents the entire document, so individual attributes should be extracted with JSON path expressions such as `$1:tenant_id`, while the raw payload can be preserved with `$1::VARIANT`.

With event notifications wired on the bucket, new files trigger the pipe automatically. The COPY is idempotent—if a file is resent with the same name and contents, it is skipped. If your upstream emits many tiny files, consider compaction or manifests to keep parallelism healthy.

7.1.3 Direct Row Ingest with Snowpipe Streaming

Here, the long-lived client (e.g., the Snowflake Kafka Connector or a bespoke service) batches rows in-memory and commits them to Snowflake with low latency. Table design mirrors the file path: keep an envelope (`tenant_id`, `event_ts`) and a `VARIANT` payload so you can evolve downstream projection without rewriting ingestion.

```
CREATE TABLE bronze.events_streaming (
  tenant_id STRING,
  event_ts  TIMESTAMP_NTZ,
  payload   VARIANT
);
-- The streaming client/connector writes rows directly into this table.
```

Practically, the important decisions live in the client: batch size vs. latency, retry/backoff policy, and how you encode idempotency (topic/partition/offset for Kafka; a source transaction id for CDC). Once rows land, Streams/Tasks pick them up exactly as they do for Snowpipe.

7.1.4 Designing for Correctness and Recovery

Concern	Practical stance	Why it works
Idempotency	Use stable file names (Snowpipe) or source offsets/keys (Streaming) and enforce keys in downstream MERGEs	Replays don't double-count; late arrivals are reconciled
Backpressure	Permit micro-batching (files or row buffers) and size a small, always-on warehouse for ingestion	Spikes are absorbed without long queues
Lateness and reordering	Process near-real-time, then **repair** a rolling window with a MERGE job	Low latency in dashboards, eventual correctness in facts
Evolution	Land lossless `VARIANT` + envelope; normalize in a staging view	Schema drift doesn't block ingestion

7.1.5 Monitoring What Matters

You can see file-level progress and errors for Snowpipe through load history; for Streaming, inspect the target table's Streams and task runs.

To inspect stream metadata, use `SHOW STREAMS` for the current state in a schema, or query `SNOWFLAKE.ACCOUNT_USAGE.STREAMS` when you want account-level or longer-retention metadata.

```
-- Pipe and load health (Snowpipe)
SELECT *
FROM TABLE(INFORMATION_SCHEMA.LOAD_HISTORY(
  START_TIME => DATEADD('hour', -6, CURRENT_TIMESTAMP())
))
ORDER BY LAST_LOAD_TIME DESC;
```

```
-- What changed for downstream consumers (current stream metadata)
SHOW STREAMS IN SCHEMA bronze;

-- Optional: filter the SHOW output in SQL
SELECT *
FROM TABLE(RESULT_SCAN(LAST_QUERY_ID()))
WHERE "table_name" ILIKE 'BRONZE%';

-- Task executions that promote deltas
SELECT *
FROM TABLE(INFORMATION_SCHEMA.TASK_HISTORY(
  SCHEDULED_TIME_RANGE_START => DATEADD('hour', -6, CURRENT_TIMESTAMP())
));
```

In practice, the two lanes often coexist. Systems that already produce curated files ride Snowpipe with minimal fuss, while event-heavy paths—clickstream, IoT, and CDC with tight SLAs—use Snowpipe Streaming to remove the file hop. The downstream contract is the same in both cases: a normalized staging view, a Stream that exposes the delta, and Tasks that merge changes into the dimensional layer.

7.2 Kafka/Confluent, Debezium, and Fivetran/Matillion Patterns

Real pipelines start with a source of truth and a transport. Kafka topics, Debezium change feeds, and SaaS ELT tools all map cleanly into Snowflake's two ingest lanes. The right pairing depends on latency tolerance, ops preferences, and how much schema drift you expect.

Source pattern	Landing approach	Notes that matter in practice
Kafka/ Confluent (events + CDC)	Kafka → (a) Snowflake Kafka Connector (Snowpipe Streaming) or (b) S3 sink → Snowpipe	The connector removes file handling and supports low-latency ingestion into Snowflake; S3 sink is simpler operationally but adds batch latency. Keep envelope columns such as `tenant_id` and `event_ts` alongside the raw `VARIANT` payload.
Debezium (database CDC)	Debezium → Kafka (or equivalent stream transport) → Snowpipe Streaming/Snowpipe	Changes arrive as creates, updates, deletes, or snapshot reads, typically with `before` and `after` images. Preserve the business key, source timestamp, transaction identifier, and log position so downstream merges remain deterministic.
Fivetran/ Matillion (managed ELT)	Tool-managed staging to cloud storage → Snowpipe (often with tool-managed pipes)	Lowest-ops path for many SaaS sources. Use the tool's delete flags and metadata columns in downstream merges, and keep source-specific metadata visible rather than flattening it away too early.
App producers (custom)	Direct SDK → Snowpipe Streaming	Useful for clickstream and IoT. Apply light validation at the edge, keep the event lossless in `VARIANT`, and store a stable envelope for pruning and lineage.

Two design habits keep all of these sane: first, encode provenance in dedicated columns or stable envelope fields, such as source system, topic, partition, offset, or file path; second, standardize the CDC envelope so downstream SQL does not care which upstream tool produced it.

Example: Querying a Debezium CDC Envelope in Snowflake

A Debezium-style payload is a good example of why this matters. In practice, the bronze layer often stores the raw message in a `VARIANT` column such as `payload`, and downstream logic extracts the fields needed for deduplication, merge logic, and auditing. Snowflake's semi-structured path syntax makes this straightforward.

```
-- Standard Debezium CDC envelope structure in Snowflake
-- Debezium wraps row changes in before/after images plus source metadata
SELECT
    payload:op::STRING AS operation,           -- c, u, d, r
    COALESCE(
        payload:after:order_id::STRING,
        payload:before:order_id::STRING
    )                                  AS order_id,
    payload:before:order_id::STRING   AS order_id_before,
    payload:after:order_id::STRING    AS order_id_after,
    TO_TIMESTAMP_NTZ(payload:source:ts_ms::NUMBER / 1000) AS source_ts,
    payload:source:txId::STRING       AS transaction_id,
    payload:source:lsn::NUMBER        AS log_sequence_number
FROM bronze.events
WHERE payload:source:db::STRING = 'orders_db';
```

For Kafka-based ingestion, Snowflake's Kafka connector can stream rows directly into Snowflake via Snowpipe Streaming; by default, tables loaded through the connector use raw content and metadata columns unless schema detection and evolution is enabled.

This is the operational reason for the table's advice. The `op` field tells you whether the row should be treated as an insert, update, delete, or snapshot read; the `before` and `after` images support Type 1 or Type 2 merge logic; and the source transaction or log-position fields help you reason about ordering and replay. Snowflake's `VARIANT` querying model is designed for exactly this kind of selective extraction from raw semi-structured records.

7.3 Change Data Capture to Dimensions and Facts

Inside Snowflake, *Streams* turn "what changed since I last looked" into a table-like interface. You can attach a stream to a landing table (row-based CDC) or to a view that normalizes tool-specific payloads into a standard shape. *Tasks* then run merges on a cadence or dependency, promoting deltas into **Type-1/Type-2 dimensions** and **append/UPSERT facts**.

7.3.1 A Small, Durable CDC Contract

Assume a normalized staging view that exposes the essentials:

```
CREATE OR REPLACE VIEW stage.orders_norm AS
SELECT
  payload:order_id::STRING            AS order_id,
  TRY_TO_NUMBER(payload:total)        AS total,
  TRY_TO_TIMESTAMP_NTZ(payload:ts)    AS event_ts,
  payload:op::STRING                  AS op, -- 'I','U','D'
  payload:src_txid::STRING            AS src_txid
FROM bronze.events
WHERE payload:type::STRING = 'order';
```

Create a stream over it:

-- This works because the view is a simple projection/filter over a native
-- table. For more complex normalization views, create the stream on the
-- base table instead.

```
CREATE OR REPLACE STREAM str_orders ON VIEW stage.orders_norm
  APPEND_ONLY = FALSE;  -- we care about updates/deletes too
```

In this example, the view is intentionally simple, so it remains compatible with Snowflake streams on views. If your normalization layer later adds unsupported constructs, move the stream to bronze.events, and perform the normalization in the task query instead. Snowflake also requires change tracking on the view and underlying tables.

Important limitation A stream can be created on a view only if the view is stream-compatible. In Snowflake, that means the view must be based on native underlying tables and use only supported operations such as projections, filters, inner/cross joins, `UNION ALL`, and `DISTINCT`. Functions in the select list must be system-defined scalar functions. Change tracking must also be enabled on the view and its underlying tables. If the normalization logic becomes more complex than those limits allow, create the stream on the base table instead, and apply the normalization in the consuming task or downstream transformation.

Type-1 Dimension: Keep Only the Latest Customer State

A Type-1 dimension stores the current version of each business entity and overwrites older attribute values when newer changes arrive. In Snowflake, a common pattern is to read incremental changes from a stream and apply them with a MERGE. Because tasks are created in the suspended state, the definition should be followed by an explicit resume step; otherwise, the schedule exists, but nothing runs. You can verify the state with SHOW TASKS or DESCRIBE TASK.

```
-- Type-1 dimension: overwrite the latest known customer state
CREATE OR REPLACE TASK t_dim_customer
  WAREHOUSE = wh_elt
  SCHEDULE = '5 MINUTE'
AS
MERGE INTO dim_customer d
USING (
  SELECT DISTINCT customer_id, email, status, event_ts
  FROM str_customers
) s
ON d.customer_id = s.customer_id
WHEN MATCHED THEN UPDATE SET
  email      = COALESCE(s.email, d.email),
  status     = s.status,
  updated_at = s.event_ts
WHEN NOT MATCHED THEN
  INSERT (customer_id, email, status, created_at, updated_at)
  VALUES (s.customer_id, s.email, s.status, s.event_ts, s.event_ts);
-- Newly created tasks do not start automatically.
ALTER TASK t_dim_customer RESUME;
-- Confirm that the task is now active.
SHOW TASKS LIKE 'T_DIM_CUSTOMER';
```

The logic is straightforward. The stream supplies changed customer rows, and the MERGE matches them on the business key, customer_id. If a target row already exists, the task overwrites the current attributes with the newest values. If no target row exists yet, the task inserts a new dimension row and stamps both created_at and updated_at from the source event timestamp. This is the defining behavior of Type-1 processing: preserve the latest state, not the full change history.

The COALESCE on email deserves a brief explanation. In many CDC feeds, not every update event carries a full after-image of the record. If an event omits email, overwriting the dimension with NULL would accidentally erase a previously known value. Using COALESCE(s.email, d.email) keeps the existing email unless the incoming event provides a replacement. The status column, by contrast, is written directly, which is appropriate when the feed guarantees that the latest status is always present.

One practical note is that SELECT DISTINCT removes exact duplicates, but it does not guarantee that only the newest event per customer is applied. If the stream can contain multiple changes for the same customer_id in one task run, a production-grade implementation should usually collapse them to the latest event, for example, with ROW_NUMBER() over PARTITION BY customer_id ORDER BY event_ts DESC. That refinement is not specific to tasks, but it often matters in real CDC pipelines.

Operationally, the important point is that the schedule alone does not activate execution. Snowflake creates tasks in the SUSPENDED state by default, so readers who copy only the CREATE TASK statement will otherwise see no runs at all until they resume the task. Snowflake's documentation explicitly recommends checking whether the task, or each task in a graph, is in the RESUMED state.

If this task later becomes part of a task graph, the root task remains the scheduled entry point for the DAG. Snowflake documents two safe ways to enable the graph: resume the child tasks you want to participate and then resume the root task, or enable the full graph from the root with SYSTEM$TASK_DEPENDENTS_ENABLE. Suspended child tasks are skipped, so they should not be left disabled accidentally.

Type-2 dimension (history via effective dating):

```
CREATE OR REPLACE TASK t_dim_customer_scd2
  WAREHOUSE = wh_elt
  AFTER t_dim_customer  -- serialize if they touch same entities
AS
MERGE INTO dim_customer_scd d
USING (
  SELECT DISTINCT customer_id, email, status, event_ts
  FROM str_customers
) s
```

```
ON d.customer_id = s.customer_id AND d.current_flag = TRUE
WHEN MATCHED AND (d.email <> s.email OR d.status <> s.status) THEN
  UPDATE SET current_flag = FALSE, valid_to = s.event_ts
WHEN NOT MATCHED THEN
  INSERT (customer_id, email, status, valid_from, valid_to, current_flag)
  VALUES (s.customer_id, s.email, s.status, s.event_ts, NULL, TRUE);
```

Facts benefit from append-first logic with repair windows. Ingest new rows as inserts; handle late corrections with a small, time-bounded MERGE that revisits recent partitions only:

```
-- Append new events
INSERT INTO fact_orders (order_id, customer_id, amount, order_ts)
SELECT order_id, customer_id, total, event_ts
FROM str_orders
WHERE op IN ('I','U')  -- Debezium upserts map to inserts here
  AND event_ts >= DATEADD('day', -2, CURRENT_TIMESTAMP()); -- guard
  for reorder

-- Periodic repair of the last N days for true updates/deletes
MERGE INTO fact_orders f
USING (
  SELECT * FROM str_orders
  WHERE event_ts >= DATEADD('day', -2, CURRENT_TIMESTAMP())
) s
ON f.order_id = s.order_id
WHEN MATCHED AND s.op = 'U' THEN
  UPDATE SET amount = s.total, order_ts = s.event_ts
WHEN MATCHED AND s.op = 'D' THEN
  DELETE;
```

7.3.2 Handling Lateness, Duplicates, and Reordering

Pain point	Practical stance	Why it works
Out-of-order events	Use an **event-time watermark** (e.g., "process up to T-5 minutes") and a rolling **repair window** in MERGEs.	Keeps latency low while guaranteeing eventual correctness.
Duplicates	Enforce idempotency with a **natural or synthetic key** plus `MERGE` semantics; consider **stream deduplication** over `(key, src_txid)`.	Protects against replay and at-least-once delivery.
Deletes (tombstones)	Carry an op code; represent deletes explicitly in streams and translate to hard deletes or **soft deletes** with a flag in dimensions.	Downstream consumers can choose how to interpret removal.
Schema drift	Land lossless `VARIANT` + envelope; normalize in views with `TRY_CAST` and defaulting; backfill via **Time Travel**.	Evolution doesn't block ingestion or break readers.

7.3.3 Observability That Keeps You Honest

The most useful signals are built-in:

```
-- What changed and what was consumed
SELECT * FROM TABLE(INFORMATION_SCHEMA.STREAMS()) WHERE TABLE_NAME ILIKE
'STR_%';

-- Task health and runtimes
SELECT * FROM TABLE(INFORMATION_SCHEMA.TASK_HISTORY(
  SCHEDULED_TIME_RANGE_START => DATEADD('hour', -6, CURRENT_TIMESTAMP())
));

-- End-to-end freshness (simple SLA probe)
SELECT MAX(event_ts) AS newest_event, CURRENT_TIMESTAMP() AS now
FROM stage.orders_norm;
```

If freshness slips, it's usually upstream backpressure (connector lag) or tiny-file churn. The fix is to widen producer batches, size warehouses to absorb spikes, or move the path to Snowpipe Streaming when files become the bottleneck.

7.3.4 Closing Thought

Streaming on Snowflake isn't a separate stack; it's the same warehouse operating on smaller, more frequent increments. Land deltas quickly with Snowpipe or its streaming sibling, standardize the CDC envelope, and promote with Streams and Tasks using merges that are safe to run again. You'll get low latency without sacrificing history, and you'll keep the option to rewind, replay, and re-publish whenever business logic changes.

CHAPTER 8

Tables, Dynamic Tables, and External Tables

Snowflake gives you three complementary ways to hold data. **Database tables** live inside Snowflake storage and behave like a classic warehouse. **Dynamic Tables** keep derived data up to date without hand-built orchestrations. **External Tables** and **Iceberg Tables** extend the platform to your lake, so you can query or even **own** data outside Snowflake while keeping SQL and governance consistent. Choosing among them is less about features and more about where the data should live, who else must use it, and how fresh it needs to be.

8.1 Permanent, Transient, and Temporary Tables

All three are "real tables" with the same SQL surface. What changes is how long Snowflake protects your data and what happens when you make mistakes.

Table class	Time Travel	Fail-safe	Common use	Notes that matter
Permanent	Configurable (typically 1–90 days)	Yes	Core facts/dims, governed marts	Highest protection and storage overhead; safest default for shared data.
Transient	Optional (0–1 day typical)	No	Staging, scratch, intermediate CTAS	Lower storage cost; no fail-safe recovery—use only for re-creatable data.
Temporary	Session only	No	Interactive work, unit tests, local temp marts	Vanishes at session end; still benefits from columnar storage and caching.

M. Hander, *Snowflake Data Warehouse Engineering*, https://doi.org/10.1007/979-8-8688-2628-3_8

A small pattern that keeps surprises away: default to **permanent** for anything shared, **transient** for reproducible intermediates, and **temporary** for developer scratch. Use explicit retention on objects that should be easy to rewind.

```
-- Permanent with 7-day Time Travel
CREATE TABLE core.fact_orders (... )
  DATA_RETENTION_TIME_IN_DAYS = 7;

-- Transient staging table (no fail-safe)
CREATE TRANSIENT TABLE stage.orders_work AS
SELECT ...;

-- Temporary scratch in a notebook/session
CREATE TEMP TABLE work.sample AS
SELECT * FROM core.dim_customer SAMPLE (1);
```

8.2 Dynamic Tables vs. Streams and Tasks

Dynamic Tables and the Streams-plus-Tasks pattern solve the same problem—keeping derived data fresh—but they do it with different philosophies. Dynamic Tables are declarative: you describe the final SELECT, set a freshness goal, and Snowflake computes the deltas. Streams and Tasks are imperative: you ask Snowflake to expose what changed, and you write the merges. Both compose into larger pipelines, both can be safe to re-run, and both can live side by side in the same platform. The choice turns on how much control you need over change semantics vs. how much orchestration you want Snowflake to absorb.

8.2.1 Two Ways to Keep Data Current

Aspect	Dynamic Tables	Streams and Tasks
Authoring model	Describe the target as a SELECT; Snowflake maintains it.	Read a Stream of changes and apply your own MERGE/INSERT/DELETE.
Freshness control	Set a target lag such as `TARGET_LAG = '5 MINUTE'`.	Use a schedule or dependency graph (`SCHEDULE = '5 MINUTE', AFTER ...`).
Incremental logic	Automatic; engine computes diffs from sources.	Explicit; you decide dedupe, repair windows, and conflict rules.
Best fit	Relational transforms (joins, filters, rollups) and multi-hop DAGs.	CDC semantics, SCD Type-2, survivorship, side outputs (rejects, audit).
Backfills and recovery	Relax the lag or force a refresh; engine recomputes safely.	Re-run tasks, widen MERGE windows, or replay Streams from offsets.
Operational feel	Less orchestration code; clearer lineage from SQL graph.	More control and observability of each promotion step.

Dynamic Tables shine when the business logic is "what rows should exist given these sources." Streams and Tasks shine when the question is "how should changes be applied to preserve history or resolve conflicts."

8.2.2 A Dynamic Table That Maintains a Mart on a Freshness Budget

The target stays within five minutes of its sources. The SQL expresses the truth of the table; Snowflake works out the incremental maintenance.

```
CREATE OR REPLACE DYNAMIC TABLE mart.orders_recent
  TARGET_LAG = '5 MINUTE'
  WAREHOUSE  = wh_elt
AS
SELECT
  o.order_id,
  o.customer_id,
```

```
  d.region,
  o.total,
  o.order_ts
FROM bronze.orders o
JOIN dim_customer d USING (customer_id)
WHERE o.order_ts >= DATEADD(day, -30, CURRENT_TIMESTAMP());
```

The dynamic table keeps a rolling 30-day slice of recent orders. Snowflake supports CURRENT_TIMESTAMP() in filter predicates for dynamic tables, so using it in the WHERE clause is valid here. The broader caution is that dynamic tables do not support every SQL construct: for example, streams cannot be used as sources, external functions are not supported, and some non-deterministic functions are restricted or disallowed.

Backfills are straightforward. If you need to re-evaluate a bigger window, temporarily broaden the WHERE clause or relax the TARGET_LAG, let the engine catch up, then restore the tighter settings.

8.2.3 The Same Outcome with Streams and Tasks for Tighter CDC Control

Here, the change feed is explicit, and the merge logic captures exactly how updates, deletes, and late arrivals should behave.

```
-- Expose deltas from the landing table
CREATE OR REPLACE STREAM str_orders ON TABLE bronze.orders;

-- Promote changes into the mart on a schedule
CREATE OR REPLACE TASK t_merge_orders
  WAREHOUSE = wh_elt
  SCHEDULE  = '5 MINUTE'
AS
MERGE INTO mart.orders_recent t
USING (
  SELECT order_id, customer_id, total, order_ts
  FROM str_orders
  WHERE order_ts >= DATEADD(day, -30, CURRENT_TIMESTAMP())
) s
```

```
ON t.order_id = s.order_id
WHEN MATCHED THEN UPDATE SET
  total    = s.total,
  order_ts = s.order_ts
WHEN NOT MATCHED THEN
  INSERT (order_id, customer_id, total, order_ts)
  VALUES (s.order_id, s.customer_id, s.total, s.order_ts);
```

For history, a second Task can maintain a Type-2 dimension with effective dating, something Dynamic Tables do not model directly:

```
-- SCD2 dimension maintenance: expire the current version, then insert the
new one.
CREATE OR REPLACE TASK t_dim_customer_scd2
  WAREHOUSE = wh_elt
  AFTER t_merge_orders
AS
BEGIN
  -- Step 1: expire the currently active version when tracked attributes
  changed.
  MERGE INTO dim_customer_scd d
  USING (
    SELECT DISTINCT customer_id, email, status, event_ts
    FROM str_customers
  ) s
  ON d.customer_id = s.customer_id
     AND d.current_flag = TRUE
  WHEN MATCHED
       AND (NVL(d.email, '∅') <> NVL(s.email, '∅')
         OR NVL(d.status, '∅') <> NVL(s.status, '∅')) THEN
    UPDATE SET
      current_flag = FALSE,
      valid_to     = s.event_ts;
  -- Step 2: insert a new current version for new or changed customers.
  INSERT INTO dim_customer_scd
    (customer_id, email, status, valid_from, valid_to, current_flag)
```

```
  SELECT
    s.customer_id,
    s.email,
    s.status,
    s.event_ts,
    NULL,
    TRUE
  FROM (
    SELECT DISTINCT customer_id, email, status, event_ts
    FROM str_customers
  ) s
  LEFT JOIN dim_customer_scd d
    ON d.customer_id = s.customer_id
   AND d.current_flag = TRUE
  WHERE d.customer_id IS NULL
     OR NVL(d.email, '∅') <> NVL(s.email, '∅')
     OR NVL(d.status, '∅') <> NVL(s.status, '∅');
END;

-- In a task graph, enable dependent tasks, then resume the root task,
-- or enable the graph recursively from the root.
SELECT SYSTEM$TASK_DEPENDENTS_ENABLE('t_merge_orders');
-- Verify state.
SHOW TASKS LIKE 'T_%CUSTOMER%';
```

> *This task maintains a slowly changing dimension of type 2. When a customer's tracked attributes change, the existing current row is expired by setting `current_flag = FALSE` and stamping `valid_to` with the incoming event timestamp. A second step then inserts the successor version as the new current row. This two-step pattern is necessary because a single `MERGE` cannot both update a matched target row and insert a successor row for that same match in one pass. In task graphs, remember that Snowflake creates tasks in the suspended state by default; suspended child tasks are skipped, so enable the graph before expecting it to run.*

8.2.4 Choosing the Right Tool for the Job

If your pipeline is primarily relational and you care most about keeping derived tables fresh with minimal orchestration, Dynamic Tables reduce boilerplate and make lineage obvious. If your pipeline encodes business-specific change rules—survivorship across sources, late-arriving corrections, delete tombstones, or compliance-grade audit trails—Streams and Tasks give you the levers to be precise.

In practice, many teams use both. Dynamic Tables build the "happy path" layers that glue sources together; Streams and Tasks handle CDC-heavy edges, correction jobs, and historical dimensions. Because both approaches read from the same landing tables and publish to the same marts, you can start declaratively and introduce imperative steps only where the data's history demands it.

8.3 External Tables and Iceberg Interoperability

Sometimes the right place for data is your lake, not Snowflake's internal storage. **External Tables** let you *query-in-place*—Snowflake projects Parquet/ORC/CSV/JSON from your bucket as SQL without ingesting it. **Iceberg Tables** go further: Snowflake reads **and writes** to data in your bucket using Apache Iceberg's table format, with ACID snapshots that other engines (Spark, Flink, Trino) also understand. The choice hinges on ownership and mutability: external for federated reads, Iceberg for shared, multi-engine **tables** with DML.

8.3.1 What Changes When You Pick One vs. the Other

Capability	External Table (files in place)	Iceberg Table (managed table in your lake)
Storage and layout	Your files (Parquet/ORC/CSV/JSON) organized by folders/partitions	Your files + Iceberg manifests, snapshots, metadata trees
Mutations from Snowflake	Read-only	Full DML: `INSERT/UPDATE/MERGE/DELETE`, time travel
Schema evolution	Implicit via file schema; projected at query time	Explicit via Iceberg schema; versioned with snapshots
Planning and pruning	Based on file paths and file/row-group stats	Adds manifest-level pruning; generally fewer files touched
Interoperability	Any engine that can read the raw files	Any **Iceberg-aware** engine shares the same table + snapshots
Catalog	N/A (Snowflake metadata over files)	Snowflake catalog or external (e.g., AWS Glue)
Best fit	Federation, gradual migration, "lake as source" analytics	Multi-engine **ownership**, ACID lakehouse, shared truth without copies

8.3.2 External Tables: Query the Lake and Keep Ownership Upstream

Define an external stage, then register an external table. Snowflake exposes a `VALUE` (VARIANT) column plus `METADATA$*` fields so you can project types and inspect files.

```
CREATE STAGE lake_sales
  URL = 's3://datalake/sales/'
  STORAGE_INTEGRATION = si_lake
  FILE_FORMAT = (TYPE = PARQUET);
CREATE EXTERNAL TABLE ext.sales (
    order_dt DATE AS TO_DATE(SPLIT_PART(METADATA$FILENAME, '/', 4),
    'YYYY-MM-DD')
)
```

```
  PARTITION BY (order_dt)
  WITH LOCATION = @lake_sales
  AUTO_REFRESH = TRUE
  FILE_FORMAT = (TYPE = PARQUET);
-- Project typed fields and prune by partition
SELECT
  value:order_id::STRING AS order_id,
  value:total::NUMBER    AS total,
  METADATA$FILENAME      AS src_file
FROM ext.sales
WHERE order_dt >= CURRENT_DATE - 1;
```

> *This minimal example works without explicit partition columns, but external tables over date-organized lake folders are usually defined with partition columns derived from* `METADATA$FILENAME`*. Filtering on those partition columns allows Snowflake to prune files before scanning them. When partitions are defined from expressions, they can still be refreshed automatically; by contrast, user-specified partitions are not compatible with automatic refresh.*

External Tables shine when another system lays down files (ETL jobs, streaming compaction) and Snowflake consumers need fast, zero-copy access. Performance tracks file hygiene: balanced ~200 MB Parquet files, consistent partitioning (e.g., `dt=/tenant=`), and `AUTO_REFRESH` or manual `ALTER EXTERNAL TABLE ... REFRESH` for discovery.

8.3.3 Iceberg Tables: One Table, Many Engines, and Real DML

Iceberg adds a table abstraction over your objects, with manifests that index data files and snapshots that version the table. Snowflake can be the writer, but Spark/Trino can read and mutate the same table through the shared catalog.

```
-- One-time infra: external volume + (optionally) an external catalog
-- Below: create an Iceberg table using Snowflake's catalog; data stays in
your bucket

CREATE ICEBERG TABLE lake_core.fact_orders
  EXTERNAL_VOLUME = my_ext_volume
  CATALOG = 'SNOWFLAKE'
```

```
AS
SELECT
  order_id,
  customer_id,
  total::NUMBER(18,2) AS total,
  order_dt::DATE      AS order_dt
FROM staging.orders_parquet;

-- ACID mutations, visible to other Iceberg engines
MERGE INTO lake_core.fact_orders t
USING staging.orders_delta s
ON t.order_id = s.order_id
WHEN MATCHED THEN UPDATE SET total = s.total
WHEN NOT MATCHED THEN INSERT (order_id, customer_id, total, order_dt)
VALUES (s.order_id, s.customer_id, s.total, s.order_dt);
```

Because Iceberg tracks snapshots, you get reliable **time travel** and lightweight rollbacks that span engines. Manifests also give Snowflake's planner more to prune than raw folder paths, which reduces file touches on selective queries.

8.3.4 Governance, Reliability, and Migration Notes

Concern	External Table stance	Iceberg stance
RBAC and policies	Apply in Snowflake at read time (object-level, masking/row access)	Same Snowflake policies at read/write; plus catalog-level ACLs for other engines
Freshness discovery	AUTO_REFRESH (event/poll) or explicit REFRESH	No discovery step; commits publish a new snapshot
Compaction and file shape	Done upstream (Spark/Glue jobs)	Can be handled by Iceberg maintenance (rewrite data files)
Backfills and replays	Swap or add folders, REFRESH, queries see both generations	New snapshot atomically replaces or appends; older snapshot remains queryable
Multi-engine writes	Risky (last-writer-wins on files)	Designed for it—snapshot isolation and conflict detection

8.3.5 Choosing Deliberately (And Switching Later)

- If the lake is a **source** and Snowflake is a **consumer**, start with **External Tables**. You get speed without moving data, and your upstream tools keep their current layout.
- If Snowflake and other engines must **co-own** a dataset with **ACID** guarantees, go **Iceberg**. It becomes your shared contract, and each engine participates through the same catalog.

You can migrate in place. A common path is to CTAS from External into Iceberg (preserving partitioning), validate row counts and query plans, then cut readers over. If an Iceberg table already exists, "register" it in Snowflake's catalog rather than copying data, so all engines converge on the same snapshots.

Whichever route you take, keep partition folders aligned with real filters (`order_dt`, `tenant_id`, `region`) and avoid tiny files. That way Snowflake can prune early, other engines stay efficient, and your lake behaves like a warehouse—without forfeiting openness.

8.3.6 Closing Thoughts

Permanent, transient, and temporary tables set the **safety level** for data that Snowflake stores. Dynamic Tables trade orchestration code for **declarative freshness**, while Streams and Tasks retain surgical control for CDC-heavy logic. External Tables and Iceberg extend Snowflake to your lake; the former favors federation, the latter shared ownership with ACID guarantees. With these choices in hand, you can place each dataset where it belongs and still serve it through one consistent SQL and governance layer.

PART III

Transformation and Compute

CHAPTER 9

SQL Engineering in Snowflake

Snowflake speaks ANSI SQL, but its engine leans hard on metadata, micro-partitions, and a services layer that makes safety features cheap to use. You write familiar queries, then reach for Snowflake's specifics—semi-structured paths, robust DDL, and orchestration-friendly SQL—when you need them. This chapter focuses on practical patterns: what differs from "plain ANSI," how to use windowing and DML to shape data incrementally, and how Time Travel and zero-copy cloning make development safe.

9.1 ANSI SQL with Snowflake Specifics

Snowflake aims to feel like standard SQL while extending the language where cloud warehousing needs it most: semi-structured data, reversible DDL, and governance that lives in SQL. You can write ordinary joins, CTEs, and window functions, then lean on Snowflake's additions—`VARIANT` paths, fail-safe casts, cloning, masking policies—when the workload demands it. The goal is familiarity with a few powerful pivots.

M. Hander, *Snowflake Data Warehouse Engineering*, https://doi.org/10.1007/979-8-8688-2628-3_9

9.1.1 What Stays ANSI vs. What's Distinct in Snowflake

Area	ANSI-familiar behavior	Snowflake specifics you'll actually use
Types	NUMBER, VARCHAR, DATE, TIMESTAMP	VARIANT, OBJECT, ARRAY; timestamp flavors: TIMESTAMP_NTZ/LTZ/TZ
Casting and parsing	CAST(x AS TYPE)	TRY_TO_NUMBER, TRY_TO_TIMESTAMP, ::TYPE shorthand (safe failure → NULL)
Semi-structured access	N/A in pure ANSI	JSON path with :/[ ], FLATTEN table function, INFER_SCHEMA
DDL and recovery	CREATE/ALTER/DROP	CLONE (zero-copy), UNDROP, data retention settings for Time Travel
Temp vs. persisted	Temp tables scoped to session	TEMP and TRANSIENT objects; retention and fail-safe configurable
Governance in SQL	GRANT/REVOKE	Hierarchical roles; masking and row-access policies bound to objects
Scripting	Procedural extensions vary	Snowflake Scripting (PL/pgSQL-like), JavaScript procs, Snowpark UDFs

9.1.2 Semi-structured Data with Safe Projection

Path notation lets you read JSON without leaving SQL. Safe casts keep bad records from aborting the run.

```
SELECT
  payload:order_id::STRING                       AS order_id,
  TRY_TO_NUMBER(payload:total)                   AS total,
   TRY_TO_TIMESTAMP_NTZ(payload:created_at) AS created_at
FROM bronze.events
WHERE payload:type::STRING = 'order_created';
```

Explode arrays with FLATTEN (a table function you join laterally):

```
SELECT
  e.payload:order_id::STRING AS order_id,
  i.value:sku::STRING        AS sku,
  i.value:qty::NUMBER        AS qty
FROM bronze.events e,
LATERAL FLATTEN(INPUT => e.payload:items) i;
```

9.1.3 Time Types You Can Reason About

Flavor	Meaning	Typical use
TIMESTAMP_NTZ	Timestamp without time zone	Warehouse-internal event time; most facts and dims
TIMESTAMP_TZ	Timestamp with stored offset	Audits crossing time zones, external interfaces
TIMESTAMP_LTZ	Normalized to session time zone	End-user display logic

Prefer NTZ for analytics; convert explicitly on ingest/egress.

```
SELECT TO_TIMESTAMP_TZ(created_ntz, 'UTC') AS created_utc
FROM core.orders;
```

9.1.4 Reversible DDL Baked into SQL

Cloning and retention settings move safety into the schema layer instead of external backups.

```
-- Instant branch of a production schema
CREATE SCHEMA prod_candidate CLONE prod;

-- Undo a mistake by swapping in a point-in-time clone
CREATE TABLE core.fact_orders_restore
  CLONE core.fact_orders
  AT (OFFSET => -600);  -- 10 minutes ago
ALTER TABLE core.fact_orders_restore SWAP WITH core.fact_orders;
```

Set retention where it matters:

```
ALTER DATABASE prod SET DATA_RETENTION_TIME_IN_DAYS = 7;
```

9.1.5 Session Scope, Transient Data, and Variables

Snowflake offers session variables and lightweight tables for scratch work without polluting shared schemas.

```
-- Session variable
SET as_of = TO_DATE('2025-09-01');

-- Temporary table lives for this session only
CREATE TEMP TABLE work_recent AS
SELECT * FROM core.fact_orders WHERE order_dt >= $as_of;
```

Choose object class deliberately:

Class	Lives	Fail-safe	Typical role
PERMANENT	Until dropped (with Time Travel)	Yes	Shared facts/dims, governed marts
TRANSIENT	Until dropped; short/zero retention	No	Staging and intermediates
TEMP	Session	No	Ad hoc exploration, unit tests

9.1.6 Governance As DDL, Not an Afterthought

Policies attach to objects, so rules travel with clones and replications.

```
-- Masking policy for PII
CREATE OR REPLACE MASKING POLICY mask_email AS (val STRING)
RETURNS STRING ->
  CASE WHEN CURRENT_ROLE() IN ('PII_READER','SYSADMIN') THEN val
       ELSE REGEXP_REPLACE(val, '(^.).+(@.+$)', '\\1***\\2') END;

ALTER TABLE dim_customer MODIFY COLUMN email SET MASKING POLICY mask_email;

-- Row access policy for tenancy
CREATE OR REPLACE ROW ACCESS POLICY tenant_guard AS (tenant_id STRING)
```

```
RETURNS BOOLEAN ->
  tenant_id = CURRENT_USER();  -- or look up from a mapping table

ALTER TABLE fact_events ADD ROW ACCESS POLICY tenant_guard ON (tenant_id);
```

9.1.7 A Few Idioms That Keep Code Clear

CTEs remain your friend for readable transformations; QUALIFY filters on windowed results without nested subqueries.

```
WITH ranked AS (
  SELECT *, ROW_NUMBER() OVER (PARTITION BY customer_id ORDER BY updated_at
DESC) rn
  FROM stage.customer_updates
)
SELECT *
FROM ranked
QUALIFY rn = 1;  -- keep latest per customer
```

And when you need parameters in pure SQL, stick to session variables so the plan stays transparent.

```
SET cutoff = DATEADD('day', -30, CURRENT_DATE());
SELECT COUNT(*) FROM core.fact_orders WHERE order_dt >= $cutoff;
```

Snowflake's twist on ANSI is small but decisive: safe casts let bad rows degrade gracefully; VARIANT and FLATTEN make semi-structured first-class; cloning and retention turn DDL into a safety mechanism; and policies live with the data they protect. Keep writing idiomatic SQL—and use these specifics where they reduce orchestration and risk.

9.2 Window Functions, MERGE, and Multi-table INSERT

Modern Snowflake SQL reduces multi-step ETL to a few tight statements. Window functions summarize without collapsing detail; MERGE expresses idempotent upserts; multi-table INSERT materializes several targets from one scan.

9.2.1 Windowing That Reads Clearly

Use partitions for business keys and orderings for time. Add frames when you need bounded calculations (e.g., rolling 7-day sums).

```
-- Rolling 7-day revenue and first/last timestamps per customer
SELECT
  customer_id,
  order_dt,
  SUM(amount) OVER (
    PARTITION BY customer_id
    ORDER BY order_dt
    RANGE BETWEEN INTERVAL '6 DAY' PRECEDING AND CURRENT ROW
  ) AS rev_7d,
  FIRST_VALUE(order_dt) OVER (PARTITION BY customer_id ORDER BY order_dt)
  AS first_order_dt,
  LAST_VALUE(order_dt)  OVER (PARTITION BY customer_id ORDER BY order_dt
                              ROWS BETWEEN UNBOUNDED PRECEDING AND
                              UNBOUNDED FOLLOWING) AS last_order_dt
FROM core.fact_orders;
```

Dense ranking and deduplication remain classics:

```
-- Keep the latest record per natural key (dedupe)
WITH ranked AS (
  SELECT *,
         ROW_NUMBER() OVER (PARTITION BY customer_id ORDER BY updated_at
DESC) AS rn
  FROM stage.customer_updates
)
SELECT * FROM ranked WHERE rn = 1;
```

9.2.2 MERGE for Idempotent Upserts

Incremental pipelines work best when the write logic is explicit and repeatable. In Snowflake, MERGE is the standard way to express that contract: match incoming rows to existing business keys, update what already exists, and insert what is missing. For Type-1 dimensions, this makes re-runs naturally idempotent as long as the source presents one authoritative row per key. Snowflake documents MERGE as a single statement that can apply WHEN MATCHED and WHEN NOT MATCHED actions based on the join between source and target.

```
MERGE INTO dim_customer d
USING stage.customer_deltas s
  ON d.customer_id = s.customer_id
WHEN MATCHED THEN
  UPDATE SET
    email      = COALESCE(s.email, d.email),
    status     = s.status,
    updated_at = s.event_ts
WHEN NOT MATCHED THEN
  INSERT (customer_id, email, status, created_at, updated_at)
  VALUES (s.customer_id, s.email, s.status, s.event_ts, s.event_ts);
```

This pattern keeps the latest known customer state in one place. Existing rows are updated in place, and new customers are inserted on first arrival. The COALESCE on email is deliberate: if an upstream delta omits an optional attribute, the merge preserves the previously known value instead of overwriting it with NULL.

Slowly changing dimensions of type 2 look similar at first glance, but the write semantics are different. A changed row must do two things: close the old current version and create a new current version. In Snowflake, a single MERGE cannot both update a matched row and then insert its successor version for that same match path, so SCD2 is better expressed as two DML steps rather than a one-statement extension of the Type-1 pattern.

```
-- Step 1: expire the current version when tracked attributes changed
MERGE INTO dim_customer_scd d
USING stage.customer_deltas s
  ON d.customer_id = s.customer_id
 AND d.current_flag = TRUE
WHEN MATCHED
 AND (
      NVL(d.email,  '∅') <> NVL(s.email,  '∅')
   OR NVL(d.status, '∅') <> NVL(s.status, '∅')
```

```
 ) THEN
  UPDATE SET
    current_flag = FALSE,
    valid_to     = s.event_ts;
-- Step 2: insert the new current version for new or changed customers
INSERT INTO dim_customer_scd
  (customer_id, email, status, valid_from, valid_to, current_flag)
SELECT
  s.customer_id,
  s.email,
  s.status,
  s.event_ts,
  NULL,
  TRUE
FROM stage.customer_deltas s
LEFT JOIN dim_customer_scd d
  ON d.customer_id = s.customer_id
 AND d.current_flag = TRUE
WHERE d.customer_id IS NULL
   OR NVL(d.email,  '∅') <> NVL(s.email,  '∅')
   OR NVL(d.status, '∅') <> NVL(s.status, '∅');
```

The first statement expires the currently active row by setting `current_flag` to `FALSE` and stamping `valid_to`. The second statement inserts the successor version as the new current row. The `NVL(..., '∅')` pattern makes the example null-safe; ordinary comparisons such as `d.email <> s.email` can miss changes when one side is `NULL`.

One practical caution applies to both Type-1 and Type-2 merges: make sure the source side is unique at the business-key level for the merge condition. Snowflake notes that when multiple source rows match the same target row, merge results can become nondeterministic for updates. In production pipelines, this usually means deduplicating or collapsing deltas before merging.

9.2.3 Multi-table INSERT from One Scan

Sometimes the same source feed must populate several downstream tables. In that case, a conditional multi-table insert can split one source scan into multiple writes, which reduces repeated I/O and keeps the outputs aligned to the same input snapshot. Snowflake supports conditional multi-table inserts through `INSERT ALL`.

```
-- Split bronze → silver in a single read
INSERT ALL
  WHEN type = 'order_created' THEN
    INTO silver.orders (order_id, customer_id, total, created_at)
    VALUES (
      payload:order_id::STRING,
      payload:customer_id::STRING,
      TRY_TO_NUMBER(payload:total),
      TRY_TO_TIMESTAMP_NTZ(payload:created_at)
    )
  WHEN type = 'order_item' THEN
    INTO silver.order_items (order_id, sku, qty)
    VALUES (
      payload:order_id::STRING,
      payload:item:sku::STRING,
      payload:item:qty::NUMBER
    )
SELECT
  payload,
  payload:type::STRING AS type
FROM bronze.events
WHERE ingest_dt = CURRENT_DATE();
```

This style is useful when one bronze event stream fans out into multiple silver targets. Each qualifying input row is evaluated once, and the `WHEN` clauses determine which target table receives it. Use the pattern carefully: the conditions should be mutually exclusive unless overlapping inserts are intentional. Snowflake's multi-table insert syntax supports exactly this "single read, multiple writes" approach.

9.3 Time Travel and Zero-Copy Cloning for Safe Dev

Snowflake's safety net has two strands. **Time Travel** lets you read or restore objects as they were at a specific point, and **zero-copy clones** give you instant branches of databases, schemas, or tables that only consume storage as they diverge. Together, they make experiments boring, rollbacks quick, and promotions atomic.

Time Travel is ideal when you need to *look back* or *undo* within the same object line. Cloning is better when you want a *parallel world*—a place to run migrations, backfills, and benchmarks without touching production. Promotions become metadata operations rather than data moves, so you can cut over and revert with the same confidence.

9.3.1 When Each Makes the Job Easier

Need	Prefer Time Travel	Prefer Clone
Inspect yesterday's facts without changing current tables	`SELECT ... AT (TIMESTAMP	OFFSET)`
Recover from an accidental `DELETE`/`MERGE` on a table	Point-in-time `CLONE` then `SWAP`	Same mechanism; clone is how you materialize the point-in-time state
Build a release candidate with production-like data	-	`CREATE DATABASE ... CLONE prod` for an isolated branch
Run high-risk backfills and measure impact	Possible but awkward	Work in a cloned schema, compare, then promote
Create per-developer sandboxes	-	Clone databases/schemas cheaply; discard when done

9.3.2 Reading and Restoring with Time Travel

```
-- Query a table as it existed at a precise time (UTC or TZ-aware)
SELECT *
FROM core.fact_orders
  AT (TIMESTAMP => TO_TIMESTAMP_TZ('2025-09-01 12:00:00 +00:00'))
WHERE order_dt >= '2025-08-01';
```

```
-- Restore damage using a point-in-time clone and an atomic swap
CREATE TABLE core.fact_orders_restore
  CLONE core.fact_orders
  AT (OFFSET => -600);  -- 10 minutes ago

ALTER TABLE core.fact_orders_restore
  SWAP WITH core.fact_orders;  -- no data copy, instant cutover
```

Time Travel is also a lens for audits and reproducibility. You can re-run a report against "last Friday's warehouse" without maintaining snapshots yourself or validate a fix by comparing present vs. past result sets.

9.3.3 Branching and Promoting with Zero-Copy Clones

```
-- Whole-environment branch for a release
CREATE DATABASE prod_candidate_2025_10_01 CLONE prod;

-- Run migrations, backfills, and tests safely
-- ...

-- Promote a finished mart atomically
ALTER SCHEMA prod.mart_sales
  SWAP WITH prod_candidate_2025_10_01.mart_sales;
```

Clones are cheap up front because Snowflake reuses the same micro-partitions. Storage accrues only for partitions that change after the clone. That makes short-lived feature branches and hotfix sandboxes practical, even for multi-terabyte databases.

9.3.4 Settings That Shape Safety and Cost

Setting	Where you set it	Why it matters
DATA_RETENTION_TIME_IN_DAYS	Table, schema, or database	Controls the Time Travel window. Longer means more rewind room and more storage retained.
Object class (Permanent/Transient/Temporary)	At create time	Only **Permanent** objects have fail-safe. Transient/Temp are cheaper but offer no last-ditch recovery.
Clone scope (DB/Schema/Table)	CREATE … CLONE … target	Cloning higher in the tree captures more dependencies (UDFs, views, roles' grants on objects).

Typical posture is a 7–14-day retention on shared databases, short retention or transient class for staging layers, and explicit retention on critical marts that need longer rewind horizons.

```
-- Safer defaults on shared data
ALTER DATABASE prod SET DATA_RETENTION_TIME_IN_DAYS = 7;

-- Transient staging with minimal retention
CREATE TRANSIENT SCHEMA stage;
ALTER SCHEMA stage SET DATA_RETENTION_TIME_IN_DAYS = 1;
```

9.3.5 A Simple, Reliable Release Rhythm

1. **Clone** production into a candidate database.
2. **Apply** migrations, backfills, and validation queries in the clone.
3. **Promote** with SWAP for atomic cutover.
4. **Revert** with another SWAP if needed; the old state remains intact.

Because these steps are metadata-only, they complete quickly and don't block readers.

9.3.6 Troubles You Can Avoid

Symptom	Likely cause	Fix in practice
Time Travel query fails	Retention window too short	Increase `DATA_RETENTION_TIME_IN_DAYS` on the relevant scope before you need it; for now, restore from a more recent point if possible.
Clone grows unexpectedly	Heavy rewrites post-clone	Limit wide `UPDATE`/`DELETE` in clones; prefer CTAS into new tables, then `SWAP`.
Cutover breaks views	Dependencies not cloned together	Clone at the **schema** or **database** level so views/UDFs stay aligned; validate with a smoke test before `SWAP`.

The habit to cultivate is simple: treat Time Travel as your *undo button* and clones as your *feature branches*. Query the past to understand, clone the present to change, and swap with confidence when you go live.

9.3.7 Closing Notes

Snowflake's SQL feels familiar on purpose, but it rewards you for using its native moves. Keep semi-structured data in `VARIANT` and project cleanly with safe casts. Reach for window functions when you need running context without losing detail. Encode your write rules with `MERGE`, and bundle related outputs using multi-table `INSERT` to make scans count. Above all, treat Time Travel and cloning as part of your daily muscle memory: branch freely, promote atomically, and recover without drama.

CHAPTER 10

Orchestration and ELT

In Snowflake, orchestration is about moving **small, safe deltas** through clear boundaries. Streams expose change, Tasks move those changes, and external orchestrators (Airflow, Prefect, dbt) coordinate when multiple systems or repos are involved. The aim is a graph that is observable, idempotent, and easy to rewind when business logic evolves.

10.1 Streams and Tasks Design Patterns

Streams provide a table-like view of "what changed since last consumption." Tasks execute SQL (or Snowpark) on a schedule or after dependencies. Together they turn ingestion into **ELT** with simple contracts: land → normalize → merge → publish.

10.1.1 Common Shapes and Where They Fit

Pattern	Shape in Snowflake	When it fits	Notes that matter
Landing → Normalization	Raw table → **view** that standardizes payloads	Semi-structured, multi-source CDC	Keep envelope columns (`tenant_id`, `event_ts`) for pruning
Delta promotion	**Stream** on table/view → MERGE in a Task	Type-1 dims, upsert facts	Streams on views help hide tool-specific payloads
Historical dims	Stream → SCD-2 **MERGE**	Regulatory history, audits	Effective dating; prefer narrow repair windows
Append + repair	INSERT now → periodic **MERGE** over last N days	High-volume facts, late corrections	Keeps latency low and correctness eventual
DAG of transforms	Chains of **Tasks** or **Dynamic Tables**	Relational "happy path" pipelines	Use `AFTER` for dependency; reserve Tasks for CDC/side effects

M. Hander, *Snowflake Data Warehouse Engineering*, https://doi.org/10.1007/979-8-8688-2628-3_10

Minimal normalization view → stream:

```
CREATE OR REPLACE VIEW stage.orders_norm AS
SELECT
  payload:order_id::STRING          AS order_id,
  payload:customer_id::STRING       AS customer_id,
  TRY_TO_NUMBER(payload:total)      AS total,
  TRY_TO_TIMESTAMP_NTZ(payload:ts)  AS event_ts,
  payload:op::STRING                AS op   -- 'I','U','D'
FROM bronze.events
WHERE payload:type::STRING = 'order';

CREATE OR REPLACE STREAM str_orders ON VIEW stage.orders_norm;
```

Minimal normalization view → stream: A stream can sit on top of a normalization view when the view remains stream-compatible. This is often a clean way to hide tool-specific payload shapes and expose a small CDC contract to downstream tasks. If the view later becomes too complex, move the stream to the base table, and apply the normalization logic in the consuming task instead. Change tracking must be enabled on the view and its underlying tables.

```
-- Fallback when the view is not a good stream source:

CREATE OR REPLACE STREAM str_orders ON TABLE bronze.events;

-- Apply the normalization/filtering logic in the task query
instead.
```

Promotion Task (idempotent upsert):

```
CREATE OR REPLACE TASK t_orders_upsert
  WAREHOUSE = wh_elt
  SCHEDULE  = '5 MINUTE'
AS
MERGE INTO mart.orders t
USING (
  SELECT order_id, customer_id, total, event_ts
  FROM str_orders
) s
```

```
ON t.order_id = s.order_id
WHEN MATCHED THEN UPDATE SET total = s.total, updated_at = s.event_ts
WHEN NOT MATCHED THEN INSERT (order_id, customer_id, total, created_at,
updated_at)
VALUES (s.order_id, s.customer_id, s.total, s.event_ts, s.event_ts);
```

SCD-2 where history matters:

```
CREATE OR REPLACE TASK t_dim_customer_scd2
  WAREHOUSE = wh_elt
  SCHEDULE  = '10 MINUTE'
AS
BEGIN
  -- Step 1: expire the current version when tracked attributes changed
  MERGE INTO dim_customer_scd d
  USING (
    SELECT DISTINCT customer_id, email, status, event_ts
    FROM str_customers
  ) s
  ON d.customer_id = s.customer_id
 AND d.current_flag = TRUE
  WHEN MATCHED
   AND (
        NVL(d.email,  '∅') <> NVL(s.email,  '∅')
     OR NVL(d.status, '∅') <> NVL(s.status, '∅')
   ) THEN
    UPDATE SET
      current_flag = FALSE,
      valid_to     = s.event_ts;
  -- Step 2: insert the new current version for new or changed customers
  INSERT INTO dim_customer_scd
    (customer_id, email, status, valid_from, valid_to, current_flag)
  SELECT
    s.customer_id,
    s.email,
```

```
    s.status,
    s.event_ts,
    NULL,
    TRUE
  FROM (
    SELECT DISTINCT customer_id, email, status, event_ts
    FROM str_customers
  ) s
  LEFT JOIN dim_customer_scd d
    ON d.customer_id = s.customer_id
   AND d.current_flag = TRUE
  WHERE d.customer_id IS NULL
     OR NVL(d.email,  '∅') <> NVL(s.email,  '∅')
     OR NVL(d.status, '∅') <> NVL(s.status, '∅');
END;
ALTER TASK t_dim_customer_scd2 RESUME;
```

In Snowflake, SCD2 is not just a “MERGE with one more clause.” A changed row must both close the existing version and create a successor version, which is why production SCD2 pipelines are typically expressed as two coordinated DML steps.

10.2 Task Graphs, Scheduling, and Retries

A Snowflake task graph is a set of small SQL units that run on a cadence or after one another. The job of orchestration is to keep those units **re-runnable**, keep runs **short and frequent**, and make failures **observable**. Streams expose deltas; Tasks promote them. When a task can run twice without harming the result, scheduling and retries become straightforward.

10.2.1 Designing Reliable Task Graphs

Concern	Practical stance	Why it works
Granularity	Prefer short units that finish in minutes	Small blast radius; faster recovery and parallelism
Ordering	Use AFTER for strict dependencies; avoid deep chains	Simpler reasoning; fewer cascading failures
Cadence	Run often (e.g., every five minutes) with small deltas	Lower latency and quicker retries
Idempotency	Express writes as MERGE/CTAS; read from Streams	Safe to retry; duplicates collapse on keys
Concurrency	Dedicated small warehouse; auto-suspend quick	Warm enough to start fast; cheap between runs
Repair	Add a periodic "windowed repair" task	Corrects late/out-of-order data without heavy backfills
Observability	Persist metrics and query TASK_HISTORY	You can prove freshness and see where time went

10.2.2 A Small Graph: Normalize ➤ Upsert Facts ➤ Maintain Dims

Root task on a schedule:

```
CREATE OR REPLACE TASK t_normalize
  WAREHOUSE = wh_elt
  SCHEDULE  = '5 MINUTE'
  USER_TASK_TIMEOUT_MS = 1200000
AS
CREATE OR REPLACE TRANSIENT TABLE work.orders_norm AS
SELECT *
FROM stage.orders_norm
WHERE event_ts >= DATEADD('minute', -10, CURRENT_TIMESTAMP());
```

The root task materializes a normalized working set that downstream tasks can consume. Because task runs do not share a common session context, a temporary table is the wrong handoff mechanism here: temporary tables are session-scoped and disappear with the session that created them. For staging between tasks, use a persistent object such as a transient table instead. Transient tables remain visible across task runs and sessions while avoiding the full permanence of standard tables.

Parallel children after normalization:

```
-- Upsert fact table from the normalized batch
CREATE OR REPLACE TASK t_fact_orders
  WAREHOUSE = wh_elt
  AFTER t_normalize
AS
MERGE INTO mart.fact_orders t
USING work.orders_norm s
ON t.order_id = s.order_id
WHEN MATCHED THEN UPDATE SET amount = s.total, order_ts = s.event_ts
WHEN NOT MATCHED THEN INSERT (order_id, customer_id, amount, order_ts)
VALUES (s.order_id, s.customer_id, s.total, s.event_ts);

-- Maintain Type-2 dim with effective dating
CREATE OR REPLACE TASK t_dim_customer_scd2
  WAREHOUSE = wh_elt
  AFTER t_normalize
AS
MERGE INTO dim_customer_scd d
USING (SELECT DISTINCT customer_id, email, status, event_ts FROM work.
orders_norm) s
ON d.customer_id = s.customer_id AND d.current_flag = TRUE
WHEN MATCHED AND (d.email <> s.email OR d.status <> s.status) THEN
  UPDATE SET current_flag = FALSE, valid_to = s.event_ts
WHEN NOT MATCHED THEN
  INSERT (customer_id, email, status, valid_from, valid_to, current_flag)
  VALUES (s.customer_id, s.email, s.status, s.event_ts, NULL, TRUE);
```

A rolling repair that revisits the last two days (handles late arrivals without heavy reprocessing):

```
CREATE OR REPLACE TASK t_fact_orders_repair
  WAREHOUSE = wh_elt
  SCHEDULE  = '1 HOUR'
AS
MERGE INTO mart.fact_orders f
USING (
  SELECT * FROM stage.orders_norm
  WHERE event_ts >= DATEADD('day', -2, CURRENT_TIMESTAMP())
) s
ON f.order_id = s.order_id
WHEN MATCHED THEN UPDATE SET amount = s.total, order_ts = s.event_ts
WHEN NOT MATCHED THEN INSERT (order_id, customer_id, amount, order_ts)
VALUES (s.order_id, s.customer_id, s.total, s.event_ts);
```

Tip Keep **all** write paths keyed (`order_id`, plus a source transaction id if available) so retries and replays converge.

10.2.3 Scheduling and Ad Hoc Kicks

Need	Statement	Notes
Start a tree now	`ALTER TASK t_normalize EXECUTE;`	Triggers children via `AFTER` links
Pause a subtree	`ALTER TASK t_normalize SUSPEND;`	Children won't run; resume with `RESUME`
Change cadence	`ALTER TASK t_normalize SET SCHEDULE = '2 MINUTE';`	Prefer frequent, small runs over big batches

10.2.4 Retries That Don't Hurt

Snowflake will re-run a task when you or an external scheduler kicks it again. Make the SQL safe to run twice:

- **Reads**: Use Streams or time windows; avoid "consume and delete" patterns.
- **Writes**: MERGE or INSERT ... ON (via MERGE semantics), never blind INSERT.
- **Temp work**: Write to TEMP/TRANSIENT objects named per run (e.g., work.orders_norm) so partial results don't bleed across runs.

If a task hard-fails (timeout, permissions), fix the cause and kick it again; the idempotent merge ensures convergence.

10.2.5 Monitoring That Answers "Are We Fresh?"

Store a tiny metrics table and rely on Task History for run state:

```
-- Emit run metrics at the end of a task
INSERT INTO ops.run_metrics(feed, max_event_ts, loaded_rows, ran_at)
SELECT 'orders', MAX(event_ts), COUNT(*), CURRENT_TIMESTAMP()
FROM work.orders_norm;

-- Task runs
SELECT name, state, scheduled_time, query_id, return_value
FROM TABLE(INFORMATION_SCHEMA.TASK_HISTORY(
  SCHEDULED_TIME_RANGE_START => DATEADD('hour', -6, CURRENT_TIMESTAMP())
))
ORDER BY scheduled_time DESC;

-- Freshness SLO
CREATE OR REPLACE VIEW ops.freshness AS
SELECT
  'orders' AS feed,
  MAX(event_ts) AS newest_event,
```

```
  TIMESTAMPDIFF('minute', MAX(event_ts), CURRENT_TIMESTAMP()) AS
minutes_lag
FROM stage.orders_norm;
```

10.2.6 Failure Modes You Can Neutralize

Symptom	Likely root cause	Fix that lasts
Duplicate rows after a retry	Non-idempotent `INSERT`	Rewrite as `MERGE` keyed on business id (and source tx id)
Long queues on the half hour	Overlapping cadences; one tiny warehouse	Stagger schedules; give tasks a dedicated small warehouse or multi-cluster
Tasks flip-flop between success/fail	Timeouts or cold starts	Set `USER_TASK_TIMEOUT_MS` sensibly; keep the ELT warehouse small but warm (short auto-suspend)
Freshness drifts silently	No SLO surfaced	Publish a freshness view and alert when lag exceeds threshold

The pattern is simple: make every task safe to run again, schedule short cycles, and keep a repair window. With that in place, retries are just re-invocations, and your graph remains calm even when upstreams wobble.

10.3 Airflow/Prefect/dbt Orchestration Topologies

External orchestrators matter when pipelines span **multiple repos or systems**, or when promotion and approvals must live outside the warehouse. The guiding principle is simple: let Snowflake do the heavy lifting (Streams, Tasks, Dynamic Tables), and use Airflow, Prefect, or dbt to **coordinate, validate, and promote**. You'll move faster when the orchestrator triggers *units of work the warehouse already knows how to run* rather than re-implementing ELT logic in Python.

10.3.1 Where Each Tool Naturally Leads

Topology	What the tool does best	Snowflake's role	When this sings
dbt-native (dbt Cloud/ Core)	Compiles SQL graphs; incremental materializations; tests and docs	Execute compiled SQL; manage warehouses, permissions, and data	Pure SQL transforms; analysts/engineers share one graph and lineage
Airflow-driven	Cross-system DAGs, SLAs, approvals, sensors	Run Tasks/Dynamic Tables; expose Streams as stable inputs	Enterprise workflows with many external steps (files, ML, APIs)
Prefect-driven	Pythonic flows, state handling, retries with lightweight infra	Same: Snowflake runs the data work	Teams preferring Python ergonomics and cloud-hosted control plane
Hybrid (dbt inside Airflow/ Prefect)	Orchestrator schedules dbt jobs plus Snowflake SQL	Warehouse remains the compute plane	One "pane of glass" for ops; mixed ecosystems

10.3.2 Contract Between Orchestrator and Warehouse

Concern	Good contract	Anti-pattern
Unit of work	"Resume/execute this Task tree" or "run this dbt job/model selection"	Python loops issuing many ad hoc queries
Idempotency	Warehouse statements are `MERGE`/CTAS and safe to retry	Orchestrator dedupes in memory only
State and lineage	Lives with SQL (Tasks, dbt manifest)	Hidden inside Python operators

10.3.3 dbt-native: The Warehouse As the Only Engine

dbt compiles your models into a DAG and runs them in Snowflake; tests and docs travel with the graph. Incremental materializations keep runs cheap; selectors restrict scope for hotfixes and backfills.

```
-- models/silver/orders.sql
{{ config(materialized='incremental', unique_key='order_id') }}

SELECT
  order_id,
  customer_id,
  total,
  order_ts
FROM {{ ref('stage_orders_norm') }}
{% if is_incremental() %}
WHERE order_ts > (SELECT COALESCE(MAX(order_ts), '1900-01-01') FROM {{
this }})
{% endif %}
```

Promotion stays outside the graph (e.g., `ALTER SCHEMA … SWAP`) so cutovers are atomic and reversible.

10.3.4 Airflow-Driven: Schedule, Gate, and Watch Snowflake Do the Work

Use Airflow to **trigger** Snowflake Tasks (or dbt jobs), gate on upstream signals, and record SLAs. Let Streams/Tasks perform the ELT.

```
from airflow import DAG
from airflow.providers.snowflake.operators.snowflake import
SnowflakeOperator
from datetime import datetime

with DAG("elt_orders", start_date=datetime(2025, 9, 1), schedule="@hourly",
catchup=False) as dag:
    resume = SnowflakeOperator(
        task_id="resume_task_tree",
```

```
        sql="ALTER TASK t_normalize RESUME; ALTER TASK t_orders RESUME;
        ALTER TASK t_items RESUME;"
    )
    run_now = SnowflakeOperator(
        task_id="kick_tasks",
        sql="ALTER TASK t_normalize EXECUTE;"
    )
    pause = SnowflakeOperator(
        task_id="pause_task_tree",
        trigger_rule="all_done",
        sql="ALTER TASK t_normalize SUSPEND; ALTER TASK t_orders SUSPEND;
        ALTER TASK t_items SUSPEND;"
    )
    resume >> run_now >> wait_for_completion >> pause
```

The task graph must not be suspended immediately after EXECUTE TASK, because child tasks execute asynchronously after the root task starts the graph run. In orchestration tools such as Airflow, insert a wait step that polls Snowflake until the graph is no longer running, for example, via CURRENT_TASK_GRAPHS or task history, and only then suspend the graph if temporary enablement is required.

Airflow sensors (S3/GCS/HTTP) work well to **gate** the start of a Task tree, while Task History in Snowflake remains the system of record for what ran.

10.3.5 Prefect-Driven: Pythonic Control and SQL at the Core

Prefect's flows keep retries and state expressive in code, while the SQL still runs in Snowflake. This is a good fit for teams that prefer minimal infra and Python ergonomics.

```
from prefect import flow, task
from snowflake.connector import connect

@task(retries=3, retry_delay_seconds=30)
def run_sql(sql):
    with connect(...) as con:
        con.cursor().execute(sql)
```

```
@flow
def elt_flow():
    run_sql.submit("ALTER TASK t_normalize EXECUTE IMMEDIATE TASK;")
    run_sql.submit("ALTER TASK t_orders EXECUTE IMMEDIATE TASK;")
    run_sql.submit("ALTER TASK t_items EXECUTE IMMEDIATE TASK;")

if __name__ == "__main__":
    elt_flow()
```

10.3.6 Environment Promotion and Secrets: Keep Them Boring

Topic	Practical posture	Why it works
Dev/Test/Prod	Separate Snowflake **accounts**; orchestrator targets env via connection blocks	Real isolation; identical code, different secrets
Cutover	Use SWAP on schemas/tables after validation	Atomic, reversible promotions
Secrets	Store creds in orchestrator's secret store; Snowflake uses **Storage Integrations** for stages	Key rotation outside SQL; least-privilege by role
Observability	dbt artifacts + Snowflake `TASK_HISTORY` + orchestrator logs	One place to debug SQL, one to debug scheduling

10.3.7 Failure Modes Worth Planning For

Symptom	Likely root	Fix that pays off
Duplicate rows after retries	Non-idempotent writes	Encode writes as `MERGE` keyed by business id (and source tx id)
Long queues at the top of the hour	Overlapping schedules; single small warehouse	Stagger starts; dedicate a tiny always-on ELT warehouse or multi-cluster
Silent staleness in dashboards	No freshness SLO wired	Publish a "freshness" view (max `event_ts` vs. now) and alert in the orchestrator
Flaky file gating	Tiny-file storms or event lag	Compact upstream or move hot paths to Snowpipe Streaming

10.3.8 Putting It Together

A resilient topology looks like this: dbt (or SQL) defines the graph of transformations; Streams and Tasks perform **change-aware** moves inside Snowflake; Airflow or Prefect coordinates cross-system steps and promotions; secrets and environments are cleanly separated; freshness is measured in SQL and surfaced in the orchestrator. Keep the orchestrator **thin** and the warehouse **authoritative**, and you'll get fast cycles, low toil, and debuggable runs.

10.3.9 Closing Notes

Let Snowflake handle **change detection** and **writes**; use Streams and Tasks for dependable, idempotent movement. Keep tasks small and frequent, add a repair window for correctness, and expose freshness as a first-class metric. Bring in Airflow, Prefect, or dbt to coordinate across systems and repos, but resist the urge to re-implement the warehouse inside the orchestrator. When each tool stays in its lane, ELT is fast, observable, and easy to fix.

CHAPTER 11

Snowpark and UDF/UDAF/UDTF

Snowpark brings a **dataframe API** to Snowflake so you can write transforms in Python, Scala, or Java that compile down to SQL and run **inside** the warehouse. When logic truly can't be expressed as relational operations, you reach for user-defined code: UDFs (row-wise), UDAFs (aggregations), and UDTFs (table-returning). The guiding idea is simple: **push work to the engine** and keep data in Snowflake. Use Snowpark for orchestration-friendly pipelines; fall back to UDF-family only for the parts SQL can't express cleanly.

11.1 Snowpark for Python/Scala/Java

Snowpark programs look like Spark-style dataframes but execute as optimized SQL in Snowflake. You write in your preferred language, create a `Session`, and build a plan; Snowflake handles query generation, security, and scaling.

M. Hander, *Snowflake Data Warehouse Engineering*, https://doi.org/10.1007/979-8-8688-2628-3_11

11.1.1 What Changes by Language

Dimension	Python	Scala	Java
Ergonomics	Fast iteration in notebooks/CI; rich Anaconda package support	Strong typing; best for big code bases on the JVM	Industrial packaging; widest enterprise tooling
Dependency model	Anaconda (Snowflake channel) + `packages=` in session; external access via integrations	JAR on a stage; `--packages` via build (SBT/Maven)	JAR on a stage; Maven/Gradle
Cold start feel	Short for DF plans; UDF cold start tied to runtime	Similar	Similar
Ideal use	Data prep, ML feature logic near data, glue around SQL	Shared libraries, strict schemas, UDAFs	Platform services, governed libraries

Regardless of language, **data stays in Snowflake**; code runs in a secure sandbox with optional, governed egress (External Access Integration + Network Rules) when you truly must call out.

11.1.2 Anatomy: Connect, Build a Plan, and Execute

Python (Snowpark)

```
from snowflake.snowpark import Session
from snowflake.snowpark.functions import col, try_to_double, sum as ssum

session = Session.builder.configs({
    "account": "<acct>",
    "user": "<user>",
    "password": "<pwd>",
    "role": "TRANSFORMER",
    "warehouse": "WH_ELT",
    "database": "CORE",
    "schema": "PUBLIC"
}).create()
```

```
orders = session.table("CORE.FACT_ORDERS")

silver = (
    orders
    .select("ORDER_ID","CUSTOMER_ID","ORDER_DT","AMOUNT")
    .with_column("AMOUNT_NUM", try_to_double(col("AMOUNT")))
    .filter(col("ORDER_DT") >= "2025-09-01")
)

by_cust = silver.group_by("CUSTOMER_ID").agg(ssum(col("AMOUNT_NUM")).
alias("REV"))
by_cust.write.save_as_table(
    "SILVER.REV_BY_CUSTOMER",
    mode="overwrite"
) # full replacement each run; simple for examples, heavier at scale
```

This example uses overwrite to keep the write path simple and deterministic: each run rebuilds the target from the current DataFrame result. That is easy to reason about, but it can become expensive as tables grow. When the target is append-only, append is often sufficient. When the target must be reconciled incrementally, use a merge against the destination table instead of rewriting it wholesale.

Scala (Snowpark)

```
import com.snowflake.snowpark._
import com.snowflake.snowpark.functions._

val session = Session.builder.configs(Map(
  "URL" -> "<acct>.snowflakecomputing.com",
  "USER" -> "user",
  "PASSWORD" -> "pwd",
  "ROLE" -> "TRANSFORMER",
  "WAREHOUSE" -> "WH_ELT",
  "DB" -> "CORE",
  "SCHEMA" -> "PUBLIC"
)).create
```

```
val orders = session.table("CORE.FACT_ORDERS")
val byCust = orders
  .filter(col("ORDER_DT") >= lit("2025-09-01"))
  .groupBy(col("CUSTOMER_ID"))
  .agg(sum(col("AMOUNT")).as("REV"))

byCust.write.saveAsTable("SILVER.REV_BY_CUSTOMER", SaveMode.Overwrite)
```

The dataframe code above compiles to SQL and benefits from pruning, caching, and warehouse sizing—no row shipping, no client-side loops.

11.1.3 When to Stay in DF/SQL vs. Drop into UDFs

Situation	Prefer DF/SQL	Reach for UDF/UDAF/UDTF
Filtering, joins, windowed calcs	✅ Native DF/SQL	-
Complex parsing, fuzzy matching, custom tokenization	⚠ Try SQL first; if awkward...	✅ UDF/UDTF with a vetted library
Aggregations and rollups	✅ SQL/DF	✅ UDAF only for custom reduce/merge logic
Explode/reshape semi-structured	✅ FLATTEN, DF explode patterns	✅ UDTF if multi-row emission is cleaner in code

11.1.4 Python UDF, UDAF, and UDTF in Practice

Row-Wise UDF (Normalize String)

For a **demo-only** version:

```
from snowflake.snowpark.functions import udf
from snowflake.snowpark.types import StringType
@udf(
    name="norm_email",
    replace=True,
    is_permanent=False,
    return_type=StringType()
)
```

```
def norm_email(s: str) -> str:
    if s is None:
        return None
    return s.strip().lower()
session.sql("""
  SELECT norm_email(email) AS email_norm
  FROM DIM_CUSTOMER
""").collect()
```

For a **production-oriented** version:

```
from snowflake.snowpark.functions import udf
from snowflake.snowpark.types import StringType
@udf(
    name="norm_email",
    replace=True,
    is_permanent=True,
    stage_location="@code_stage",
    return_type=StringType(),
    packages=[]
)
def norm_email(s: str) -> str:
    if s is None:
        return None
    return s.strip().lower()
```

UDAF (Median with t-Digest as Illustration—Simplified Sketch)

```
from snowflake.snowpark.types import FloatType
from snowflake.snowpark.functions import udaf

class TDigestUDAF:
    def __init__(self): self.values = []
    def accumulate(self, x: float):
        if x is not None: self.values.append(float(x))
    def merge(self, other):
        self.values.extend(other.values)
    def finish(self) -> float:
```

```
        vs = sorted(self.values)
        if not vs: return None
        mid = len(vs)//2
        return (vs[mid] if len(vs)%2==1 else (vs[mid-1]+vs[mid])/2.0)

median_udaf = udaf(TDigestUDAF, return_type=FloatType(), name="median_
udaf", replace=True)

session.sql("SELECT CUSTOMER_ID, median_udaf(AMOUNT) AS med FROM FACT_
ORDERS GROUP BY CUSTOMER_ID").collect()
```

UDTF (Emit Multiple Rows per Input)

```
from typing import Iterable, Tuple
from snowflake.snowpark.types import StructType, StructField, StringType,
IntegerType
from snowflake.snowpark.functions import udtf

@udtf(
  output_schema=StructType([StructField("TOKEN", StringType()),
  StructField("LEN", IntegerType())]),
  name="split_tokens",
  replace=True,
  is_permanent=False
)
class SplitTokens:
    def process(self, text: str) -> Iterable[Tuple[str, int]]:
        if not text: return []
        for t in text.split():
            yield (t, len(t))

session.sql("SELECT * FROM TABLE(split_tokens('a quick snowflake'))").
collect()
```

Secure Egress (Only If Absolutely Needed)

Snowpark code runs in a sandbox without an outbound network by default. If you must call an external API (e.g., reference data), request **External Access Integration** + **Network Rule** and keep calls minimal and cached in tables.

11.1.5 Packaging and Dependencies Without Drama

Topic	Python	Scala/Java
Third-party libraries	Prefer Anaconda packages from Snowflake's channel; for others, use **custom wheels** on a stage and `packages=['<pkg>']`	Build a **fat JAR** (shaded) and stage it; reference in session `imports`
Versioning	Pin in code; store package manifest in repo; test in lower envs	Pin in Maven/Gradle; CI builds immutable artifacts
Deploy target	Stage (internal) + `IMPORTS`/`PACKAGES` on session/SP/UDF	Stage JAR + `IMPORTS` on session/SP/UDF

11.1.6 Practical Guardrails

Concern	Guidance	Why it holds up
Row-by-row logic	Avoid for bulk transforms; prefer DF/SQL	Vectorized plans prune/parallelize; UDFs are per-row
`collect()` misuse	Don't pull big frames client-side	Keep data in Snowflake; write to tables/views
Caching	Rely on Snowflake caches; keep warehouses warm for bursts	You don't manage executors; the engine does
Testing	Unit-test UDF handler logic as plain Python; integration-test in a dev account	Catches runtime/env surprises early

```
# Keep the core logic as a normal Python function
def norm_email_logic(s: str) -> str:
    if s is None:
        return None
    return s.strip().lower()
# Unit test locally without Snowflake
def test_norm_email_logic():
    assert norm_email_logic("  User@Example.COM  ") == "user@example.com"
    assert norm_email_logic(None) is None
```

```
# Register the same logic as a Snowpark UDF
from snowflake.snowpark.functions import udf
from snowflake.snowpark.types import StringType
@udf(name="norm_email", replace=True, is_permanent=False, return_
type=StringType())
def norm_email(s: str) -> str:
    return norm_email_logic(s)
```

For pure handler logic, local unit tests are fast and sufficient. For Snowpark DataFrame transformations, Snowpark's local testing framework can emulate DataFrame operations without a live Snowflake connection; then a dev-account integration test can validate packaging, permissions, and runtime behavior.

11.1.7 UDF, UDAF, and UDTF: Picking the Right Shape

Kind	Contract	Great for	Keep in mind
UDF	Input row → single value	Cleansing, small transforms, custom math	Inline with SELECTs; avoid heavy I/O
UDAF	Many rows → one value	Custom aggregates (percentiles, sketches)	Define `accumulate`/`merge`/`finish` carefully
UDTF	One row → many rows	Tokenization, structured parsing, mini-explodes	Compose with `TABLE()` and lateral joins

When in doubt, try the relational path first. If it turns into contortions, a small, tested UDF/UDTF often clarifies intent while keeping execution inside Snowflake.

11.2 UDFs, UDAFs, and Stored Procedures (JavaScript vs. Snowpark)

Snowflake gives you two pragmatic paths for custom logic. **Snowpark** lets you author UDFs/UDAFs/UDTFs and procedures in Python/Scala/Java that run inside secure sandboxes and compile dataframe plans down to SQL. **JavaScript stored procedures** run on Snowflake's native JavaScript runtime and are great for orchestration, control flow, and utility work that isn't expressible as a single query. Both live *in* the warehouse, close to the data, and can be governed and versioned like any other object.

11.2.1 Picking the Right Mechanism

Need	UDF (row-wise)	UDAF (aggregate)	UDTF (table-returning)	Stored procedure: JavaScript	Stored procedure: Snowpark
Per-row transform or validation	✅ Fast, inline with SELECT	-	-	-	-
Custom aggregate (median, sketches)	-	✅	-	-	-
Emit multiple rows per input (tokenize, explode)	-	-	✅	-	-
Multi-step control flow, DDL/DML orchestration	-	-	-	✅ Excellent	✅ Excellent
Heavy numeric/text libs, ML feature code	⚠ (small logic ok)	✅	✅	-	✅ Strong (Python/Java deps)
Simplicity, no dependency management	✅	✅	✅	✅ Strong	⚠ Manage packages/JARs

Rule of thumb: keep data-parallel math and parsing in UDF/UDAF/UDTF; use procedures for *process* (loops, branching, retries, DDL). Reach for Snowpark procedures when you need rich language ecosystems; stick to JavaScript when you want tiny, portable controllers with zero external deps.

11.2.2 Authoring and Deployment Model in Practice

Dimension	JavaScript procedure	Snowpark (Python/Scala/Java) UDF/Proc
Runtime	Built-in JS engine	Language runtime in a secure sandbox
Dependencies	None (ships with runtime)	Packages/JARs staged or Anaconda channel
Cold start	Very fast	Short for DF plans; longer for first load of runtime/deps
External calls	No egress by default (same for Snowpark)	Both require **External Access Integration** + Network Rules if calling out
Best use	Control flow, DDL/DML orchestration, lightweight utilities	Data science-ish transforms, complex parsing, custom aggregates

11.2.3 Compact Examples

Python UDF (Snowpark): Permanent Deployment for Reusable Cleansing

A temporary UDF is useful in notebooks and ad hoc sessions. For reusable warehouse logic, register the function as a permanent UDF so it remains available across sessions and can be called directly from SQL.

```
from snowflake.snowpark.functions import udf
from snowflake.snowpark.types import StringType
def norm_email_logic(s: str) -> str:
    if s is None:
        return None
    return s.strip().lower()
@udf(
    name="util.norm_email",
    replace=True,
    is_permanent=True,
    stage_location="@code_stage",
    return_type=StringType(),
    packages=[]
)
```

```
def norm_email(s: str) -> str:
    return norm_email_logic(s)
```

Invoke it from SQL like any other scalar function:

```
SELECT
  customer_id,
  util.norm_email(email) AS email_norm
FROM dim_customer;
```

This is the useful production distinction: the function is no longer just a Snowpark session artifact; it becomes a reusable database object.

Python UDAF (Snowpark): Robust Aggregate with Null Handling

Instead of repeating the same illustrative median, make the aggregate example slightly more operational. The version below computes an average order value while ignoring nulls and safely handling empty groups.

```
from snowflake.snowpark.functions import udaf
from snowflake.snowpark.types import FloatType
class AvgNonNullUDAF:
    def __init__(self):
        self.total = 0.0
        self.count = 0
    def accumulate(self, x):
        if x is not None:
            self.total += float(x)
            self.count += 1
    def merge(self, other):
        self.total += other.total
        self.count += other.count
    def finish(self):
        if self.count == 0:
            return None
        return self.total / self.count
```

```
avg_non_null = udaf(
    AvgNonNullUDAF,
    return_type=FloatType(),
    name="avg_non_null",
    replace=True
)
```

Invoke it in SQL:

```
SELECT
  customer_id,
  avg_non_null(amount) AS avg_amount
FROM fact_orders
GROUP BY customer_id;
```

This keeps the example compact while showing the real structure of a Snowpark aggregate: accumulate, merge, finish.

JavaScript Stored Procedure: Controlled Promotion with Retry

Stored procedures are useful when an operation is procedural rather than relational. Promotion steps are a good example: they are usually short, imperative, and sensitive to transient failures.

```
CREATE OR REPLACE PROCEDURE ops.swap_mart(src_schema STRING, tgt_
schema STRING)
RETURNS STRING
LANGUAGE JAVASCRIPT
AS
$$
var sqlCmd = "ALTER SCHEMA " + src_schema + " SWAP WITH " + tgt_schema;
var attempts = 0;
while (attempts < 3) {
  try {
    snowflake.execute({sqlText: sqlCmd});
    return "SWAP OK";
  } catch (e) {
    attempts++;
```

```
      snowflake.execute({sqlText: "CALL system$wait(5)"});
      if (attempts == 3) {
        throw e;
      }
    }
  }
$$;
```

Call it from a task or orchestrator:

```
CALL ops.swap_mart('prod_candidate.mart_sales', 'prod.mart_sales');
```

This example earns its place because it shows something SQL alone does not express cleanly: retry logic around a controlled administrative action.

Snowpark Stored Procedure: Parameterized Backfill

A stored procedure becomes useful when a job needs several coordinated steps or runtime parameters. A backfill is a natural example because the repair window changes from run to run.

```
from snowflake.snowpark import Session
from snowflake.snowpark.types import StringType
from snowflake.snowpark.functions import sproc
@sproc(
    name="ops.backfill_orders",
    replace=True,
    is_permanent=True,
    stage_location="@code_stage",
    return_type=StringType(),
    packages=[]
)
def backfill_orders(session: Session, days: int) -> str:
    if days <= 0:
        return "days must be positive"
    session.sql(f"""
        MERGE INTO mart.fact_orders t
```

```
        USING (
          SELECT *
          FROM stage.orders_norm
          WHERE event_ts >= DATEADD('day', -{days}, CURRENT_TIMESTAMP())
        ) s
        ON t.order_id = s.order_id
        WHEN MATCHED THEN
          UPDATE SET
            amount   = s.total,
            order_ts = s.event_ts
        WHEN NOT MATCHED THEN
          INSERT (order_id, customer_id, amount, order_ts)
          VALUES (s.order_id, s.customer_id, s.total, s.event_ts)
    """).collect()
    return f"Backfilled {days} days"
```

Invoke it like this:

```
CALL ops.backfill_orders(7);
```

Governance, Performance, and Safety Notes

Concern	Practical stance	Why it helps
Idempotency	Encode writes in procedures as MERGE or SWAP	Safe retries from Tasks/ orchestrators
Security	Use roles narrowly; mark UDFs/UDTFs **IMMUTABLE** where true; avoid external calls unless necessary	Tighter execution guarantees; simpler reviews
Dependencies	Prefer Snowflake Anaconda channel (Python) or shaded JARs (JVM)	Reproducible builds and quicker cold starts
Observability	Write procedure logs/metrics to small ops tables; surface errors via return values	Easier production triage than console-only logs

Choosing Deliberately (And Mixing Safely)

- Use **UDF/UDAF/UDTF** when the question is *"how do I compute this value from the data?"*
- Use **procedures** when the question is *"what series of steps should run, in what order, with what guards?"*
- Prefer **JavaScript procedures** for tiny, dependency-free controllers and **Snowpark procedures** when your logic benefits from rich libraries or shared code across data apps.

Both styles live happily in the same pipeline: UDFs/UDAFs handle custom math and parsing inside SELECTs; a thin JavaScript (or Snowpark) procedure coordinates a `SWAP`, kicks Tasks, and records run metadata.

11.3 Packaging, Dependencies, and Secure Execution

Snowpark code should feel like part of your schema: versioned, reproducible, and safe to run in production. Packaging and dependency choices determine how quickly functions start, whether builds are portable across environments, and how tightly you can restrict outbound access. Aim for small, deterministic artifacts and prefer Snowflake-native dependency channels so cold starts stay short and reviews are simple.

11.3.1 Packaging Models That Age Well

Language	Preferred packaging	How dependencies arrive	Why this works
Python (Snowpark)	Pin to Snowflake's Anaconda channel; add only missing libs as wheels on a stage	`packages=[...]` and `imports=['@stage/wheels/*.whl']` in UDF/SP/session	Pre-built binaries → fast cold starts; minimal artifact drift
Scala/Java (Snowpark)	Single **shaded/fat JAR** per library or service	`imports=['@stage/jars/app-all.jar']`	One file to version; no runtime classpath hunting

Language	Preferred packaging	How dependencies arrive	Why this works
JavaScript procedures	Inline source (no external deps)	N/A	Zero packaging, instant cold starts; great for control flow

Python Session with Packages and Imports

```
from snowflake.snowpark import Session
sess = Session.builder.configs({
  "account": "...", "user": "...", "password": "...",
  "warehouse": "WH_ELT", "role": "TRANSFORMER",
  "database": "CORE", "schema": "PUBLIC",
  # Arrow result format can speed up large result transfers and works
    well with
  # Arrow/pandas-oriented workflows; ensure the client environment
    includes the
  # needed Arrow dependencies. Omit this if you want to keep the example
    minimal.
  "PYTHON_CONNECTOR_QUERY_RESULT_FORMAT": "arrow"
}).create()

# Package hints when creating UDFs/SPs:
# @udf(..., packages=["numpy==1.26.4"], imports=["@code.whl/myutil-0.1-py3-
none-any.whl"])
```

JVM Import (Scala/Java)

```
val session = Session.builder
  .configs(Map( /* conn props */ ))
  .imports(Seq("@code.jar/app-all.jar"))
  .create
```

11.3.2 Versioning and Promotion Without Surprises

Concern	Practice	Commit point
Immutability	Never mutate artifacts in-place; publish new wheel/JAR names with version suffix	`@code.whl/myutil-0.2.whl`, `@code.jar/app-1.4.3-all.jar`
Referencing	Bind versions at object creation time (UDF/SP `imports/packages`)	Object DDL is the "lockfile"
Promotion	Recreate UDFs/procs in higher envs referencing the same artifact names	CI deploy step per env
Rollback	Keep the previous artifact staged; `CREATE OR REPLACE` back to v-1	Metadata-only change

11.3.3 Secrets and Configuration

Keep credentials out of code. Bind configuration to roles, stages, and integrations so execution is environment-agnostic.

Item	Snowflake-native home	Example
Object storage creds	**Storage Integrations**	`CREATE STORAGE INTEGRATION ...` then reference on stages
Network egress policy	**Network Rules** + **External Access Integration**	Restrict domains/ports for outbound HTTP
App secrets	**Secrets** (if available) or key/value tables with restricted roles	UDF/SP reads via role-limited view
Connection fan-out	Roles + GRANTs	Procedures run under the invoker/owner's role as designed

11.3.4 Secure Execution and Outbound Calls

Snowpark and JavaScript procedures run in a sandbox with **no outbound network** by default. If you must call an API, make it explicit, minimal, and governable.

```
-- Allow only specific HTTPS domains
CREATE NETWORK RULE ext_refdata
  TYPE = HOST_PORT
  MODE = EGRESS
  VALUE_LIST = ('api.example.com:443');

CREATE EXTERNAL ACCESS INTEGRATION eai_refdata
  ALLOWED_NETWORK_RULES = (ext_refdata)
  ENABLED = TRUE;

-- Bind to a procedure
CREATE OR REPLACE PROCEDURE refdata.refresh()
RETURNS STRING
LANGUAGE PYTHON
RUNTIME_VERSION = '3.10'
EXTERNAL_ACCESS_INTEGRATIONS = (eai_refdata)
PACKAGES = ('requests')
AS $$
# call out to api.example.com and write results to a table
$$;
```

Principle: import data once, then reuse it in tables. Avoid per-row egress in UDFs; if you must reach out, do it in a scheduled procedure and cache results.

11.3.5 Cold Start and Performance Hygiene

Symptom	Root cause	Fix that lasts
Long first-call latency	Large wheels/JARs; many small imports	Consolidate to a single artifact; prune unused packages; size the warehouse appropriately for Snowpark-heavy workloads
Unpredictable failures	Native libs not available in Snowflake runtimes	Prefer pure-Python/Java libs or Snowflake-channel builds
Slow row-by-row UDFs	Logic belongs in DF/SQL	Push to dataframe/SQL; keep UDFs narrow and stateless
Memory pressure	Heavy in-memory ops in UDF/UDAF	Use CTAS + window functions; for UDAFs, keep state compact

Large dependency bundles are a common cold-start cost. Snowflake explicitly recommends Snowpark-optimized warehouses for workloads with larger memory needs; in practice, teams also often reduce latency by avoiding tiny warehouses for dependency-heavy UDFs and by keeping warehouses from suspending too aggressively, but that is more of an operational tuning pattern than a hard platform rule.

11.3.6 Reproducible Builds in CI/CD

Step	Python (Snowpark)	Scala/Java
Build	Resolve from Snowflake Anaconda first; vendor extras	Build shaded JAR with pinned deps
Stage	`PUT file://dist/myutil-0.2.whl @code.whl AUTO_COMPRESS=FALSE;`	`PUT file://target/app-1.4.3-all.jar @code.jar;`
Deploy	`CREATE OR REPLACE FUNCTION ... imports=['@code.whl/myutil-0.2.whl']`	`CREATE OR REPLACE PROCEDURE ... imports=['@code.jar/app-1.4.3-all.jar']`
Verify	Smoke query + integration test in a cloned schema	Same, plus UDF/UDAF correctness tests

11.3.7 Governance Baked into Code Objects

Control	How to apply	Effect
IMMUTABLE UDFs	`CREATE FUNCTION ... IMMUTABLE`	Planner can cache; guarantees no side effects
SECURE UDF/VIEW	`CREATE SECURE FUNCTION/VIEW`	Definition hidden; safer sharing
Ownership and roles	Separate **object owner** and **runtime role**	Least privilege at execution time
Auditing	Log to an ops table from procedures	Traceability without external systems

A good mental model is: **artifacts are data.** You stage versioned wheels/JARs, bind them in object DDL, and promote by metadata. Network access is denied unless explicitly allowed; secrets live in managed integrations. With those constraints, Snowpark UDFs and procedures run predictably, start quickly, and remain easy to audit months later.

CHAPTER 12

ML and Advanced Analytics Workloads

Snowflake can be your entire ML data plane: you build features where the facts live, keep point-in-time guarantees with Time Travel and SCDs, and publish training/serving views that stay consistent across experiments and production. You don't have to move rows to engineer features—windowed SQL, semi-structured projection, and Snowpark cover nearly everything. When you do need custom math, a small UDF or Snowpark routine runs in-database, so the data never leaves.

12.1 Feature Engineering in Snowflake

The goal is simple: **produce leakage-free features at the entity-time grain you will serve**. Features should be reproducible (same query + same cutoff → same values), cheap to refresh, and easy to audit. Most of that comes from how you **join history**, **window events**, and **materialize** outputs—not from exotic code.

M. Hander, *Snowflake Data Warehouse Engineering*, https://doi.org/10.1007/979-8-8688-2628-3_12

12.1.1 What "Good" Looks Like

Concern	Practical stance	Why it holds up in Snowflake
Grain	Key every row by `(entity_id, as_of_ts)`	Features align with how models receive data at inference time
Leakage control	Join with **time bounds**; use SCD/Time Travel to see "only what we knew then"	Past-state reconstruction is first-class
Freshness	Incremental recompute: Streams/Tasks or Dynamic Tables	Small deltas keep costs and SLAs predictable
Auditability	Persist feature queries and inputs; log snapshot ids	Rebuild any training set byte-for-byte
Serving parity	Training view ≈ Serving view (same SQL, only cutoff differs)	Eliminates train/serve skew

12.1.2 Point-in-Time Joins (No Leakage)

Two safe ways to "travel back" to the dimension state: SCD Type-2 effective dating or a point-in-time clone with Time Travel. SCD-2 scales better day-to-day; Time Travel is perfect for audits.

```
-- Fact grain: (customer_id, event_ts)
-- Dim SCD2: valid_from, valid_to, current_flag
SELECT
  f.customer_id,
  f.event_ts AS as_of_ts,
  d.segment,
  d.status
FROM fact_events f
JOIN dim_customer_scd d
  ON d.customer_id = f.customer_id
 AND d.valid_from <= f.event_ts
 AND COALESCE(d.valid_to, '9999-12-31') > f.event_ts;
```

Point-in-time with Time Travel (ad hoc audit):

```
-- What the dim looked like exactly at a past cutoff
SELECT *
FROM dim_customer
  AT (TIMESTAMP => TO_TIMESTAMP_TZ('2025-09-01 00:00:00 +00:00'));
```

12.1.3 Windowed Features That Serve and Train

Classic behavioral features are rolling counts, sums, and recency. Keep them **bounded** to your serving horizon.

```
-- 7d, 30d behaviors at (customer_id, as_of_ts) grain
WITH b AS (
  SELECT customer_id, event_ts, amount
  FROM fact_events
)
SELECT
  customer_id,
  event_ts AS as_of_ts,

  COUNT_IF(amount > 0)
    OVER (PARTITION BY customer_id
          ORDER BY event_ts
          RANGE BETWEEN INTERVAL '7 DAY' PRECEDING AND CURRENT ROW)
          AS cnt_7d,
  SUM(amount)
    OVER (PARTITION BY customer_id
          ORDER BY event_ts
          RANGE BETWEEN INTERVAL '30 DAY' PRECEDING AND CURRENT ROW) AS
          amt_30d,
  DATEDIFF('day',
           MAX(event_ts) OVER (PARTITION BY customer_id
                               ORDER BY event_ts
                               RANGE BETWEEN UNBOUNDED PRECEDING AND
                               CURRENT ROW),
           event_ts) AS days_since_last
FROM b;
```

Tip For daily scoring, pre-bucket `as_of_ts` to midnight and compute windows on day buckets; you'll get stronger pruning and cheaper refresh.

12.1.4 Categorical and Text Features

Use conformed dimensions for high-cardinality categories and keep **encodings** in tables, not notebooks.

```
-- Target encoding with leakage guard: fit on t-1 window, apply to day t
WITH label as (
  SELECT order_id, customer_id, target, DATE_TRUNC('day', event_ts) as day
  FROM training_labels
),
fit AS (
  SELECT brand_id, day,
         AVG(target) AS te
  FROM label
  GROUP BY brand_id, day
),
apply AS (
  SELECT l.order_id, l.day, l.target, p.brand_id
  FROM label l
  JOIN dim_product p USING (order_id)
)
SELECT a.order_id, a.target,
       f_prev.te AS brand_te_prevday
FROM apply a
JOIN fit f_prev
  ON f_prev.brand_id = a.brand_id
 AND f_prev.day      = DATEADD('day', -1, a.day);
```

Text embeddings are most convenient when stored once and reused many times. In accounts where Snowflake's VECTOR type and vector similarity functions are available, store embeddings in a VECTOR column and compare them with native vector functions such as VECTOR_COSINE_SIMILARITY. In environments where VECTOR is not available, store the embedding as an array in VARIANT, and perform similarity calculations in Snowpark or another client-side processing layer.

Preferred pattern when VECTOR **is available:**

```
CREATE OR REPLACE TABLE nlp.product_text_features (
  product_id NUMBER,
  embedding VECTOR(FLOAT, 768)
);
SELECT
  product_id,
  VECTOR_COSINE_SIMILARITY(
    embedding,
    [/* query embedding values */]::VECTOR(FLOAT, 768)
  ) AS similarity
FROM nlp.product_text_features
ORDER BY similarity DESC;
```

Snowflake's vector functions operate on VECTOR values of matching type and dimension, and VECTOR_COSINE_SIMILARITY returns a float in [-1, 1].

Fallback pattern when VECTOR **is not available:**

```
CREATE OR REPLACE TABLE nlp.product_text_features (
  product_id NUMBER,
  embedding VARIANT
);
```

If native vector support is unavailable, persist embeddings in a VARIANT column as arrays, and perform similarity calculation in Snowpark or another application layer. This preserves the storage model while avoiding dependence on VECTOR-specific SQL functions.

```
-- Example layout
CREATE TABLE nlp.product_text_features (
  product_id STRING PRIMARY KEY,
  updated_at TIMESTAMP_NTZ,
  embedding  VECTOR(768)  -- or VARIANT if VECTOR not available
);
```

Similarity search (cosine distance):

```
-- Given a query vector :qv
SELECT product_id
FROM nlp.product_text_features
ORDER BY VECTOR_COSINE_DISTANCE(embedding, :qv)
LIMIT 50;
```

12.1.5 Materializing Features (Incremental, Auditable)

A clean pattern is "**bronze → silver → feature**" with **Dynamic Tables** for relational transforms and **Tasks** where CDC semantics or repair windows matter.

Layer	Contents	Good materialization
Bronze	Raw facts + envelopes (`tenant_id`, `event_ts`)	Snowpipe/Streaming to permanent tables
Silver	Normalized, SCD-aware dims; tidy facts	Streams+Tasks or Dynamic Tables
Feature	`(entity_id, as_of_ts, f1...fn)`	Dynamic Table with `TARGET_LAG` or Task-driven incremental MERGE

Dynamic Table example (kept within ten minutes of source):

```
CREATE OR REPLACE DYNAMIC TABLE feat.customer_daily
TARGET_LAG = '10 MINUTE'
WAREHOUSE  = wh_elt
AS
SELECT
  c.customer_id,
  DATE_TRUNC('day', e.event_ts) AS as_of_ts,
  COUNT_IF(e.amount > 0)        AS cnt_pos_7d,
  SUM(e.amount)                 AS amt_30d
FROM fact_events e
JOIN dim_customer_scd c
  ON c.customer_id = e.customer_id
 AND c.valid_from <= e.event_ts
```

```
 AND COALESCE(c.valid_to, '9999-12-31') > e.event_ts
WHERE e.event_ts >= DATEADD(day, -30, CURRENT_DATE())
GROUP BY 1, 2;
```

Task-driven increment with a repair window (late data):

```
CREATE OR REPLACE TASK t_feat_customer_daily
  WAREHOUSE = wh_elt
  SCHEDULE  = '15 MINUTE'
AS
MERGE INTO feat.customer_daily t
USING (
  SELECT /* recompute last 2 days only */
    c.customer_id,
    DATE_TRUNC('day', e.event_ts) AS as_of_ts,
    COUNT_IF(e.amount > 0)        AS cnt_pos_7d,
    SUM(e.amount)                 AS amt_30d
  FROM fact_events e
  JOIN dim_customer_scd c
    ON c.customer_id = e.customer_id
   AND c.valid_from <= e.event_ts
   AND COALESCE(c.valid_to,'9999-12-31') > e.event_ts
  WHERE e.event_ts >= DATEADD(day, -2, CURRENT_TIMESTAMP())
  GROUP BY 1,2
) s
ON t.customer_id = s.customer_id AND t.as_of_ts = s.as_of_ts
WHEN MATCHED THEN UPDATE SET cnt_pos_7d = s.cnt_pos_7d, amt_30d = s.amt_30d
WHEN NOT MATCHED THEN INSERT (customer_id, as_of_ts, cnt_pos_7d, amt_30d)
VALUES (s.customer_id, s.as_of_ts, s.cnt_pos_7d, s.amt_30d);
```

12.1.6 Training/Serving Parity and Snapshots

Keep one **feature view** that both training and serving use; change only the cutoff predicate.

```
CREATE OR REPLACE VIEW feat.v_customer_features AS
SELECT * FROM feat.customer_daily;
```

```
-- Training set for a historical window (labels joined separately)
SELECT *
FROM feat.v_customer_features
WHERE as_of_ts BETWEEN '2025-06-01' AND '2025-08-31';

-- Serving for the most recent day
SELECT *
FROM feat.v_customer_features
WHERE as_of_ts = DATE_TRUNC('day', CURRENT_DATE());
```

Snapshot your inputs so experiments are reproducible:

Snapshot lever	How to use it	What you get
Time Travel	Query sources/feature tables with AT (TIMESTAMP => ...)	Exact past state without exporting
Zero-copy clone	CREATE DATABASE train_2025_09_01 CLONE prod;	Frozen copy to build and share datasets
Version tags	Tag feature views/tables with a release id	Traceability in catalogs and notebooks

12.1.7 Operational Guardrails You'll Use

Symptom	Likely cause	Fix that lasts
Feature drift vs. training	Training used a different SQL than serving	Single source of truth view; unit tests on aggregates
Slow refreshes	Too many tiny files or unbounded windows	Compact upstream; bound windows; use repair-window MERGE
Leakage in join	Joined to current dim state	Enforce SCD join or Time Travel in linters/reviews
Inconsistent encodings	Notebook-only mappings	Persist encoders as tables; join by id at both train/serve

12.1.8 Snowpark When SQL Gets Awkward

For the ~10% of logic that's cleaner in code—e.g., custom tokenization, robust outlier handling, or sketch-based aggregations—use small, tested Snowpark UDFs or UDAFs and keep them **pure**.

Example: Robust z-Score As a UDF (Illustrative)

```
from snowflake.snowpark.functions import udf
from snowflake.snowpark.types import FloatType
@udf(
    name="feat.winsor",
    replace=True,
    is_permanent=True,
    stage_location="@code_stage",
    return_type=FloatType(),
    packages=[]
)
def winsor(x: float, lo: float, hi: float) -> float:
    if x is None:
        return None
    return max(lo, min(hi, x))
```

> Examples in this section use permanent deployment semantics. Temporary Snowpark UDFs exist only for the creating session; production UDFs that are called from tasks, procedures, or shared SQL should be registered with `is_permanent=True` and a valid `stage_location`.

Use it inside your feature SELECT; the heavy lifting (scans, joins, windows) stays in SQL, and the UDF just tidies edge cases.

Deployment note In Snowpark, named Python UDFs, UDAFs, UDTFs, and stored procedures are temporary by default. That is convenient for notebooks and ad hoc sessions, but not for production pipelines. Whenever an object is intended to survive the session and be callable from tasks, procedures, or shared SQL, register it as permanent and provide a non-temporary internal `stage_location`.

12.1.9 Closing Note

Great features in Snowflake are mostly **great SQL**: point-in-time joins, bounded windows, and sober materialization. Snowpark and UDFs round off the corners, Time Travel keeps you honest, and Streams/Tasks or Dynamic Tables make freshness mundane. Keep the contract—`(entity_id, as_of_ts)`—front and center, and your models will be both accurate and easy to operate.

12.2 Model Scoring Patterns (Batch/Streaming)

Scoring in Snowflake boils down to two questions: **how fresh must predictions be** and **where does the model live**. Once you fix those, everything else is plumbing—join the right features, call the model, and write back results in an idempotent way. In practice, you'll use the same feature views for both training and serving; the difference is whether you evaluate them on a schedule (batch) or as events land (streaming/micro-batch).

12.2.1 Picking the Lane and the Runtime

Dimension	Batch scoring (scheduled)	Streaming/micro-batch scoring
Typical SLA	Minutes → hours	Seconds → minutes
Trigger	Task/dbt run/orchestrator	Stream over landing table; Snowpipe/ Snowpipe Streaming
Feature source	Daily/hourly **feature tables/views**	Latest **feature tables** + event payload; sometimes "on-the-fly" windows
Model location	In-database UDF/UDTF (Snowpark), External Function, or SQL rules	Same options; prefer UDFs to avoid per-row egress
Writeback	Upsert into `PREDICTIONS_*` with `(entity_id, as_of_ts, model_version)`	Append + periodic merge/repair over recent window
Observability	Post-run metrics table, drift checks	Per-window counters, error DLQ, freshness view

12.2.2 In-Database Scoring with a Snowpark UDF (Batch)

Package the model once, keep the data in Snowflake, and run on a cadence. The UDF loads the serialized model from an import (wheel/JAR or artifact on a stage) and returns a scalar (or a small OBJECT) per row.

```
# Python (conceptual): score_probability UDF

from snowflake.snowpark.functions import udf
from snowflake.snowpark.types import FloatType
@udf(
    name="ml.score_probability",
    replace=True,
    is_permanent=True,
    stage_location="@code_stage",
    return_type=FloatType(),
    packages=[]
)
def score_probability(v1: float, v2: float, v3: float) -> float:
    # model.predict_proba(...) -> probability of class 1
    return float(0.5 * v1 + 0.3 * v2 + 0.2 * v3)  # placeholder
```

> Because scoring functions are typically reused across sessions and often called from scheduled pipelines, register them as permanent UDFs. In Snowpark, permanent Python UDFs require `is_permanent=True` and a non-temporary internal `stage_location`.

Schedule the upsert with a Task; key by `(entity_id, as_of_ts, model_version)` so retries converge:

```
CREATE OR REPLACE TASK t_score_daily
  WAREHOUSE = wh_ml
  SCHEDULE  = 'DAILY USING CRON 5 1 * * * UTC'
AS
MERGE INTO pred.churn_daily t
USING (
  SELECT
    f.customer_id  AS entity_id,
    f.as_of_ts,
```

```
    'churn_v7'     AS model_version,
    ml.score_probability(f.cnt_pos_7d, f.amt_30d, f.days_since_last)
    AS p_churn
  FROM feat.customer_daily f
  WHERE f.as_of_ts = DATE_TRUNC('day', CURRENT_DATE() - 1)
) s
ON t.entity_id = s.entity_id
AND t.as_of_ts = s.as_of_ts
AND t.model_version = s.model_version
WHEN MATCHED THEN
  UPDATE SET
    p_churn   = s.p_churn,
    scored_at = CURRENT_TIMESTAMP()
WHEN NOT MATCHED THEN
  INSERT (entity_id, as_of_ts, model_version, p_churn, scored_at)
  VALUES (s.entity_id, s.as_of_ts, s.model_version, s.p_churn, CURRENT_
  TIMESTAMP());
ALTER TASK t_score_daily RESUME;
```

Operational note Scheduled tasks are created suspended and must be resumed explicitly. Also, any Python UDF called from a scheduled task must be registered as a permanent UDF; temporary Snowpark UDFs are session-scoped and disappear when the creating session ends.

Why this ages well: the model stays close to the data, you avoid network egress, and idempotent merges make retries safe.

12.2.3 Micro-batch Scoring with Streams and Tasks

Use a Stream on the normalized event feed, enrich with the latest features, score, and append predictions. A periodic repair task revisits a short window to heal late arrivals.

```
CREATE OR REPLACE DYNAMIC TABLE feat.customer_latest
  TARGET_LAG = '5 MINUTE'
  WAREHOUSE  = wh_elt
```

```
AS
SELECT *
FROM feat.customer_daily
QUALIFY ROW_NUMBER() OVER (
  PARTITION BY customer_id
  ORDER BY as_of_ts DESC
) = 1;

-- 1) Expose deltas
CREATE OR REPLACE STREAM str_events ON VIEW stage.events_norm;
-- 1a) Keep only the latest feature snapshot per customer
CREATE OR REPLACE DYNAMIC TABLE feat.customer_latest
  TARGET_LAG = '5 MINUTE'
  WAREHOUSE  = wh_elt
AS
SELECT *
FROM feat.customer_daily
QUALIFY ROW_NUMBER() OVER (
  PARTITION BY customer_id
  ORDER BY as_of_ts DESC
) = 1;
-- 2) Near-real-time scoring task
CREATE OR REPLACE TASK t_score_stream
  WAREHOUSE = wh_ml
  SCHEDULE  = '5 MINUTE'
AS
MERGE INTO pred.events_online t
USING (
  SELECT
    e.customer_id AS entity_id,
    e.event_ts,
    'churn_v7' AS model_version,
    score_probability(
      f.cnt_pos_7d,
      f.amt_30d,
      f.days_since_last
```

```
    ) AS score,
    CURRENT_TIMESTAMP() AS scored_at
  FROM str_events e
  JOIN feat.customer_latest f
    ON f.customer_id = e.customer_id
   -- Optionally guard for "ready" features:
   -- AND f.as_of_ts >= e.event_ts - INTERVAL '2 MINUTE'
) s
ON  t.entity_id     = s.entity_id
AND t.event_ts      = s.event_ts
AND t.model_version = s.model_version
WHEN MATCHED THEN
  UPDATE SET
    score     = s.score,
    scored_at = s.scored_at
WHEN NOT MATCHED THEN
  INSERT (entity_id, event_ts, model_version, score, scored_at)
  VALUES (s.entity_id, s.event_ts, s.model_version, s.score, s.scored_at);
```

Repair window (late data, out-of-order):

```
CREATE OR REPLACE TASK t_score_repair
  WAREHOUSE = wh_ml
  SCHEDULE  = 'HOURLY'
AS
MERGE INTO pred.events_online t
USING (
  SELECT
    e.customer_id AS entity_id,
    e.event_ts,
    'churn_v7'    AS model_version,
    score_probability(f.cnt_pos_7d, f.amt_30d, f.days_since_last) AS score
  FROM stage.events_norm e
  JOIN feat.customer_latest f ON f.customer_id = e.customer_id
  WHERE e.event_ts >= DATEADD('hour', -2, CURRENT_TIMESTAMP())
) s
```

```
ON (t.entity_id, t.event_ts, t.model_version) = (s.entity_id, s.event_ts,
s.model_version)
WHEN MATCHED THEN UPDATE SET score = s.score, scored_at = CURRENT_
TIMESTAMP()
WHEN NOT MATCHED THEN
  INSERT (entity_id, event_ts, model_version, score, scored_at)
  VALUES (s.entity_id, s.event_ts, s.model_version, s.score, CURRENT_
  TIMESTAMP());
```

12.2.4 Calling an External Model Endpoint (When You Must)

Sometimes, the model lives behind an HTTPS endpoint (third-party or your own). Prefer **External Functions** for HTTPS calls; they use an API Integration, handle auth, and keep SQL as the contract. Use them in **batch** (not row-by-row interactive UDF calls) and cache results.

Concern	Good posture	Rationale
Transport	External Function with API Integration	Governed egress, retries, observability
Grain	Batch inputs (arrays/JSON) per call if supported	Fewer round-trips
Caching	Persist `(entity_id, as_of_ts, model_version) → score`	Avoids duplicate calls on retries
Fallback	Default/last-good score on outage	Keeps downstream healthy

(DDL for External Functions varies by environment; bind it once, then call in your scoring SQL exactly like a UDF.)

12.2.5 Choosing Model Placement

Option	When to prefer	Trade-offs
UDF in Snowflake (Snowpark)	Data is large; latency is tight; dependencies fit Snowflake runtimes	Package management; cold starts if artifacts are big
External Function (HTTPS)	Centralized inference service; specialized hardware; A/B outside Snowflake	Network egress, rate limits; design for batching/caching
SQL rules	Baselines, guardrails, feature flags	Easy, transparent—but limited expressiveness

12.2.6 A/B, Canaries, and Rollbacks

Predictions should carry **model metadata** so you can mix versions safely and compare outcomes.

```
-- Tag predictions with version and policy
SELECT entity_id, as_of_ts,
       CASE WHEN rand() < 0.1 THEN 'churn_v8' ELSE 'churn_v7' END AS model_
       version,
       score_probability(...) AS p
FROM feat.customer_daily;
```

Promote by swapping schemas or by flipping a small routing view; revert the same way. Keep both versions writing so you can analyze lift and drift without re-scoring.

12.2.7 Observability That Matters

Signal	Where it lives	Why you care
Freshness	View that compares `MAX(as_of_ts)` (features) and latest `scored_at` (predictions)	Alerts when serving lags features
Coverage	Count of entities scored vs. expected	Finds join gaps and data skew

(*continued*)

Signal	Where it lives	Why you care
Stability	Summary stats of features and scores per window	Flags drift or busted encoders
Errors	DLQ for failed rows/external calls	Retries and forensics without blocking the pipeline

Example freshness probe:

```
CREATE OR REPLACE VIEW ops.pred_freshness AS
SELECT
  'churn' AS model,
  (SELECT MAX(as_of_ts) FROM feat.customer_daily)      AS newest_features,
  (SELECT MAX(scored_at) FROM pred.churn_daily)        AS newest_scores,
  TIMESTAMPDIFF('minute',
    (SELECT MAX(as_of_ts) FROM feat.customer_daily),
    (SELECT MAX(scored_at) FROM pred.churn_daily)
  ) AS minutes_lag;
```

12.2.8 Putting It Together

Keep **features authoritative** in Snowflake, score with **UDFs** when possible, and reserve **external calls** for cases that truly need them—always with batching and caching. Batch for cheap, predictable throughput; stream when the business needs immediacy; and in both cases, write predictions with a stable key and model version so retries, A/B tests, and audits are trivial.

12.3 Integration with External ML Platforms

Snowflake plays well with outside ML stacks when you keep the **data contract** simple and put governance first. The core flows are predictable: export features or labels for training, register models and metadata where your MLOps stack expects them, and bring predictions back with an auditable key. When you need online inference, call a managed endpoint through a governed interface. The pieces change—SageMaker, Vertex AI, Azure ML, Databricks, Hugging Face Inference, your own service—but the pattern doesn't.

12.3.1 Which Path to Use (And Why)

Scenario	Data movement	Inference call site	Best fit	Why it works
External training, in-Snowflake scoring	Export features once; keep features authoritative in Snowflake	Snowpark UDF/ UDTF (model artifact imported)	Large data, tight SLAs, minimal egress	Data stays put, cheap batch/stream scoring, simple rollback
External training and external online scoring	Features join in Snowflake; send compact payloads	**External Function** to HTTPS endpoint	Real-time APIs, centralized serving infra	Governed egress, one SQL surface, model team owns runtime
External training and batch scoring outside	Export features → external job scores → re-ingest predictions	Round-trip via stage (S3/ADLS/ GCS) or Iceberg	Heavy GPU jobs, specialty runtimes	Parallel offline scoring; robust with large models
Cross-engine lakehouse	Share tables via Iceberg or External Tables	Either side	Mixed Spark/ Trino + Snowflake shops	One table, many engines, no copies

12.3.2 Clean Contracts for Training ↔ Serving

Use stable IDs and timestamps so you can stitch runs together across systems.

Contract piece	Shape	Example
Primary key	`(entity_id, as_of_ts)`	`customer_id='C123', as_of_ts='2025-09-01'`
Feature view	SQL view/table in Snowflake	`feat.v_customer_features`
Model version	String tag carried with predictions	`churn_v7`
Prediction table	Upsertable with id + version	`pred.churn_daily(entity_id, as_of_ts, model_version, score, scored_at)`

12.3.3 External Training: Ship Bytes Out, Not the Warehouse

Export features or training sets with explicit schema and snapshot lineage; train externally; record artifacts in your registry; bring metrics back for lineage.

```
-- Snapshot training set (point-in-time safe)
CREATE OR REPLACE TABLE train.churn_2025_09_01 AS
SELECT f.*, l.label
FROM feat.v_customer_features f
JOIN labels.churn l
  ON l.customer_id = f.customer_id
 AND l.as_of_ts    = f.as_of_ts
WHERE f.as_of_ts BETWEEN '2025-06-01' AND '2025-08-31';

-- Deliver to the platform bucket
COPY INTO @ml_exports/churn_2025_09_01/
FROM train.churn_2025_09_01
FILE_FORMAT = (TYPE = PARQUET);
```

On return, store model metadata so audits live next to data:

```
CREATE TABLE ops.model_registry (
  model_name STRING, version STRING, trained_at TIMESTAMP_NTZ,
  data_snapshot STRING, metrics VARIANT, artifact_uri STRING
);
```

For illustration, the following example shows what a registry entry might look like:

```
INSERT INTO ops.model_registry VALUES
('churn','v7',CURRENT_TIMESTAMP(),'train.churn_2025_09_01',
 PARSE_JSON('{"auc":0.816,"pr_auc":0.44}'),
 's3://ml-models/churn/v7/');
```

In production, this row should not be inserted with hard-coded metric values. Instead, the training or deployment pipeline should write the registry entry automatically, using metrics and artifact locations emitted by the external training platform or orchestration layer, such as CI/CD workflows, MLflow, or a managed ML service.

12.3.4 Online Inference Through Governed Endpoints (External Functions)

External Functions let SQL call HTTPS with an **API Integration**. You keep calls batched and cache results in tables to avoid duplicate requests.

```
-- (Admin) API integration bound to your serving gateway
CREATE API INTEGRATION ai_infer
  API_PROVIDER = aws_api_gateway
  API_AWS_ROLE_ARN = 'arn:aws:iam::123456789012:role/snowflake_extfn'
  API_ALLOWED_PREFIXES = ('https://api.myserve.company/')
  ENABLED = TRUE;

-- External Function: send feature vector; get score
CREATE OR REPLACE EXTERNAL FUNCTION ml.score_churn(v VARIANT, model_
version STRING)
RETURNS FLOAT
API_INTEGRATION = ai_infer
HEADERS = ( 'x-model-version' = model_version )
URL = 'https://api.myserve.company/churn/score';

-- Use it sparingly and batch where possible
INSERT INTO pred.churn_events (entity_id, event_ts, model_version, score,
scored_at)
SELECT
  e.customer_id,
  e.event_ts,
  'churn_v7',
  ml.score_churn(OBJECT_CONSTRUCT(
      'cnt_pos_7d', f.cnt_pos_7d, 'amt_30d', f.amt_30d, 'days_since_last',
      f.days_since_last
  ), 'churn_v7') AS score,
  CURRENT_TIMESTAMP()
FROM stage.events_norm e
JOIN feat.customer_latest f ON f.customer_id = e.customer_id;
```

Good hygiene: throttle via your API gateway, batch inputs if your service supports it, and persist (entity_id, as_of_ts, model_version) → score so retries never double-call.

12.3.5 Batch Scoring Outside and Round-Tripping Predictions

When GPU runtimes or proprietary ops require external batch scoring, exchange Parquet through a stage (or write/read the same **Iceberg** table).

```
-- Export feature window
COPY INTO @ml_exports/churn_score_batch_2025_09_25/
FROM (
  SELECT entity_id, as_of_ts, cnt_pos_7d, amt_30d, days_since_last
  FROM feat.customer_daily
  WHERE as_of_ts = '2025-09-25'
)
FILE_FORMAT - (TYPE = PARQUET);

-- (External job writes results back to @ml_imports ...)

-- Ingest predictions idempotently
MERGE INTO pred.churn_daily t
USING (
  SELECT $1:entity_id::STRING AS entity_id,
         $1:as_of_ts::TIMESTAMP_NTZ AS as_of_ts,
         'churn_v7' AS model_version,
         $1:score::FLOAT AS score
  FROM @ml_imports/churn_score_batch_2025_09_25/ (FILE_FORMAT => (TYPE =
  PARQUET))
) s
ON (t.entity_id, t.as_of_ts, t.model_version) = (s.entity_id, s.as_of_ts,
s.model_version)
WHEN MATCHED THEN UPDATE SET score = s.score, scored_at = CURRENT_TIMESTAMP()
WHEN NOT MATCHED THEN INSERT (entity_id, as_of_ts, model_version, score,
scored_at)
VALUES (s.entity_id, s.as_of_ts, s.model_version, s.score, CURRENT_
TIMESTAMP());
```

12.3.6 Working with SageMaker, Vertex AI, and Azure ML (At a Glance)

Platform	Train data ingress	Serve back into Snowflake	Notes
AWS SageMaker	S3 from `COPY INTO`; or Iceberg table read via Spark	External Function to API Gateway/Lambda → SageMaker Endpoint; or batch transform round-trip to S3	Keep roles minimal with **Storage Integrations**; batch to reduce calls
Google Vertex AI	GCS from `COPY INTO`; BigQuery-adjacent patterns via External Tables	External Function to Cloud Run/API Gateway → Vertex endpoint	Prefer same-region GCS; cache scores in Snowflake
Azure ML	ADLS from `COPY INTO`; Synapse/Spark on External/Iceberg	External Function to Azure Functions/APIM → AML endpoint	Use **Private Link** where required; stage Parquet for batch

12.3.7 Governance and Networking That Won't Keep You Up

Concern	Snowflake control	Posture
Egress restriction	**API Integration** (External Functions) or **External Access Integration** (Snowpark)	Allow only specific domains; log usage
Secrets and auth	Integration-managed credentials; header injection	No secrets in SQL code; rotate outside the warehouse
Data residency	Same-region stages; avoid cross-region endpoints	Export and inference stay local to the account region
Lineage	Tables for snapshots, registry, prediction writes	One place to audit end-to-end runs

12.3.8 Making Drift and Freshness Visible

Keep small, queryable signals in Snowflake; have your external platform push metrics back after each train/score.

```
CREATE OR REPLACE TABLE ops.model_metrics (
  model_name STRING, version STRING, run_id STRING,
  metric_name STRING, metric_value FLOAT, logged_at TIMESTAMP_NTZ
);

-- Example freshness SLO: features vs. predictions
CREATE OR REPLACE VIEW ops.pred_freshness AS
SELECT
  'churn' AS model,
  (SELECT MAX(as_of_ts) FROM feat.customer_daily) AS newest_features,
  (SELECT MAX(scored_at) FROM pred.churn_daily)   AS newest_scores,
  TIMESTAMPDIFF('minute',
    (SELECT MAX(as_of_ts) FROM feat.customer_daily),
    (SELECT MAX(scored_at) FROM pred.churn_daily)
  ) AS minutes_lag;
```

12.3.9 Putting It Together

Train where it's convenient, but keep **features, snapshots, and predictions** authoritative in Snowflake with stable keys and model versions. For online serving, prefer **External Functions** over ad hoc egress so calls are governed and observable. For batch, exchange **Parquet** (or share **Iceberg** tables), and merge results idempotently. With those seams in place, you can swap platforms, roll models forward or back, and still answer the only question that matters later: *What did we know, which model did we use, and what did it predict?*

PART IV

Performance and Cost

CHAPTER 13

Query Performance Tuning

Snowflake performance comes from **reading less**, **doing less**, and **not waiting**. Reading less is pruning—good micro-partition shapes, selective predicates, and avoiding unnecessary columns. Doing less is the optimizer choosing cheaper joins, pushing filters early, and reusing caches and results. Not waiting is workload design: right-sized warehouses, burst handling with multi-cluster, and avoiding cross-team contention. Tuning, then, isn't about mysterious knobs; it's about confirming which of those three levers failed for a slow query and fixing that, not everything at once.

A practical workflow is (1) profile what actually ran, (2) diagnose if time was spent **queuing, scanning, shuffling, or outputting**, and (3) change either the **query shape** or the **environment**. Most wins arrive by tightening predicates, shrinking the scanned data set, and isolating bursty workloads to their own warehouses.

13.1 Profiling with QUERY_HISTORY and Query Profile

Start by **observing**. Snowflake records every query with rich metadata (timings, rows, bytes, warehouse, role). Snowsight Query Profile shows what actually happened at runtime, including where execution time accumulated across operators; `EXPLAIN`, by contrast, shows the logical plan Snowflake would use before execution. Reading these two together tells you whether the problem is **before execution** (queueing/compilation), **during execution** (scan volume, bad join order), or **after** (massive result sets, network).

M. Hander, *Snowflake Data Warehouse Engineering*, https://doi.org/10.1007/979-8-8688-2628-3_13

13.1.1 What the Two Views Each Tell You

Lens	**Answers quickly**	**Use it to decide**
INFORMATION_ SCHEMA.QUERY_ HISTORY*	Near-real-time query history for active troubleshooting: who ran what, when, on which warehouse, and whether time was spent queued or executing. Use this first during live incidents. SNOWFLAKE.ACCOUNT_USAGE. QUERY_HISTORY	Historical query history for reporting, trending, and broader retrospective analysis. Use this when you need longer retention, but remember that Account Usage data can lag by roughly 45 minutes to 3 hours, depending on the view.
Snowsight **Query Profile** (operator tree)	Which steps dominated time; which tables scanned how much; join order; partition pruning effectiveness	Do we need better predicates, fewer columns, a different join strategy, or clustering/search optimization for a stubborn filter?

For live debugging, prefer the Information Schema query-history functions. Use SNOWFLAKE.ACCOUNT_USAGE.QUERY_HISTORY for historical analysis and dashboards, not as your primary source during an active incident because Account Usage data is delayed.

*Rule of thumb: if the **queue** bars dominate, fix the warehouse and workload isolation. If **scan**/operator bars dominate, fix the **query shape** and **table layout**.*

13.1.2 Quickly Finding the Worst Offenders

Top queries by elapsed time (last six hours), with simple attribution:

```
SELECT
  QUERY_ID,
  START_TIME,
  USER_NAME,
  WAREHOUSE_NAME,
  TOTAL_ELAPSED_TIME/1000.0 AS elapsed_s,
  EXECUTION_TIME/1000.0     AS exec_s,
```

```
  (QUEUED_OVERLOAD_TIME + QUEUED_PROVISIONING_TIME)/1000.0 AS queued_s,
  ROWS_PRODUCED,
  QUERY_TEXT
FROM TABLE(INFORMATION_SCHEMA.QUERY_HISTORY(
  END_TIME_RANGE_START => DATEADD('hour', -6, CURRENT_TIMESTAMP()),
  RESULT_LIMIT => 2000
))
ORDER BY TOTAL_ELAPSED_TIME DESC
LIMIT 50;
```

If **queued_s** is large relative to **elapsed_s**, you're starved for compute or concurrency. If **exec_s** dominates, dive into the profile.

Find "expensive scans" (big execution time **and** high output) to prioritize review:

```
SELECT
  QUERY_ID, START_TIME, WAREHOUSE_NAME,
  EXECUTION_TIME/1000.0 AS exec_s,
  ROWS_PRODUCED,
  QUERY_TEXT
FROM TABLE(INFORMATION_SCHEMA.QUERY_HISTORY(
  END_TIME_RANGE_START => DATEADD('hour', -6, CURRENT_TIMESTAMP()),
  RESULT_LIMIT => 2000
))
WHERE EXECUTION_STATUS = 'SUCCESS'
ORDER BY exec_s DESC
LIMIT 50;
```

Spot warehouse pressure and prove it's not a single rogue query by correlating with load history:

```
SELECT *
FROM TABLE(INFORMATION_SCHEMA.WAREHOUSE_LOAD_HISTORY(
  WAREHOUSE_NAME => 'WH_BI',
  START_TIME => DATEADD('hour', -6, CURRENT_TIMESTAMP())
))
ORDER BY START_TIME;
```

13.1.3 Reading the Query Profile Like an Engineer

Open the query in Snowsight and click **Profile**. Work left-to-right over the operator tree:

- **Scan operators**: Check partitions/bytes scanned vs. output rows. If it reads a lot to return a little, you have a **pruning** problem (missing/ loose predicates, selecting too many columns, or data not co-located for common filters).
- **Join operators**: Look for skew—one input huge, the other tiny—and confirm the join keys are selective. If a broadcast-style pattern re-reads a large dimension, consider materializing or pre-aggregating.
- **Sort/aggregate**: Large sorts imply too-wide `SELECT` lists or late filters. Push projections/filters up.
- **Result**: Millions of output rows to the client will be slow, even if the compute is fine. Persist to a table or limit earlier.

A minimal rewrite flow is: **drop unused columns**, **push predicates into the earliest CTE** that can satisfy them, and **filter by scalar envelope columns** (date/tenant) before digging into `VARIANT`.

13.1.4 Common Symptoms → What to Measure → First Move

Symptom in history/ profile	What to measure	First move you can ship today
Big **queued_s** bars; long waits at the top of the hour	Warehouse load history; concurrent statements; queued time	Give BI/ELT their own warehouses; enable multi-cluster for bursts; stagger schedules
Scans read **many partitions** for simple date filters	Partition ranges in profile; predicate selectivity	Ensure filters hit scalar columns (`order_dt`, `tenant_id`); load by narrow date windows; cluster only if selective filters still under-prune

(continued)

Symptom in history/ profile	What to measure	First move you can ship today
Join node dominates; one input huge	Cardinality of join keys; null/ duplicate rates	Pre-filter early; coerce keys to common types; consider pre-agg on large side
Heavy sort/aggregate after filters	Column list width; placement of `WHERE/QUALIFY`	Project only needed columns; move filters into inner CTEs; add `QUALIFY` to avoid nested subqueries
Fast re-runs; first run slow	Result/local cache effects	Keep a small always-on warehouse for dashboards; avoid constant suspend/resume thrash during peak windows

13.1.5 Proving a Data-Layout Issue (And Fixing It)

If the profile shows poor pruning on a large fact, quantify and decide whether to introduce a **clustering key**:

```
-- How well layout aligns with (event_dt, user_id) filters?
SELECT SYSTEM$CLUSTERING_INFORMATION('FACT_EVENTS');
```

Good pruning and shallow depth? Don't cluster. Poor pruning on your most common filters? Nudge layout:

```
ALTER TABLE fact_events CLUSTER BY (event_dt, user_id);
```

Then run the same query and compare **scanned partitions** and **exec_s**.

13.1.6 Catching "Obvious" Query-Shape Issues with a Quick Checklist

Check	Why it matters	Fix
Selecting `SELECT *` from wide tables	More IO, bigger shuffles	Project only needed columns
Filtering inside JSON only	Pruning can't engage	Add envelope columns (`event_dt`, `tenant_id`) and filter there first
Mixing types on join keys	Forces casts, blocks stats	Cast once in a CTE; align types in tables
Late `DISTINCT` for dedupe	Full sort on big sets	Use window + `QUALIFY rn=1` on a partitioned order
Giant client fetch	Network + serialization cost	`CREATE TABLE AS SELECT` ... then sample/limit for inspection

13.1.7 A Few Commands You'll Reach for Repeatedly

Inspect the optimizer's **estimated logical plan** without running the statement:

```
EXPLAIN USING TEXT
SELECT ...;
```

Use EXPLAIN to understand planned operators, join shape, and estimated scans before execution. Do not treat it as runtime evidence: EXPLAIN does not execute the query, and its partition and byte figures are estimates rather than actual observed execution metrics. For what really happened, use Query History and Snowsight Query Profile.

Focus on your own recent work:

```
SELECT *
FROM TABLE(INFORMATION_SCHEMA.QUERY_HISTORY_BY_SESSION(RESULT_LIMIT
=> 200))
ORDER BY START_TIME DESC;
```

Tie a slow dashboard tile to its warehouse and queued time:

```
SELECT QUERY_ID, START_TIME, WAREHOUSE_NAME,
       (QUEUED_OVERLOAD_TIME + QUEUED_PROVISIONING_TIME)/1000.0 AS
       queued_s,
       EXECUTION_TIME/1000.0 AS exec_s,
       QUERY_TEXT
FROM TABLE(INFORMATION_SCHEMA.QUERY_HISTORY(
  END_TIME_RANGE_START => DATEADD('hour', -1, CURRENT_TIMESTAMP()),
  RESULT_LIMIT => 500
))
WHERE USER_NAME = 'BI_SERVICE_USER'
ORDER BY START_TIME DESC;
```

13.1.8 Bringing It Together

Performance tuning in Snowflake is a **measurement discipline**. Let QUERY_HISTORY tell you if you're waiting or working; let the **Query Profile** show where work piled up. If you're waiting, fix the **workload** (warehouses, schedules, isolation). If you're working too hard, fix the **query** (predicates, projections, join order) and, only when necessary, the **layout** (load cadence, clustering). Do the smallest thing that makes the next run scan less and wait less—and confirm it with the same tools that sent you here.

13.2 Micro-partition Pruning and Clustering Strategies

Snowflake is fast when it **doesn't read**. Micro-partition pruning is the primary accelerator: each partition carries column min/max and other stats, so a good predicate lets the optimizer skip whole chunks. Clustering is a *nudge* you add only when natural write order fights your most common filters. Tuning is therefore mostly about shaping data and predicates, so pruning is ruthless—and only then deciding whether a clustering key is worth the ongoing maintenance.

13.2.1 Diagnose First, Then Choose the Lightest Fix

Symptom in Query Profile	Likely cause	Lightest effective change
Many partitions scanned for simple date filters	Loads land in wide, mixed time windows	Ingest by narrow time slices (hour/day) so `WHERE dt BETWEEN ...` prunes naturally
Tenant/entity filters still scan broadly	Data interleaves tenants/entities within each load batch	Add `tenant_id` (or entity) to load order/path; if still broad, consider `(dt, tenant_id)` clustering
Selective equality on a high-cardinality key (e.g., `user_id = ?`)	Key values spread randomly across partitions	If queries are consistently "entity over time," a composite clustering key such as `(dt, user_id)` can help; otherwise, consider Search Optimization for sparse point lookups. Keep in mind that Snowflake generally recommends ordering multi-column clustering keys from lower cardinality to higher cardinality.
Filters live inside JSON only	Pruning can't use `VARIANT` paths alone	Keep scalar envelope columns (e.g., `event_dt`, `tenant_id`, `type`) and filter on them first
Good predicates, still lots of scans	Columns are wide; projecting `SELECT *`	Project only needed columns; push `WHERE` into the earliest CTE

13.2.2 Measure Pruning: Don't Guess

```
-- How well the table aligns to (event_dt, user_id) filters?
SELECT SYSTEM$CLUSTERING_INFORMATION('FACT_EVENTS') AS info;

-- Storage/partition footprint
SELECT *
FROM INFORMATION_SCHEMA.TABLE_STORAGE_METRICS
WHERE TABLE_NAME = 'FACT_EVENTS';
```

Read Query Profile's **"partitions scanned"** and **bytes scanned** before/after changes; that's the ground truth.

13.2.3 Practical Layout Patterns That Prune Well

Situation	Layout that wins	Why it helps
Time-series facts	Land files per **hour/day**; keep `event_dt` as a scalar column	Date predicates exclude whole ranges early
Multi-tenant SaaS	Keep **tenant_id** prominent in file paths or order of inserts	Tenant filters skip most partitions
"Entity over time" lookups	Cluster on `(event_dt, entity_id)` *only* if natural loads don't align	Co-locates hot entity ranges enough for pruning
Semi-structured logs	Store `event_dt`, `tenant_id`, `type` beside `payload VARIANT`	Pruning decides before JSON parsing
External tables (Parquet)	Partition folders by `dt=/tenant=` with ~200 MB files	Planner prunes by folder and row-group stats

13.2.4 When Clustering Is Worth It (And When It Isn't)

Clustering is justified when **all** are true:

- You have **stable, selective predicates** (e.g., `WHERE dt BETWEEN … AND ... AND user_id = ?`).
- Natural ingest order can't produce good locality.
- Query volume justifies **ongoing maintenance cost.**

If that's your case:

```
ALTER TABLE fact_events CLUSTER BY (event_dt, user_id);
```

For multi-column clustering keys, order columns from lower cardinality to higher cardinality. In many "entity over time" cases, a date column is placed before a more granular identifier for exactly that reason.

Snowflake maintains the layout in the background; you can pause/resume if needed:

```
ALTER TABLE fact_events SUSPEND RECLUSTER;  -- pause maintenance
ALTER TABLE fact_events RESUME RECLUSTER;   -- resume later
```

If clustering info shows **shallow depth** and high pruning already, **do nothing**—extra maintenance won't make scans meaningfully smaller.

13.2.5 Alternatives to Clustering for Needle-in-Haystack Filters

Need	Option	When to use it
Point/equality lookups on sparse keys; occasional wildcards	**Search Optimization** on specific columns	When queries remain selective but data resists locality (especially strings, semi-structured paths)
Repeatable complex filters	**Materialized views** over narrowed ranges	When a subset is hot and stable enough to maintain
Many tiny predicates from BI	**Result cache** + dedicated BI warehouse	When the same query text repeats frequently

```
-- Estimate cost before enabling Search Optimization
SELECT SYSTEM$ESTIMATE_SEARCH_OPTIMIZATION_COSTS('RAW.EVENTS');
```

Use Search Optimization surgically. Before enabling it, estimate build, storage, and maintenance cost with `SYSTEM$ESTIMATE_SEARCH_OPTIMIZATION_COSTS`, because the service adds ongoing cost that depends on the number of optimized columns, their distinct values, and how much the table changes over time. Cost estimates are best-effort and actual costs can vary from the estimate.

13.2.6 A Small, Repeatable Tuning Loop

1. Profile a slow query using `QUERY_HISTORY` and the Snowsight Query Profile (see 13.1) → note queued vs. execution time, partitions and bytes scanned, and the operators that dominate.
2. **Tighten the query** → add scalar filters, drop unused columns, align types on joins.
3. **Shape ingest** → narrower time slices; order rows to group by tenant/region.

4. **Re-measure** → if scans remain broad on hot predicates, add a clustering key.
5. **Validate cost** → watch automatic recluster metrics and warehouse/maintenance spend.

13.2.7 Example: From Broad Scans to Sharp Pruning

Before (poor locality, filters inside JSON):

```
SELECT COUNT(*)
FROM raw.events
WHERE payload:type::STRING = 'click'
  AND payload:user_id::STRING = 'U123'
  AND payload:ts::DATE BETWEEN '2025-09-01' AND '2025-09-07';
```

After (envelope + optional clustering):

```
-- Land envelope columns at ingest:
-- (tenant_id, event_dt, type, payload VARIANT)

SELECT COUNT(*)
FROM raw.events
WHERE type = 'click'
  AND user_id = 'U123'              -- extracted once into a scalar
  AND event_dt BETWEEN '2025-09-01' AND '2025-09-07';

-- If pruning remains weak for (event_dt, user_id), add:
ALTER TABLE raw.events CLUSTER BY (event_dt, user_id);
```

Re-run and compare **partitions scanned**; if they collapse to a small fraction, you've won without oversizing compute.

13.2.8 Common Traps and Quick Exits

Trap	Why it hurts	Safer move
Clustering "just in case"	Ongoing maintenance with no scan reduction	Prove need with SYSTEM$CLUSTERING_INFORMATION first
Clustering on too many columns	Dilutes locality; increases maintenance	1–2 columns that mirror real predicates
Filtering only inside VARIANT	Pruning can't engage	Add scalar envelope columns and filter them
Huge SELECT * on wide tables	Reads unnecessary columns and bytes	Project only what you use, and earlier in the plan
Ignoring file layout in lakes	Planner can't prune folders/row groups	Partition folders sensibly; compact to balanced Parquet

Bottom line: make pruning cheap with envelope columns and disciplined ingest; measure before and after; and reserve clustering for the few tables and predicates where natural locality can't keep up.

13.3 Result Cache, Warehouse Sizing, and Concurrency

Snowflake feels fast when repeated work doesn't run again and new work doesn't wait. The first is the **result cache** (and warm micro-partition caches). The second is **right-sized warehouses** with enough lanes for concurrent queries. Tuning here is mostly operational: make queries cache-friendly; separate ELT from BI; and pick scale-up vs. scale-out based on the kind of pressure you see.

13.3.1 How the Caches Actually Help

Cache	What it holds	Lives where	When you benefit	How you lose it
Result cache	Persisted query results can be reused when a matching query is submitted and the referenced results have not been invalidated by data or object changes.	Global (typically available for up to 24 hours), independent of a running warehouse	Repeated queries against unchanged underlying data	Underlying data or dependent objects change, session settings alter semantics, privileges differ, or the submitted statement is no longer a matching query for persisted-result reuse. Do not focus on whitespace normalization as a tuning tactic.
Warehouse (local) cache	Recently read micro-partitions	On each running warehouse cluster	Iterative exploration; repeated scans during refresh windows	Warehouse suspends/ scales to zero; different warehouse; heavy churn spins caches away
Metadata/ plan cache	Table stats, partition ranges, compiled plans	Services layer	Stable schemas and predicates	DDL churn; volatile filters; primitives that force re-planning

Practical takeaways: keep BI query patterns stable, parameterize carefully, avoid gratuitous `SELECT *`, and do not bounce the BI warehouse during peak periods because warm local caches reduce repeated scan cost. Stable query shape matters more than formatting trivia such as whitespace.

13.3.2 Result-Cache Eligibility at a Glance

Condition	Must be true for a cache hit
SQL text and session state	The submitted statement must be a matching query for persisted-result reuse, and session settings must not change query semantics. In practice, aim for stable, repeatable query text, but do not treat whitespace normalization as the key tuning lever.
Data stability	None of the referenced objects changed since the last run (no INSERT/UPDATE/MERGE/DDL that affects results)
Privileges	Caller still has rights to all objects the original query touched

If you can't rely on result caching (fast-changing tables), you can still hit **local caches** by keeping the same warehouse alive across dashboard refreshes.

13.3.3 Warehouse Sizing: Scale Up vs. Scale Out

Workload pattern	Tell-tale signs	Better move	Why
BI, many short queries	Queue spikes, small per-query scans	**Scale out** (multi-cluster)	More lanes; each query stays on a small, cheap cluster
Heavy ELT step (windowed joins, large aggregations)	Long execution time, little queuing	**Scale up** (larger size, one cluster)	More cores per query; finish sooner, then scale back
Mixed BI + ELT on one warehouse	BI waits behind ELT	**Separate warehouses**	Isolation keeps caches and SLAs stable

Baseline Posture

- Give BI a dedicated **multi-cluster** warehouse (small size, MIN=1, MAX=~3–5; `SCALING_POLICY = STANDARD`).
- Give ELT one or more **single-cluster** warehouses sized for the step; schedule to avoid BI peaks; use auto-suspend to limit idle cost.

```
-- BI: small, elastic, cache-friendly
CREATE WAREHOUSE wh_bi
  WAREHOUSE_SIZE = 'SMALL'
  MIN_CLUSTER_COUNT = 1
  MAX_CLUSTER_COUNT = 4
  SCALING_POLICY = 'STANDARD'
  AUTO_SUSPEND = 60
  AUTO_RESUME = TRUE
  INITIALLY_SUSPENDED = TRUE;
```

In Snowflake, each warehouse cluster has a MAX_CONCURRENCY_LEVEL, which defaults to 8. When that level is reached, additional statements queue on a single-cluster warehouse; for a multi-cluster warehouse in auto-scale mode, Snowflake can start additional clusters instead. If dashboard bursts still overwhelm the warehouse, readers may see rising queued time or statement cancellations if STATEMENT_QUEUED_TIMEOUT_IN_SECONDS is set. In practice, the fixes are to raise MAX_CLUSTER_COUNT, split BI and ELT workloads, or tune concurrency expectations explicitly.

```
-- ELT: bursty, bigger muscles, no contention with BI
CREATE WAREHOUSE wh_elt
  WAREHOUSE_SIZE = 'LARGE'
  MIN_CLUSTER_COUNT = 1
  MAX_CLUSTER_COUNT = 1
  AUTO_SUSPEND = 300
  AUTO_RESUME = TRUE;
```

> *Rule of thumb: if **queued time** dominates, you need more **lanes** (scale out or split workloads). If **execution time** dominates, you need more **horsepower** (scale up) or better pruning.*

13.3.4 Reading Pressure vs. Waiting (So You Size the Right Thing)

```
-- Spot queuing vs. execution for recent queries
SELECT
  QUERY_ID, WAREHOUSE_NAME,
  TOTAL_ELAPSED_TIME/1000 AS elapsed_s,
```

```
  (QUEUED_OVERLOAD_TIME + QUEUED_PROVISIONING_TIME)/1000 AS queued_s,
  EXECUTION_TIME/1000 AS exec_s,
  ROWS_PRODUCED
FROM TABLE(INFORMATION_SCHEMA.QUERY_HISTORY(
  END_TIME_RANGE_START => DATEADD('hour', -2, CURRENT_TIMESTAMP()),
  RESULT_LIMIT => 1000
))
ORDER BY elapsed_s DESC;

-- Warehouse load shape over time
SELECT *
FROM TABLE(INFORMATION_SCHEMA.WAREHOUSE_LOAD_HISTORY(
  WAREHOUSE_NAME => 'WH_BI',
  START_TIME => DATEADD('hour', -2, CURRENT_TIMESTAMP())
))
ORDER BY START_TIME;
```

If queued_s swells during dashboard refresh minutes, raise MAX_CLUSTER_COUNT or stagger the BI scheduler. If exec_s rules, revisit predicates and micro-partition layout before buying bigger compute.

13.3.5 Concurrency: Making "More Lanes" Work for You

Setting	Value to prefer	Effect
MIN_CLUSTER_COUNT	1	Keeps at least one warm lane for caches
MAX_CLUSTER_COUNT	3–5 (BI), 1 (ELT)	Elasticity for bursts vs. single strong lane
SCALING_POLICY	STANDARD for BI; ECONOMY when bursts can wait a little	STANDARD adds clusters quickly; ECONOMY consolidates to save cost
Auto-suspend/resume	Short suspend for BI (30–120s)	Preserves caches through short lulls; avoids thrash

Multi-cluster doesn't make a single query faster; it prevents queries from **waiting** behind each other. Pair it with small warehouse sizes so each lane is cheap.

13.3.6 Resource Monitors and Spend Safety

Set hard or soft caps that match your operating model.

```
CREATE OR REPLACE RESOURCE MONITOR rm_bi
  WITH CREDIT_QUOTA = 500
  FREQUENCY = MONTHLY
  START_TIMESTAMP = IMMEDIATELY
  TRIGGERS ON 80 PERCENT DO NOTIFY
           ON 95 PERCENT DO SUSPEND
           ON 100 PERCENT DO SUSPEND_IMMEDIATE;

ALTER WAREHOUSE wh_bi SET RESOURCE_MONITOR = rm_bi;
```

Monitors won't tune performance, but they keep experiments and runaway dashboards from surprising your budget. One additional lever worth noting is Snowflake's Query Acceleration Service (QAS): for eligible outlier queries, especially those with large scans combined with selective filters or aggregations, Snowflake can offload part of the execution to serverless compute rather than relying only on warehouse sizing. Because QAS is billed separately from warehouse credits, it should be evaluated deliberately with tools such as `SYSTEM$ESTIMATE_QUERY_ACCELERATION` and the `QUERY_ACCELERATION_ELIGIBLE` view.

13.3.7 Troubles You Can Fix Quickly

Symptom	Likely cause	Smallest effective change
Dashboards fast once, slow later	Warehouse suspends between tiles; local caches lost	Lengthen BI AUTO_SUSPEND; keep MIN_CLUSTER=1 during peak windows
Repeated dashboard tiles do not reuse persisted results	The query is no longer a matching query for persisted-result reuse, underlying data changed, or semantics differ because of session settings or privileges	Parameterize carefully, keep query patterns stable, and accept that fresh data invalidates persisted results; do not treat whitespace alignment as the main remedy.
Queue spikes at the top of the hour	Many jobs start together	Stagger schedules; raise MAX_CLUSTER_COUNT; separate BI/ELT
ELT still slow after scaling up	Scanning too much	Fix pruning (date/tenant filters, fewer columns) before throwing compute
BI cost drift	Idle multi-cluster lingering	Use ECONOMY scaling off-peak; tighten AUTO_SUSPEND; cap with a resource monitor

13.3.8 Putting It Together

Make repeated work **cache-eligible**, not just bigger-warehouse-tolerant. Give BI its own small, elastic warehouse so concurrency never blocks, and give ELT a strong but isolated lane so heavy steps don't evict BI caches. Watch **queued vs. execution** time to decide whether to add lanes or horsepower, and confirm changes with the same history views that sent you tuning in the first place.

CHAPTER 14

Warehouse Design and Workload Management

Snowflake performance isn't only about query shape; it's about **where queries run** and **who they run with**. A great design treats warehouses as lanes on a highway. BI dashboards shouldn't idle behind nightly ELT. Data science experiments shouldn't evict caches that keep finance reports snappy. The goal is simple: give each workload a lane sized for its behavior, keep the lanes warm just enough to hit caches, and control spend with guardrails instead of heroics.

Warehouses are cheap to create and safe to delete. That makes *isolation* the default move: separate by workload class, sometimes by team or domain, and only share when behaviors are truly compatible. Everything else—caches, result reuse, concurrency—gets easier once the lanes are tidy.

14.1 Multi-warehouse Strategy (XL vs. Many Smalls)

There are two broad ways to buy speed: one **very large** warehouse (scale up) or **several small** warehouses (scale out). Both can be right; the choice depends on whether you're fighting **long single queries** or **a crowd of small ones** and whether isolation matters more than raw horsepower.

M. Hander, *Snowflake Data Warehouse Engineering*, https://doi.org/10.1007/979-8-8688-2628-3_14

14.1.1 How the Strategies Trade Off

Dimension	One XL warehouse	Many small warehouses (per workload)
Query pattern fit	Heavy ELT steps, big joins/ aggregations finish faster on more cores	BI dashboards, concurrent ad hoc exploration, micro-batch jobs
Concurrency	Many queries still **wait** behind each other	More lanes; fewer waits; each lane stays cheap
Cache behavior	One big cache; but ELT churn can evict BI locality	Each lane retains its own warm partitions; BI caches survive ELT runs
Blast radius	Noisy neighbor can lag everyone	Incidents contained to one lane
Cost control	Simple to meter; hard to apportion by team	Tag/monitor per lane; budgets and alerts by warehouse
Tuning levers	"Scale up" during heavy windows, then down	"Scale out" via multi-cluster for BI; right-size per lane
Operability	Fewer objects to manage	Clearer SLOs and ownership, slightly more objects

> *Rule of thumb: if **execution time** dominates and queries are individually heavy, an **XL** (or temporarily larger) warehouse is the lever. If **queued time** dominates and queries are small but many, **many Smalls** (often with multi-cluster) win.*

14.1.2 A Sensible Lane Plan

Most organizations land here:

Lane	Purpose	Sizing and elasticity	Policy that keeps it calm
BI	Dashboards, analysts, apps	SMALL, multi-cluster (MIN=1, MAX=3–5), SCALING_POLICY=STANDARD	Stagger schedules; keep AUTO_SUSPEND modest (30–120s) so caches persist
ELT	Transforms, merges, windowed jobs	LARGE (or step-specific), single cluster	Run on cadence, not during BI peaks; use Tasks; auto-suspend between waves
Ad hoc	Notebooks, exploration	XSMALL/SMALL, single cluster per team	Soft quotas via resource monitors; query tags for attribution
Data Science	Feature builds, sampling	SMALL/MEDIUM, bursty	Encourage CTAS/results in tables; isolate experiments
Ingestion	Snowpipe Streaming/ Snowpipe helpers	XSMALL always-on	Keep steady; tiny but warm to avoid cold-start latency

14.1.3 Routing, Guardrails, and Attribution

Make the path of least resistance the **right** warehouse. Bind roles to default warehouses; tag queries for chargeback; cap spend with monitors. Default warehouses improve routing, but effective access still depends on role hierarchy inheritance and direct grants.

```
-- Role defaults steer users to the right lane
ALTER ROLE analyst SET DEFAULT_WAREHOUSE = wh_bi;
ALTER ROLE elt_runner SET DEFAULT_WAREHOUSE = wh_elt;

-- Query tagging for attribution
ALTER SESSION SET QUERY_TAG = 'team=finance;purpose=dashboard';

-- Spend guardrails
CREATE OR REPLACE RESOURCE MONITOR rm_bi
  WITH CREDIT_QUOTA = 500
  FREQUENCY = MONTHLY
  TRIGGERS ON 80 PERCENT DO NOTIFY
           ON 95 PERCENT DO SUSPEND
```

```
        ON 100 PERCENT DO SUSPEND_IMMEDIATE;
ALTER WAREHOUSE wh_bi SET RESOURCE_MONITOR = rm_bi;
```

14.1.4 Choosing Scale-up vs. Scale-out for a Bottleneck

Symptom (from QUERY_HISTORY/load history)	Best next move	Why it works
High **queued** time, low execution time	Increase **lanes**: multi-cluster on BI or split workloads	Reduces wait without overspending on cores you don't use
High **execution** time, queries CPU-bound	Temporarily **scale up** ELT warehouse; fix pruning next	More cores per query shorten critical steps while you improve the layout
BI tiles fast once, slow later	Keep the BI lane warm longer; don't bounce during peak	Preserves local caches for repeat scans
BI slowed by ELT	Separate warehouses (and schedules)	Isolation protects BI caches and SLOs

14.1.5 Concrete, Boring Defaults That Hold Up

```
-- BI: elastic lanes, small cores, warm enough for caches
CREATE WAREHOUSE wh_bi
  WAREHOUSE_SIZE = 'SMALL'
  MIN_CLUSTER_COUNT = 1
  MAX_CLUSTER_COUNT = 4
  SCALING_POLICY = 'STANDARD'
  AUTO_SUSPEND = 60
  AUTO_RESUME = TRUE;

-- ELT: muscle for heavy steps, no contention with BI
CREATE WAREHOUSE wh_elt
  WAREHOUSE_SIZE = 'LARGE'
  MIN_CLUSTER_COUNT = 1
  MAX_CLUSTER_COUNT = 1
  AUTO_SUSPEND = 300
  AUTO_RESUME = TRUE;
```

When a monthly close or a backfill needs more power, scale **ELT** up for the window, not forever. When a product launch drives BI bursts, raise **MAX_CLUSTER_COUNT** temporarily.

14.1.6 What to Watch So You Don't Guess

Signal	Where to read it	Decision it informs
Queued vs. execution time	`INFORMATION_SCHEMA.QUERY_HISTORY*`	Add lanes or add horsepower
Warehouse pressure over time	`WAREHOUSE_LOAD_HISTORY`	Stagger schedules; tune MAX_CLUSTER_COUNT
Credits by lane/team	`METERING_HISTORY`, query tags	Budgets, chargeback, right-sizing
Cache behavior	First-run vs. re-run latencies	Adjust auto-suspend; keep BI warm during peak windows

14.1.7 Closing Guidance

Start with **many Smalls**: one BI lane with multi-cluster, one ELT lane sized for the heaviest step, and a small lane for ad hoc. Promote lanes to **larger** only when profiling shows execution, not queuing, is the enemy. Keep warehouses warm just long enough to benefit from caches; keep them separate so a noisy neighbor cannot steal your good day. When in doubt, try the smallest change that reduces **waiting** or **reading**—and verify it with the same history views that sent you tuning in the first place.

14.2 Auto-suspend/resume and Scaling Policies

Warehouses should be **awake when needed, asleep when idle, and elastic only when bursts arrive**. Auto-suspend/resume and scaling policies do that work. The art is picking suspend times that preserve caches during peak windows and a scaling policy that trades speed vs. spend the way your workload needs.

14.2.1 What the Settings Actually Control

Setting	What it does	Why it matters
AUTO_SUSPEND = <seconds>	Idles a warehouse after N seconds with no queries	Short values save credits but drop local caches; longer values keep dashboards snappy
AUTO_RESUME = TRUE/ FALSE	Starts a suspended warehouse on demand	Turn it on for human/BI workloads; off only for tightly scheduled batch
MIN_CLUSTER_ COUNT/MAX_CLUSTER_ COUNT	Bounds the number of concurrent clusters (multi-cluster)	More clusters = more lanes; doesn't speed a single query
`SCALING_POLICY = STANDARD	ECONOMY`	Determines how aggressively clusters spin up/down

14.2.2 STANDARD vs. ECONOMY at a Glance

Policy	Spin-up behavior	Scale-down behavior	Best for
STANDARD	Minimizes queuing by favoring starting additional clusters	Keeps responsiveness high during bursts	BI peaks, scheduled reports, ad hoc spikes
ECONOMY	Starts additional clusters only when Snowflake estimates enough load to keep them busy for at least six minutes	Shuts down lightly loaded clusters when less than about six minutes of work remain	Overnight jobs, steady but latency—tolerant workloads

In practice, ECONOMY can leave short BI bursts queued longer than STANDARD, because it is intentionally conservative about starting extra clusters.

14.2.3 Practical Defaults by Lane

Lane	Size and clusters	Suspend	Policy	Rationale
BI	SMALL, MIN=1, MAX=3–5	60–120 s	STANDARD	Keeps one warm lane for caches; bursts get extra lanes fast
ELT	LARGE, MIN=1, MAX=1	300 s	ECONOMY or omit	One strong lane; suspend between waves to save credits
Ad hoc	XSMALL/SMALL, MIN=1, MAX=1	60 s	ECONOMY	Cheap, quick to wake; cost-biased when idle
Ingestion helpers	XSMALL, MIN=1, MAX=1	N/A (keep on)	ECONOMY	Avoid cold-start latency on frequent micro-loads

14.2.4 Boring, Good SQL to Set the Posture

```
-- BI: elastic, cache-friendly
CREATE WAREHOUSE wh_bi
  WAREHOUSE_SIZE = 'SMALL'
  MIN_CLUSTER_COUNT = 1
  MAX_CLUSTER_COUNT = 4
  SCALING_POLICY = 'STANDARD'
  AUTO_SUSPEND = 90
  AUTO_RESUME = TRUE;

-- ELT: muscle for heavy steps; no multi-cluster
CREATE WAREHOUSE wh_elt
  WAREHOUSE_SIZE = 'LARGE'
  MIN_CLUSTER_COUNT = 1
  MAX_CLUSTER_COUNT = 1
  AUTO_SUSPEND = 300
  AUTO_RESUME = TRUE
  SCALING_POLICY = 'ECONOMY';
```

Adjust on the fly without recreate:

```
ALTER WAREHOUSE wh_bi SET AUTO_SUSPEND = 120;
ALTER WAREHOUSE wh_bi SET MAX_CLUSTER_COUNT = 6;
ALTER WAREHOUSE wh_bi SET SCALING_POLICY = 'ECONOMY'; -- off-peak
```

14.2.5 Picking Suspend Times That Don't Fight Caches

Symptom	What's happening	Smallest effective change
Dashboards fast once, then sluggish	Local caches vanish between tiles	Increase AUTO_SUSPEND to cover the refresh ripple (e.g., 60→120s); keep MIN_CLUSTER=1 during peak
Cost drift overnight	Warehouses linger warm with no queries	Shorten AUTO_SUSPEND off-peak; switch BI to ECONOMY after hours via schedule
Queues at the top of the hour	Many jobs start together	Keep STANDARD, raise MAX_CLUSTER_COUNT, or stagger schedules

14.2.6 When to Scale Up vs. Scale Out (And What Policies Do)

Bottleneck	Move	Why
Long **execution** time on single heavy queries	Scale **up** size; keep MAX_CLUSTER=1	More cores per query; policy doesn't matter
High **queued** time with many small queries	Scale **out** MAX_CLUSTER>1; STANDARD during peaks	More lanes; STANDARD adds them quickly
Steady trickle, cost sensitive	ECONOMY	Consolidates lanes sooner, fewer idle minutes

14.2.7 Observability Loop (So You Tune with Evidence)

```
-- Is time spent waiting or working?
SELECT
  WAREHOUSE_NAME,
  AVG((QUEUED_OVERLOAD_TIME + QUEUED_PROVISIONING_TIME)/1000) AS avg_
  queued_s,
  AVG(EXECUTION_TIME/1000) AS avg_exec_s
FROM TABLE(INFORMATION_SCHEMA.QUERY_HISTORY(
  END_TIME_RANGE_START => DATEADD('hour', -2, CURRENT_TIMESTAMP())
))
GROUP BY 1
ORDER BY avg_queued_s DESC;

-- Did scaling actually add lanes when needed?
SELECT *
FROM TABLE(INFORMATION_SCHEMA.WAREHOUSE_LOAD_HISTORY(
  WAREHOUSE_NAME => 'WH_BI',
  START_TIME => DATEADD('hour', -2, CURRENT_TIMESTAMP())
))
ORDER BY START_TIME;
```

Read these alongside user complaints. If **queued** rises while active clusters stay at **1**, raise MAX_CLUSTER_COUNT or move to STANDARD. If **exec** dominates, fix pruning or size up temporarily, then right-size back down.

14.2.8 Safety Rails You'll Want

Control	Example	What it prevents
Resource monitor per lane	Credits + suspend triggers	Surprise spend during bursts
Time-based policy flips	Scheduled ALTER WAREHOUSE ... SCALING_POLICY='ECONOMY' at night	Paying for elasticity you don't need off-peak
Default warehouses by role	ALTER ROLE analyst SET DEFAULT_WAREHOUSE=wh_bi	Misrouted queries landing on ELT or vice versa

Bottom line: keep BI **warm and elastic**, keep ELT **strong and solitary**, and let STANDARD absorb peaks while ECONOMY trims idle cost off-peak. Tune with history—not hunches—and favor the smallest change that reduces **waiting** without inflating **work**.

14.3 Resource Monitors and Workload Isolation

Two levers keep Snowflake calm on busy days and predictable on the last day of the month: **resource monitors** to cap or warn on credits and **workload isolation** so one team's spike never sinks another's SLA. Treat monitors as guardrails, not brakes; they should notify early and only suspend the lane that's misbehaving. Resource monitors are guardrails for **warehouse** spend, not for all compute spend. Serverless features follow separate cost views and governance controls. Isolation does the rest: dedicated warehouses per class of work, clear routing via roles and tags, and simple attribution.

14.3.1 What Resource Monitors Actually Do

Important: resource monitors govern **user-managed warehouses only**. They do not track or suspend spend from serverless features, including serverless tasks. For serverless task cost visibility, use `SNOWFLAKE.ACCOUNT_USAGE.SERVERLESS_TASK_HISTORY`; for governance, use budgets rather than warehouse resource monitors.

Concept	What it controls	Where it applies	Typical posture
Credit quota	Total credits a monitor will watch in a period	Account or one/many warehouses	Monthly quotas per lane (BI, ELT, DataSci)
Triggers	Actions at % thresholds	Per monitor	80% notify, 95% suspend, 100% suspend immediate
Frequency	Reset the cadence of the meter	Monthly/Weekly/Daily	Match the finance reporting cycle
Scope	Which warehouses count	`ALTER WAREHOUSE ... SET RESOURCE_MONITOR=`	Bind each lane to its own monitor

Create and bind (separate lanes, separate budgets):

```
-- BI lane: warn early, suspend before runaway
CREATE OR REPLACE RESOURCE MONITOR rm_bi
  WITH CREDIT_QUOTA = 500
  FREQUENCY = MONTHLY
  START_TIMESTAMP = IMMEDIATELY
  TRIGGERS ON 80 PERCENT DO NOTIFY
           ON 95 PERCENT DO SUSPEND
           ON 100 PERCENT DO SUSPEND_IMMEDIATE;

ALTER WAREHOUSE wh_bi SET RESOURCE_MONITOR = rm_bi;

-- ELT lane: bigger budget, same pattern
CREATE OR REPLACE RESOURCE MONITOR rm_elt
  WITH CREDIT_QUOTA = 1200
  FREQUENCY = MONTHLY
  TRIGGERS ON 85 PERCENT DO NOTIFY
           ON 100 PERCENT DO SUSPEND;
ALTER WAREHOUSE wh_elt SET RESOURCE_MONITOR = rm_elt;
```

14.3.2 Workload Isolation That Survives Real Life

Isolation dimension	How to apply it	Why it helps
Warehouses by class of work	BI, ELT, Ad hoc, Ingestion each get a warehouse (sizes/policies from §14.1–14.2)	Noisy neighbors can only harm their own lane
RBAC routing	Default warehouses per role; grant USAGE only where intended	Users and Tasks "fall into" the right lane
Tags and query tags	Tag warehouses/queries for chargeback; require a QUERY_TAG in prod	Clear attribution, faster root cause
Task pinning	`CREATE TASK ... WAREHOUSE = wh_elt`	ELT never steals BI capacity
Time fences	Schedule heavy ELT off BI peaks	Caches and SLAs stay stable

Concrete routing:

```
-- Default lanes by role
ALTER ROLE ANALYST    SET DEFAULT_WAREHOUSE = wh_bi;
ALTER ROLE ELT_RUNNER SET DEFAULT_WAREHOUSE = wh_elt;

-- Primary boundaries: grant warehouse access only to the intended roles,
-- and verify role hierarchy inheritance so access is not reintroduced
through parent roles.
GRANT USAGE ON WAREHOUSE wh_bi  TO ROLE ANALYST;
REVOKE USAGE ON WAREHOUSE wh_bi FROM ROLE ELT_RUNNER;
```

Note REVOKE only restricts the specific role. If ELT_RUNNER is granted to a higher role (e.g., SYSADMIN) that also has USAGE on wh_bi, isolation is not enforced at that higher role level. Audit role hierarchies and effective grants with SHOW GRANTS to confirm that boundaries actually hold. Snowflake documents that privileges granted to a role are inherited by higher-level roles in the role hierarchy.

```
-- Set attribution in apps, drivers, and schedulers; Snowflake does not
natively enforce QUERY_TAG presence
ALTER SESSION SET QUERY_TAG = 'team=finance;purpose=dashboard';
```

Note QUERY_TAG is useful for attribution, but Snowflake does not provide a native control to make query tags mandatory at the warehouse or account level, and session policies do not enforce QUERY_TAG. In practice, tags must be applied through application, driver, orchestration, or scheduler configuration, and missing tags create silent attribution gaps. Set QUERY_TAG at connection time where possible, and audit for untagged queries in query history.

Optional cost tags on objects (for automated chargeback):

```
CREATE TAG cost_center ALLOWED_VALUES ('fin','mkt','ops');
ALTER WAREHOUSE wh_bi SET TAG cost_center = 'fin';
```

14.3.3 Observability and Chargeback in Two Views

Question	Query to use	What to look for
Who spent credits?	ACCOUNT_USAGE.WAREHOUSE_METERING_HISTORY (group by warehouse/tag)	Spikes against monitors; drift vs. budget
Were caps hit?	SHOW RESOURCE MONITORS + monitor notifications	Trigger times and actions taken
How much did serverless tasks cost?	SNOWFLAKE.ACCOUNT_USAGE.SERVERLESS_TASK_HISTORY	Serverless task credits that will not appear as warehouse-metering spend

Examples:

```
-- Credits by warehouse (and tag) in the last 30 days
SELECT
  WAREHOUSE_NAME,
  TO_DATE(START_TIME) AS day,
  SUM(CREDITS_USED)   AS credits
FROM SNOWFLAKE.ACCOUNT_USAGE.WAREHOUSE_METERING_HISTORY
WHERE START_TIME >= DATEADD(day, -30, CURRENT_TIMESTAMP())
GROUP BY 1,2
ORDER BY 2,1;

-- Attribution by query tag
SELECT
  SPLIT_PART(QUERY_TAG, ';', 1) AS team_kv,
  SUM(CREDITS_USED_CLOUD_SERVICES + CREDITS_USED_COMPUTE) AS credits
FROM SNOWFLAKE.ACCOUNT_USAGE.QUERY_HISTORY
WHERE START_TIME >= DATEADD(day, -30, CURRENT_TIMESTAMP())
GROUP BY 1
ORDER BY credits DESC;
```

14.3.4 Failure Modes You Can Prevent

Symptom	Likely cause	Smallest effective change
BI stalls mid-day	ELT shares the same warehouse and spikes	Split warehouses; bind ELT Tasks to `wh_elt`; add a BI monitor
Sudden suspension during close	Monitor bound to the **account** instead of the BI lane	Scope monitors to warehouses; raise BI quota with Notify at 80%
Spend surprises	Missing tags/mixed workloads in one lane	Enforce `QUERY_TAG` in prod; tag warehouses; add per-lane monitors
"Why is cache cold?"	Frequent suspend/resume across teams in one lane	Isolate lanes; lengthen BI `AUTO_SUSPEND` to keep caches through bursts

14.4 Bringing It Together

Warehouse design works best when it becomes an operating habit rather than a one-time setup exercise. This chapter's lane model, scaling controls, and resource guardrails all point to the same practice: isolate incompatible workloads, size each lane for its behavior, and tune the warehouse before blaming the SQL. In most environments, the stable rhythm looks like this:

1. **Isolate by class of work**: BI, ELT, Ad hoc, and Ingestion should run in separate lanes by default.
2. **Bind each lane to its own guardrails**: Resource monitors for warehouse-backed workloads, and separate visibility for any serverless features.
3. **Route intentionally**: Use default warehouses, minimal grants, and consistent `QUERY_TAG` conventions in applications and schedulers.
4. **Review and adjust monthly**: Use metering history, query history, and warehouse load history to tune quotas, suspend times, and cluster limits before rewriting queries that are not actually execution-bound.

When these practices hold, the platform becomes predictable. BI stays responsive, ELT finishes without blocking analysts, and cost discussions happen in scheduled reviews instead of during incidents. That is the real goal of workload management: not maximum complexity, but calm and repeatable control.

CHAPTER 15

FinOps in Snowflake

FinOps in Snowflake is the discipline of **making performance a cost-conscious default**. The platform already gives you strong levers—elastic warehouses, usage telemetry, and reversible data retention—so most savings come from **clear workload lanes**, **right-sized warehouses**, and **time-bound storage** rather than heroics. Treat cost as an engineering signal you can query, tag, and regress-test.

15.1 Cost Drivers: Storage, Compute, and Data Transfer

At a high level, you pay for (1) **storage at rest**, (2) **compute credits** burned by warehouses and cloud services, and (3) **movement across boundaries** (egress, replication, external calls). Good FinOps starts by measuring each separately, then shaping workloads to keep spend proportional to value.

M. Hander, *Snowflake Data Warehouse Engineering*, https://doi.org/10.1007/979-8-8688-2628-3_15

15.1.1 What Actually Drives the Bill

Driver	What it includes	How it grows	Knobs you control	Quick wins that last
Compute	Warehouse execution, automatic services tied to queries	Concurrency spikes, large scans, long ELT	Warehouse size and cluster count; auto-suspend; scaling policy; workload isolation	Many small BI lanes (multi-cluster) instead of one XL; prune scans; use `MERGE` over blind inserts; keep BI warm just enough to hit caches
Cloud Services (overhead)	Planning, metadata, result cache management	Explodes with *tons* of tiny queries or long compiles	Query shape; avoid chatter; batch where possible	Parameterize dashboards; reduce duplicate/chatty queries; favor CTAS over row-by-row loops
Storage	Table micro-partitions, Time Travel, Fail-safe; stages; clones' deltas	Long retention on hot datasets; wide tables; many small files landing	`DATA_RETENTION_TIME_IN_DAYS`; permanent vs transient; file compaction; VARIANT envelope discipline	Set 7–14 day defaults; transient for intermediates; compact lakes to ~200 MB Parquet; drop unused columns early
Data transfer	Cross-region/ Cloud egress, replication, reader accounts, external functions/egress	Replicating big DBs; cross-region queries; model APIs per-row	Keep data and consumers in-region; cache external results; batch calls	Co-locate BI with data; stage training/ score batches once; call external models via External Functions in batches and cache outputs

Mindset: ***compute is today's lever, storage is a slow drift,*** *and* ***data transfer punishes architecture mistakes.*** *Measure each one with a simple, boring query and adjust the smallest knob that changes behavior.*

15.1.2 Measuring the Three Pillars (Ready-to-Run Queries)

Compute Credits by Warehouse (Last 30 Days)

```
SELECT
  warehouse_name,
  TO_DATE(start_time) AS day,
  SUM(credits_used)   AS credits
FROM SNOWFLAKE.ACCOUNT_USAGE.WAREHOUSE_METERING_HISTORY
WHERE start_time >= DATEADD(day, -30, CURRENT_TIMESTAMP())
GROUP BY 1,2
ORDER BY 2,1;
```

> ***Serverless costs need separate coverage.*** *Warehouse metering does not capture all compute spend. If your account uses Automatic Clustering, Search Optimization, Snowpipe, Query Acceleration Service, Dynamic Tables, or serverless Tasks, review their dedicated usage views as well. Warehouse-only reporting understates the total compute cost in those environments.*

Useful serverless views include

- AUTOMATIC_CLUSTERING_HISTORY
- SEARCH_OPTIMIZATION_HISTORY
- PIPE_USAGE_HISTORY
- QUERY_ACCELERATION_HISTORY
- DYNAMIC_TABLE_REFRESH_HISTORY
- SERVERLESS_TASK_HISTORY

Storage Footprint by Database (Current)

```
SELECT
  database_name,
  active_bytes/POWER(1024,3)  AS active_gb,
  time_travel_bytes/POWER(1024,3) AS tt_gb,
  failsafe_bytes/POWER(1024,3)    AS fs_gb,
```

```
  (active_bytes+time_travel_bytes+failsafe_bytes)/POWER(1024,3) AS total_gb
FROM SNOWFLAKE.ACCOUNT_USAGE.DATABASE_STORAGE_USAGE_HISTORY
QUALIFY ROW_NUMBER() OVER (PARTITION BY database_name ORDER BY usage_date
DESC) = 1
ORDER BY total_gb DESC;
```

Data Transfer by Day and Transfer Type (Last 30 Days)

```
SELECT
  TO_DATE(START_TIME) AS day,
  TRANSFER_TYPE,
  SOURCE_CLOUD,
  SOURCE_REGION,
  TARGET_CLOUD,
  TARGET_REGION,
  SUM(BYTES_TRANSFERRED)/POWER(1024,3) AS transferred_gb
FROM SNOWFLAKE.ACCOUNT_USAGE.DATA_TRANSFER_HISTORY
WHERE START_TIME >= DATEADD(day, -30,   CURRENT_TIMESTAMP())
GROUP BY 1,2,3,4,5,6
ORDER BY 1,2;
```

(Object names can vary by edition/privileges; if a view is empty in your account, use its INFORMATION_SCHEMA counterpart where available.)

> ***Serverless costs are separate from warehouse costs.*** *In addition to warehouse metering, review serverless usage through AUTOMATIC_CLUSTERING_HISTORY, SERVERLESS_TASK_HISTORY, SEARCH_OPTIMIZATION_HISTORY, PIPE_USAGE_HISTORY, QUERY_ACCELERATION_HISTORY, and DYNAMIC_TABLE_REFRESH_HISTORY where relevant. Warehouse views alone will understate total compute spend.*

Cloud services billing nuance: *Snowflake bills cloud services usage only when daily cloud services consumption exceeds 10% of daily virtual warehouse usage. This adjustment is calculated daily in UTC. Serverless compute does not count toward that 10% threshold, so cloud services cost analysis and total compute-cost analysis should be interpreted separately.*

15.1.3 Turning Measurement into Action

Signal you observe	Likely root	Smallest effective change
High credits during dashboard hours with a big queued time	Not enough lanes	Raise BI MAX_CLUSTER_COUNT; keep size small; stagger report starts
High credits from long ELT steps, low queue	Heavy scans/under-pruned queries	Push scalar filters (date/tenant) earlier; drop SELECT *; consider narrow clustering on 1–2 columns after proving need
Storage growth mostly in Time Travel	Long retention on mutable tables	Set DATA_RETENTION_TIME_IN_DAYS to 7–14 on shared DBs; 1–2 on staging; use **transient** for intermediates
Rising stage/storage from tiny files	File explosion from upstream	Compact upstream; target ~200 MB Parquet; use manifests or partitioned COPY
Data-transfer spikes	Cross-region reads or chatty external calls	Co-locate compute; replicate only needed schemas; batch External Function calls and cache results

15.1.4 Tagging and Chargeback (Make Cost Attributable)

Attach simple tags and use QUERY_TAG so finance and engineering see the same ledger.

```
CREATE TAG cost_center ALLOWED_VALUES ('fin','mkt','ops','eng');

ALTER WAREHOUSE wh_bi_fin SET TAG cost_center = 'fin';
-- Set in apps and schedulers; audit for missing tags because Snowflake
does not natively enforce QUERY_TAG presence
ALTER SESSION SET QUERY_TAG = 'team=fin;purpose=dashboard';
```

Roll Up by Tag

```
SELECT
SPLIT_PART(qh.QUERY_TAG, ';', 1) AS team_kv,
SUM(qah.CREDITS_ATTRIBUTED_COMPUTE) AS compute_credits
FROM SNOWFLAKE.ACCOUNT_USAGE.QUERY_ATTRIBUTION_HISTORY qah
```

```
JOIN SNOWFLAKE.ACCOUNT_USAGE.QUERY_HISTORY qh
ON qh.QUERY_ID = qah.QUERY_ID
WHERE qh.START_TIME >= DATEADD(day, -30, CURRENT_TIMESTAMP())
GROUP BY 1
ORDER BY compute_credits DESC;
```

Note QUERY_ATTRIBUTION_HISTORY attributes **warehouse compute** to individual queries. It does not include storage, data transfer, cloud services billing adjustments, or serverless-feature costs, so use it as one part of the FinOps picture rather than a full bill replica.

15.1.5 Storage Hygiene That Doesn't Hurt Reliability

Topic	Recommendation	Reason
Retention	Default 7–14 days on prod DBs; 1–2 on staging/transient	Enough rewind for safety without hoarding
Object class	Permanent for shared facts/dims; transient for intermediates; temp for scratch	Fail-safe where it matters; savings where it doesn't
Clones	Use for branches; drop promptly after promotion	Clones are cheap until divergent writes accumulate
VARIANT payloads	Keep small scalar envelopes outside JSON	Better pruning → lower compute and smaller retention deltas
External stages	Periodically clear orphaned/export artifacts	Stages contribute to storage if you forget them

```
-- Set safer, finite retention
ALTER DATABASE prod SET DATA_RETENTION_TIME_IN_DAYS = 7;
ALTER SCHEMA stage  SET DATA_RETENTION_TIME_IN_DAYS = 1;
```

15.1.6 Compute Discipline in One Page

Lever	Good default	Why
BI	Small, **multi-cluster** (MIN=1, MAX=3–5), STANDARD, 60–120s auto-suspend	Keeps caches and adds lanes under burst
ELT	Single cluster, size to step; 300s auto-suspend; scheduled off BI peaks	Muscle without evicting BI caches
Ad hoc	XSmall/Small per team; soft quotas via resource monitors	Experiments can't starve prod
Result cache	Stable query text; parameterize in app	Free wins for repeated dashboards

Resource monitor example:

```
CREATE OR REPLACE RESOURCE MONITOR rm_bi
  WITH CREDIT_QUOTA = 500
  FREQUENCY = MONTHLY
  TRIGGERS ON 80 PERCENT DO NOTIFY
           ON 95 PERCENT DO SUSPEND
           ON 100 PERCENT DO SUSPEND_IMMEDIATE;
ALTER WAREHOUSE wh_bi SET RESOURCE_MONITOR = rm_bi;
```

15.1.7 Data Transfer Realities (Keep Data Close, Calls Rare)

Pattern	Risk	Safer pattern
Cross-region querying/ replication by default	Surprise egress	Keep analytics and data in the same region; replicate only hot schemas and only at the needed cadence
Per-row external scoring calls	Many small egress hits + latency	Use **External Functions**; batch inputs; cache `(entity_id, as_of_ts, model_version)` → `score`
Reader accounts for BI across clouds	Egress if regions differ	Use same-region shares or Stage → local copy for heavy consumers

15.1.8 A Lightweight FinOps Operating Rhythm

1. **Budget by lane.** Each warehouse gets a monitor and a tag; BI/ELT/Ad hoc are separate.
2. **Review weekly.** Credits by warehouse, top 20 queries by total runtime and bytes scanned, storage growth by database.
3. **Fix the smallest thing.** More lanes (BI) or better pruning (ELT) before more size. Reduce retention before chasing compaction projects.
4. **Codify improvements.** Warehouse DDL in IaC; retention and monitors in migrations; tags in orchestrator defaults.
5. **Prove it.** Re-run the same metering queries after changes; store deltas in an `ops.finops_kpis` table for trend lines.

FinOps works best when it's boring. Make costs intelligible with tags and one-page reports, change **one** knob at a time, and keep your lanes tidy. The same habits that make Snowflake fast—pruning hard, isolating workloads, and sizing sanely—make it economical, too.

15.2 Cost Guardrails and Governance

A warehouse-only view of FinOps is incomplete in Snowflake. Serverless features such as Automatic Clustering, Search Optimization, Snowpipe, Query Acceleration Service, Dynamic Table refreshes, and serverless Tasks have their own usage views and should be reviewed separately when those features are in use. Cloud services consumption also needs interpretation: Snowflake bills cloud services only when daily cloud services usage exceeds 10% of daily warehouse usage, and serverless compute does not count toward that threshold.

Good FinOps isn't "cutting spend"; it's **making the right behavior the default.** Guardrails nudge teams toward efficient warehouses, measured data retention, and attributable usage; governance turns those nudges into policy you can audit. The trick is to attach controls to the same places you already manage—roles, warehouses, databases—and to surface simple, trusted numbers that product and finance both believe.

15.2.1 The Guardrail Toolbox (What to Use, Where to Put It)

Guardrail	Attach to	What it enforces	Why it sticks
Resource monitors	Warehouses (per lane)	Notify/suspend at quotas	Limits the blast radius; budgets map to org units
Default warehouses by role	RBAC	Routing to the right lane	"Falling into" the cheapest correct path
Scaling posture	Warehouse DDL	Auto-suspend, min/max clusters, policy	Encodes elasticity vs. savings
Retention policy	DB/Schema/Table	`DATA_RETENTION_TIME_IN_DAYS`	Storage drift slows without heroics
Query tags and object tags	Sessions/ Warehouses	Attribution for chargeback and anomaly detection	Cost is explainable and debuggable
Change control	IaC/Migrations	Warehouse sizes, monitors, retention in code	Reviews catch expensive drift before it ships

15.2.2 Opinionated Defaults You Can Ship Today

```
-- 1) Route users to the correct lane
ALTER ROLE ANALYST     SET DEFAULT_WAREHOUSE = wh_bi;
ALTER ROLE ELT_RUNNER SET DEFAULT_WAREHOUSE = wh_elt;

-- 2) BI lane: elastic, cache-friendly, with a budget
CREATE OR REPLACE RESOURCE MONITOR rm_bi
  WITH CREDIT_QUOTA = 500
  FREQUENCY = MONTHLY
  TRIGGERS ON 80 PERCENT DO NOTIFY
           ON 95 PERCENT DO SUSPEND;

ALTER WAREHOUSE wh_bi SET
  WAREHOUSE_SIZE = 'SMALL',
  MIN_CLUSTER_COUNT = 1,
```

```
  MAX_CLUSTER_COUNT = 4,
  SCALING_POLICY = 'STANDARD',
  AUTO_SUSPEND = 90,
  AUTO_RESUME = TRUE,
  RESOURCE_MONITOR = rm_bi;

-- 3) Storage hygiene: reasonable rewind windows
ALTER DATABASE prod  SET DATA_RETENTION_TIME_IN_DAYS = 7;
ALTER SCHEMA   stage SET DATA_RETENTION_TIME_IN_DAYS = 1;  -- transient/
intermediate
```

15.2.3 Attribution and Policy-as-Data

Control	How to apply it	What you gain
Query tags (required in prod)	`ALTER SESSION SET QUERY_TAG = 'team=fin;purpose=dashboard';`	Chargeback and anomaly grouping
Cost center tags	`CREATE TAG cost_center ALLOWED_VALUES ('fin','mkt','ops');` then `ALTER WAREHOUSE wh_bi SET TAG cost_center='fin';`	Warehouse-level budget mapping
Governed roles	Limit USAGE on non-home warehouses	Prevents accidental spillover to expensive lanes
IaC (Terraform/YAML)	Warehouse/monitor/retention definitions in code	Diff and review before costs change

Snowflake does not provide a native control that makes QUERY_TAG mandatory at the warehouse or account level. In practice, standardize it through application, driver, orchestration, and scheduler configuration, then audit for missing tags in query history. Session policies do not serve as QUERY_TAG enforcement.

15.2.4 Budget Checks You Can Automate (Drop-in Queries)

Monthly Credits by Lane/Tag

```
WITH wh_tags AS (
  SELECT
    object_name AS warehouse_name,
    tag_value AS cost_center
  FROM SNOWFLAKE.ACCOUNT_USAGE.TAG_REFERENCES
  WHERE domain = 'WAREHOUSE'
    AND tag_name = 'COST_CENTER'
)
SELECT
  TO_DATE(m.start_time) AS day,
  m.warehouse_name,
  COALESCE(t.cost_center, 'unTagged') AS cost_center,
  SUM(m.credits_used) AS credits
FROM SNOWFLAKE.ACCOUNT_USAGE.WAREHOUSE_METERING_HISTORY m
LEFT JOIN wh_tags t
ON m.warehouse_name = t.warehouse_name
GROUP BY 1,2,3
ORDER BY 1,2;
```

Note TAG_REFERENCES records only **direct** object-to-tag relationships. If your governance model relies on inherited tags, this query will not reflect inherited tag values unless you model that separately.

Warehouse Compute Attributed to Queries by Team (From Query Tags)

```
SELECT
  SPLIT_PART(qh.QUERY_TAG, ';', 1) AS team_kv,
  SUM(qah.CREDITS_ATTRIBUTED_COMPUTE) AS compute_credits
FROM SNOWFLAKE.ACCOUNT_USAGE.QUERY_ATTRIBUTION_HISTORY qah
JOIN SNOWFLAKE.ACCOUNT_USAGE.QUERY_HISTORY qh
```

```
ON qh.QUERY_ID = qah.QUERY_ID
WHERE qh.START_TIME >= DATEADD(day, -30, CURRENT_TIMESTAMP())
GROUP BY 1
ORDER BY compute_credits DESC;
```

Storage Drift by Database (Current Snapshot)

```
SELECT database_name,
       (active_bytes+time_travel_bytes+failsafe_bytes)/POWER(1024,3) AS
       total_gb
FROM SNOWFLAKE.ACCOUNT_USAGE.DATABASE_STORAGE_USAGE_HISTORY
QUALIFY ROW_NUMBER() OVER (PARTITION BY database_name ORDER BY USAGE_
DATE DESC)=1
ORDER BY total_gb DESC;
```

15.2.5 Guardrail Playbook (Signal → Smallest Safe Change)

Signal	Likely root	Smallest change that holds
BI credits spike at :00 with queueing	Not enough lanes	Raise MAX_CLUSTER_COUNT on wh_bi; keep size small; stagger report jobs
ELT burns credits with long exec times	Under-pruned scans	Push scalar filters; drop SELECT *; consider narrow clustering on 1–2 columns after measuring
Storage swelling in Time Travel	Overlong retention on mutable tables	Set 7–14 days on prod DBs; 1–2 days on staging/transient layers
Unattributed spend	Missing warehouse tags or missing QUERY_TAG values	Standardize QUERY_TAG in applications and schedulers, audit for missing query tags in query history, and tag warehouses with cost_center through object tags
Surprise suspension mid-close	Monitor scoped too broadly	Bind monitors per warehouse (lane), not account-wide

15.2.6 Governance Rhythms That Keep Costs Boring

Cadence	Activity	Outcome
Weekly	Review credits by warehouse; top 20 queries by total runtime/bytes	Targeted fixes (lanes or pruning), not blanket resizing
Biweekly	Storage snapshot and growth delta	Adjust retention where drift appears
Monthly	Monitor thresholds vs. actuals; tag coverage report	Budget alignment and clean attribution
Per change	IaC PR for warehouse size/monitor/ retention edits	Prevents silent cost regressions

15.2.7 Escapes, Hatches, and Safety Nets

- **Pause or cap spend** with monitor triggers; only at the *lane* level, so other workloads continue.
- **Back off elasticity** off-hours: flip BI to `ECONOMY`, lower `MAX_CLUSTER_COUNT`, shorten `AUTO_SUSPEND`.
- **Quota sandboxes** for ad hoc: tiny warehouses + soft monitors keep experiments cheap.

Bottom line: put a budget on every lane, route work there by default, and write the policies (warehouses, monitors, retention, tags) as code. When cost has owners, tags, and weekly numbers everyone trusts, optimization turns into a couple of small, predictable tweaks—not a fire drill.

15.3 Chargeback/Showback with Tags and Views

Chargeback works when **every query and warehouse carries a label** and finance can pull the same numbers engineers see. Snowflake already records credits and storage by object; you add a thin layer of **tags + query tags** and publish a few **views** that roll up spend by team, product, or cost center. Keep the scheme boring: one canonical tag on warehouses/databases, one `QUERY_TAG` convention for apps and schedulers, and a small set of reporting views.

15.3.1 The Attribution Contract

Surface	How you label it	Example
Warehouse (lane)	Object **TAG** (e.g., cost_center, env)	ALTER WAREHOUSE wh_bi SET TAG cost_center='fin';
Query/job	QUERY_TAG with key=value;key2=value2	ALTER SESSION SET QUERY_TAG='team=fin;purpose=dashboard';
Database/schema (storage)	Object **TAG** for ownership	ALTER DATABASE prod SET TAG cost_center='fin';

*Pick **one** tag key for money ("cost_center" or "owner") and use it everywhere. Keep QUERY_TAG short and structured so you can split on ; later.*

15.3.2 Create Tags and Set Defaults

```
-- One-time: define cost center and environment tags
CREATE TAG cost_center ALLOWED_VALUES ('fin','mkt','ops','eng');
CREATE TAG environment  ALLOWED_VALUES ('prod','stg','dev');

-- Label warehouses (BI/ELT lanes)
ALTER WAREHOUSE wh_bi  SET TAG cost_center='fin', environment='prod';
ALTER WAREHOUSE wh_elt SET TAG cost_center='eng', environment='prod';

-- Label storage ownership at the database level
ALTER DATABASE prod SET TAG cost_center='fin', environment='prod';
```

Apps/orchestrators should set a QUERY_TAG per run:

```
ALTER SESSION SET QUERY_TAG='team=fin;purpose=dashboard';  --
enforce in prod
```

15.3.3 Canonical Reporting Views (Drop-in)

These views live in an ops/schema that finance can read. They normalize the raw ACCOUNT_USAGE tables into friendly rollups.

Note ACCOUNT_USAGE views are historical and can lag behind operational activity. Latency varies by view; for example, QUERY_HISTORY may lag by up to about 45 minutes, WAREHOUSE_METERING_HISTORY by up to three hours, and QUERY_ATTRIBUTION_HISTORY by several hours. Use these views for reporting and chargeback, not for real-time incident response.

Credits by Warehouse and Cost Center

```
CREATE OR REPLACE VIEW ops.v_credits_by_warehouse AS
WITH wh_tags AS (
  SELECT
    OBJECT_NAME AS warehouse_name,
    TAG_VALUE   AS cost_center
  FROM SNOWFLAKE.ACCOUNT_USAGE.TAG_REFERENCES
  WHERE DOMAIN = 'WAREHOUSE'
    AND TAG_NAME = 'COST_CENTER'
)
SELECT
  TO_DATE(m.START_TIME) AS day,
  m.WAREHOUSE_NAME,
  COALESCE(t.cost_center, 'unTagged') AS cost_center,
  SUM(m.CREDITS_USED) AS credits
FROM SNOWFLAKE.ACCOUNT_USAGE.WAREHOUSE_METERING_HISTORY m
LEFT JOIN wh_tags t
  ON t.warehouse_name = m.WAREHOUSE_NAME
GROUP BY 1,2,3;
```

This view reports **warehouse-level credits**. It does not include serverless-feature spend, and tag values come from direct TAG_REFERENCES relationships rather than inherited tags.

Credits by Query Tag (Team) from Query History

```
CREATE OR REPLACE VIEW ops.v_compute_by_team AS
SELECT
  TO_DATE(qh.START_TIME) AS day,
  SPLIT_PART(qh.QUERY_TAG, ';', 1) AS team_kv,   -- e.g., 'team=fin'
```

```
  SUM(qah.CREDITS_ATTRIBUTED_COMPUTE) AS compute_credits
FROM SNOWFLAKE.ACCOUNT_USAGE.QUERY_ATTRIBUTION_HISTORY qah
JOIN SNOWFLAKE.ACCOUNT_USAGE.QUERY_HISTORY qh
  ON qh.QUERY_ID = qah.QUERY_ID
GROUP BY 1,2;
```

Storage by Database with Owner Tag

```
CREATE OR REPLACE VIEW ops.v_storage_by_db AS
WITH latest AS (
  SELECT *,
         ROW_NUMBER() OVER (PARTITION BY DATABASE_NAME ORDER BY USAGE_DATE
         DESC) AS rn
  FROM SNOWFLAKE.ACCOUNT_USAGE.DATABASE_STORAGE_USAGE_HISTORY
),
db_tags AS (
  SELECT
    OBJECT_NAME AS database_name,
    TAG_VALUE   AS cost_center
  FROM SNOWFLAKE.ACCOUNT_USAGE.TAG_REFERENCES
  WHERE DOMAIN = 'DATABASE'
    AND TAG_NAME = 'COST_CENTER'
)
SELECT
  l.DATABASE_NAME,
  COALESCE(t.cost_center, 'unTagged') AS cost_center,
  (l.ACTIVE_BYTES + l.TIME_TRAVEL_BYTES + l.FAILSAFE_BYTES)/POWER(1024,3)
  AS total_gb,
  l.USAGE_DATE
FROM latest l
LEFT JOIN db_tags t
  ON t.database_name = l.DATABASE_NAME
WHERE l.rn = 1;
```

This view joins current storage history to direct database tag references. If ownership is tagged at the schema level or inherited, extend the model explicitly rather than assuming those tags appear automatically here.

Data Transfer (Egress) Daily

```
CREATE OR REPLACE VIEW ops.v_data_transfer AS
SELECT
  TO_DATE(START_TIME) AS day,
  TRANSFER_TYPE,
  SOURCE_CLOUD,
  SOURCE_REGION,
  TARGET_CLOUD,
  TARGET_REGION,
  SUM(BYTES_TRANSFERRED)/POWER(1024,3) AS transferred_gb
FROM SNOWFLAKE.ACCOUNT_USAGE.DATA_TRANSFER_HISTORY
GROUP BY 1,2,3,4,5,6;
```

15.3.4 Rollups Finance Can Paste into a Memo

Question	Query against the views
"How much did Finance BI spend last month?"	`SELECT day, SUM(compute_credits) FROM ops.v_credits_by_warehouse WHERE cost_center='fin' AND day >= DATEADD(month,-1,CURRENT_DATE()) GROUP BY day;`
"What warehouse compute did the ELT team generate this week by job?"	`SELECT day, team_kv, SUM(compute_credits) FROM ops.v_compute_by_team WHERE day >= DATEADD(day,-7,CURRENT_DATE()) GROUP BY day, team_kv;`
"Which DBs are our storage hotspots?"	`SELECT * FROM ops.v_storage_by_db ORDER BY total_gb DESC;`
"Are we paying egress?"	`SELECT * FROM ops.v_data_transfer WHERE TRANSFER_TYPE IN ('REPLICATION', 'EXTERNAL_FUNCTION', 'DATA_LAKE', 'COPY', 'COPY_FILES', 'INTERNAL') ORDER BY day;`

15.3.5 Guardrails to Keep Attribution Honest

Symptom	Likely cause	Smallest effective fix
"unTagged" credits in reports	Warehouse missing `cost_center` tag	Tag the warehouse; re-run views—historical rows still group by warehouse name
Lots of "(null)" team_kv	Apps or schedulers are not setting QUERY_TAG consistently	Standardize QUERY_TAG in production entrypoints, validate it in deployment/config review, and audit query history for missing values
BI credits spike with queueing	Many small queries in one lane	Raise MAX_CLUSTER_COUNT for BI; keep size small; stagger triggers
Storage charge drifts	Retention too long	Set prod DBs to 7–14 days; staging to 1–2 days; prefer transient for intermediates

15.3.6 Optional: Currency View (Simple FX)

If finance needs currency, store a daily FX rate and join:

```
CREATE OR REPLACE TABLE ops.fx_daily (day DATE PRIMARY KEY, usd_per_credit
NUMBER(10,4));

CREATE OR REPLACE VIEW ops.v_cost_by_warehouse_usd AS
SELECT v.day, v.WAREHOUSE_NAME, v.cost_center,
       v.compute_credits, f.usd_per_credit, v.compute_credits * f.usd_per_
       credit AS cost_usd
FROM ops.v_credits_by_warehouse v
JOIN ops.fx_daily f USING(day);
```

15.3.7 Operating Rhythm (Lightweight)

- **Tag once, enforce always.** Warehouses/DBs carry `cost_center`; prod sessions carry `QUERY_TAG`.
- **Publish the views.** Finance and engineering read the same tables.
- **Review weekly.** Credits by warehouse and team; storage top-N; any egress.
- **Fix the smallest thing.** Add lanes (BI) or prune scans (ELT); adjust retention where drift appears.

Chargeback/showback works best when it's boring: a couple of tags, a couple of views, and everyone reading the same ledger.

PART V

Security, Governance, and Reliability

CHAPTER 16

Security Foundations

Snowflake security assumes **everything is denied by default**, and you *delegate the minimum* a user or job needs to do its work. That posture is practical because RBAC, data masking, and network controls are all expressed as SQL. You model *who can do what, where,* and then you can review it like you review schema changes.

16.1 RBAC (Roles, Grants, Least Privilege)

Snowflake's RBAC keeps three ideas separate: **users** authenticate, **roles** authorize, and **objects** enforce. Users can hold many roles, but only the *active* role (and roles it inherits) apply to a session. Ownership sits with a *single* role per object; owners delegate with grants instead of sharing ownership.

16.1.1 What the Model Gives You

Concept	Meaning in Snowflake	Why it matters
Users vs. Roles	Users are identities; roles are bundles of privileges	Rotate people without rewriting permissions; service users act via roles
Role hierarchy	Roles can grant to roles (inheritance)	Define "team" or "domain" roles; avoid per-user grants
Object ownership	Every object has exactly one owning role	Only the owner (or SECURITYADMIN/ ACCOUNTADMIN) can grant further privileges
Privilege surfaces	Warehouse, database, schema, object-level grants	Control "can run compute" separately from "can read table"
Future grants	Schema/database-level defaults for new objects	Enforce least privilege *by default*, not by convention

M. Hander, *Snowflake Data Warehouse Engineering*, https://doi.org/10.1007/979-8-8688-2628-3_16

16.1.2 A Small, Durable Role Layout

Create **platform** roles only once; create **domain** roles per data area. Bind users to domain roles—not to SYSADMIN.

Role class	Examples	Scope	Notes
Platform (ops)	SYSADMIN, SECURITYADMIN, USERADMIN, ACCOUNTADMIN	Admin surfaces	Reserved for admins/automation; keep rare and audited
Platform (compute)	ROLE_WH_BI, ROLE_WH_ELT	USAGE/OPERATE on warehouses	Lets many teams share a lane safely
Domain (data)	ROLE_FIN_READER, ROLE_FIN_WRITER, ROLE_FIN_OWNER	DB/Schema/Table	A team's day-to-day access; owners create/drop inside the domain
Service roles	ROLE_ELT_TASKS, ROLE_BI_APP	Least privilege for jobs	Attached to service users or tasks

Create the Bones

```
-- Platform compute roles
CREATE ROLE ROLE_WH_BI;
CREATE ROLE ROLE_WH_ELT;

-- Domain roles for Finance
CREATE ROLE ROLE_FIN_OWNER;
CREATE ROLE ROLE_FIN_WRITER;
CREATE ROLE ROLE_FIN_READER;

-- Inheritance (readers/writers inherit compute lane if desired)
GRANT ROLE ROLE_WH_BI   TO ROLE ROLE_FIN_READER;
GRANT ROLE ROLE_WH_ELT  TO ROLE ROLE_FIN_WRITER;

-- Bind to users (via USERADMIN)
GRANT ROLE ROLE_FIN_READER TO USER alice;
GRANT ROLE ROLE_FIN_WRITER TO USER elt_runner;
```

16.1.3 Grants You Actually Use (And Where)

Surface	Minimal useful grants	What it enables
Warehouse	USAGE (and sometimes OPERATE) on a specific warehouse	Session can start/run queries on that lane
Database	USAGE	Visibility into schemas (not objects yet)
Schema	USAGE, plus CREATE TABLE/VIEW for writers	Ability to see or create objects in the schema
Tables/Views	SELECT, INSERT, UPDATE, DELETE, REFERENCES	Row-level access and DML
Secure views/ masking	APPLY MASKING POLICY, APPLY ROW ACCESS POLICY (to owners)	Attach governance objects
Ownership	OWNERSHIP (single role)	Full control including re-granting

Wire a Domain with Least Privilege

```
-- Finance database and schema owned by ROLE_FIN_OWNER
CREATE DATABASE FINANCE;

GRANT OWNERSHIP ON DATABASE FINANCE TO ROLE ROLE_FIN_OWNER COPY
CURRENT GRANTS;
```

In clean-slate environments, REVOKE CURRENT GRANTS is useful because it resets access during ownership transfer. In production, prefer COPY CURRENT GRANTS to preserve downstream access, then review and prune inherited grants explicitly afterward. COPY CURRENT GRANTS preserves existing access; REVOKE CURRENT GRANTS strips it.

```
GRANT USAGE ON DATABASE FINANCE TO ROLE_FIN_READER, ROLE_FIN_WRITER;

CREATE SCHEMA FINANCE.MART;
GRANT OWNERSHIP ON SCHEMA FINANCE.MART TO ROLE_FIN_OWNER REVOKE
CURRENT GRANTS;
GRANT USAGE ON SCHEMA FINANCE.MART TO ROLE_FIN_READER, ROLE_FIN_WRITER;
```

```
-- Readers can SELECT; writers can DML
GRANT SELECT ON ALL TABLES IN SCHEMA FINANCE.MART TO ROLE_FIN_READER;
GRANT SELECT, INSERT, UPDATE, DELETE
  ON ALL TABLES IN SCHEMA FINANCE.MART TO ROLE_FIN_WRITER;

-- Future grants keep it true for new tables/views
GRANT SELECT ON FUTURE TABLES IN SCHEMA FINANCE.MART TO ROLE_FIN_READER;
GRANT SELECT, INSERT, UPDATE, DELETE
  ON FUTURE TABLES IN SCHEMA FINANCE.MART TO ROLE_FIN_WRITER;
```

16.1.4 Least-Privilege Patterns That Survive Audits

Situation	Practice	Why it holds up
Day-to-day analyst work	Users get the domain **reader**; elevate via break-glass if needed	Most users never need write/own
ELT pipelines	Service user runs under **writer** (or narrower) + ELT warehouse role	Jobs can't read unrelated domains or change grants
Object creation	Only the **owner** role creates schemas/tables; writers insert into them	Ownership ≠ broad read; clean lineage and grants
Sensitive columns	Masking policies bound to columns; row access bound to tenant	Security travels with data; no per-query filters
Compute control	Warehouse roles separate from data roles	A reader can't choose an ELT lane by mistake

Attaching Policies at the Data Edge

```
-- Masking example
CREATE OR REPLACE MASKING POLICY mask_email AS (val STRING) RETURNS
STRING ->
  CASE WHEN CURRENT_ROLE() IN ('ROLE_FIN_OWNER','SECURITYADMIN') THEN val
       ELSE REGEXP_REPLACE(val, '(^.).+(@.+$)', '\\1***\\2') END;

ALTER TABLE FINANCE.MART.DIM_CUSTOMER
  MODIFY COLUMN EMAIL SET MASKING POLICY mask_email;
```

```
-- Row access example

CREATE OR REPLACE TABLE SECURITY.TENANT_ROLE_MAP (
  ROLE_NAME   STRING,
  TENANT_ID   STRING
);
CREATE OR REPLACE ROW ACCESS POLICY tenant_guard
AS (tenant_id STRING) RETURNS BOOLEAN ->
  EXISTS (
    SELECT 1
    FROM SECURITY.TENANT_ROLE_MAP m
    WHERE m.ROLE_NAME = CURRENT_ROLE()
      AND m.TENANT_ID = tenant_id
  );

ALTER TABLE FINANCE.MART.FACT_SALES
  ADD ROW ACCESS POLICY tenant_guard ON (TENANT_ID);
```

For tenant isolation, do not key row access off CURRENT_USER() unless each Snowflake user maps one-to-one to a tenant and no shared service identities exist. In production, the safer pattern is to map CURRENT_ROLE() to allowed tenants through a governance table. That keeps policy behavior stable for BI tools, service users, and delegated execution paths.

16.1.5 Warehouse Access Stays Independent

Bind compute rights to explicit roles, not to data roles implicitly.

```
GRANT USAGE ON WAREHOUSE WH_BI  TO ROLE ROLE_WH_BI;
GRANT USAGE ON WAREHOUSE WH_ELT TO ROLE ROLE_WH_ELT;

-- Optionally restrict OPERATE (resume/suspend) to ops-only
GRANT OPERATE ON WAREHOUSE WH_ELT TO ROLE SYSADMIN;
```

16.1.6 Working Sessions and Context

Task	Command	Notes
Switch role	`USE ROLE ROLE_FIN_READER;`	Only privileges of the active role (and its parents) apply
See enabled roles	`SELECT CURRENT_ROLE(), CURRENT_AVAILABLE_ROLES();`	Helps debug "why can't I see X?"
Inspect grants	`SHOW GRANTS TO ROLE ROLE_FIN_READER;`	Start with the role, not the user
Default context	`ALTER ROLE … SET DEFAULT_WAREHOUSE=…;`	Route users into the right lane automatically

16.1.7 Ownership Transfers (Without Widening Access)

When a team changes, move objects by **transferring ownership** to the new domain owner role; do not add broad grants to paper over ownership.

```
GRANT OWNERSHIP ON SCHEMA FINANCE.MART TO ROLE ROLE_FIN_OWNER REVOKE
CURRENT GRANTS;
GRANT OWNERSHIP ON TABLE FINANCE.MART.FACT_SALES TO ROLE ROLE_FIN_OWNER
REVOKE CURRENT GRANTS;
```

Ownership transfer is the correct way to move administrative control between roles, but `REVOKE CURRENT GRANTS` removes all existing grants on the object. Use it only when you intentionally want to reset downstream access. In shared environments, inventory and recreate required grants first, or perform the transfer during a controlled change window.

16.1.8 Auditing You'll Actually Run

Question	Where to look	Example
Who can read table X?	INFORMATION_SCHEMA.OBJECT_PRIVILEGES	SELECT * FROM … WHERE OBJECT_NAME='FACT_SALES' AND PRIVILEGE='SELECT';
What does a role have?	SHOW GRANTS TO ROLE <role>	Review before granting more
Who has a role?	SHOW GRANTS OF ROLE <role>	Confirm minimal membership
Who changed what?	ACCOUNT_USAGE.ACCESS_HISTORY, LOGIN_HISTORY	Tie access to the session context and client

16.1.9 Putting It Together

Model **roles, not users**. Keep compute rights independent from data rights. Make **future grants** reflect the least privilege you intend, so new objects are safe by default. Attach masking and row-access policies to the data itself so protection travels with clones and shares. And when something changes—people, teams, products—transfer **ownership** between roles instead of piling on ad hoc grants. In Snowflake, clean RBAC isn't a ceremony; it's how you keep both velocity and safety.

16.2 Authentication, SSO, and Network Policies

Snowflake's front door has two locks: **who you are** (authentication) and **where you're coming from** (network policy). Keep user sign-in behind your identity provider (IdP) with SSO and MFA; keep service accounts on key pairs or OAuth; and fence the perimeter with IP allow-lists or private connectivity. The controls live in SQL, so they can be reviewed and promoted like schema changes.

16.2.1 Choosing the Right Sign-in for Each Actor

Actor	Best auth mode	Why this fit lasts	Notes that matter
Employees and analysts (UI, notebooks)	**SAML SSO** to your IdP (Okta, Azure AD, ADFS) with **MFA**	Centralized lifecycle and MFA; JIT/SCIM provisioning; fewer Snowflake-local secrets	Use a **SAML security integration**; map the default role/warehouse via attributes
BI tools/apps (server-side)	**External OAuth** (IdP or API gateway as AS)	Short-lived tokens; easy rotation; scopes	Create **EXTERNAL OAUTH** integration; pass default role in token claims
Service users (Tasks, batch ELT)	**Key-pair auth** (RSA) for headless users	No passwords; rotation by public key; works everywhere (JDBC/ODBC/SnowSQL/Snowpark)	Store private keys in your secret store; rotate on a schedule
Partner/desktop tools	**Snowflake OAuth** or **Browser SSO**	User experience + least secrets on laptops	Prefer OAuth over stored passwords; enforce MFA via IdP

Rule of thumb: humans use **SSO + MFA**; services use **OAuth** or **key pairs**. Avoid passwords entirely in production paths.

16.2.2 Enabling SSO and Token-Based Access (Compact DDL)

SAML SSO (Okta/Azure AD, etc.)

```
CREATE OR REPLACE SECURITY INTEGRATION idp_saml
  TYPE = SAML2
  ENABLED = TRUE
  SAML2_ISSUER = 'https://idp.example.com/app/...'
  SAML2_SSO_URL = 'https://idp.example.com/sso/...'
  SAML2_X509_CERT = '-----BEGIN CERTIFICATE-----...'
  -- Optional: map IdP attributes to Snowflake session context
  SAML2_SP_INITIATED_LOGIN_PAGE_LABEL = 'Corporate SSO';
```

Bind users to SSO by default (and require MFA at the IdP). Use **SCIM** from your IdP to provision/disable users and group-to-role mappings automatically.

External OAuth (Apps, Gateways)

```
CREATE OR REPLACE SECURITY INTEGRATION ext_oauth_app
  TYPE = EXTERNAL_OAUTH
  ENABLED = TRUE
  EXTERNAL_OAUTH_ISSUER = 'https://auth.example.com'
  EXTERNAL_OAUTH_ALLOWED_ROLES_LIST = ('ROLE_BI_APP','ROLE_FIN_READER');
```

External OAuth integrations validate access tokens issued by an external authorization server. The exact DDL depends on the IdP and token shape, so treat the integration as provider-specific configuration. At a minimum, define the issuer and explicitly restrict which Snowflake roles may be assumed from the token. Do not reuse Snowflake OAuth client parameters in an External OAuth integration.

Clients present the access token to Snowflake; the integration verifies issuer, client, and scopes/roles.

Key-Pair Auth (Headless)

```
Generate a keypair once per service user;
```

Upload the **public** key:

```
ALTER USER elt_runner SET RSA_PUBLIC_KEY='MIIBIjANBgkqh...';
```

Rotate by setting `RSA_PUBLIC_KEY_2`, then swapping and removing the old key.

16.2.3 Network Policies: The Second Lock on the Door

Network policies gate connections by IP/CIDR across **web UI, JDBC/ODBC, SnowSQL, and Snowpark**. Apply one at the account and override for specific users when needed.

```
CREATE OR REPLACE NETWORK POLICY corp_office
  ALLOWED_IP_LIST = ('203.0.113.0/24')
  BLOCKED_IP_LIST = ('203.0.113.45');
```

In Snowflake network policies, `ALLOWED_IP_LIST` is the primary control boundary. Use `BLOCKED_IP_LIST` only for narrower exclusions within otherwise allowed space, not as a blanket "deny all" companion to the allow-list. `ALTER ACCOUNT SET NETWORK_POLICY = corp_office;`

```
-- Exceptions (e.g., a jump host or a vendor user)
ALTER USER vendor_user SET NETWORK_POLICY = NULL; -- or a narrower policy
```

For private connectivity, use your cloud's private service (**AWS PrivateLink**, **Azure Private Link**, **GCP Private Service Connect**) with Snowflake's private endpoints so traffic never crosses the public Internet. Pair this with a network policy that allows only your VPC/VNet egress IPs.

16.2.4 Putting Auth and Network Together (Minimal, Strong Defaults)

Control	Default to set	Effect
Primary auth	SAML SSO for humans; External OAuth for apps; key-pairs for batch	No passwords at rest; MFA central
Session posture	Disable non-SSO login for SSO users; short session/token lifetimes	Shrinks the blast radius of stolen cookies/tokens
Network perimeter	Account-level allow-list; private endpoints for server-to-server	Removes the Internet as a dependency
Lifecycle	SCIM-backed provisioning; roles via IdP groups; automated deprovision	Least privilege follows joiners/movers/leavers
Secrets handling	Private keys + OAuth client secrets in a vault; no creds in code	Reproducible, auditable rotation

16.2.5 Operational Checks You'll Actually Run

Who Is Bypassing SSO?

Search for password/keypair logins among human roles and close them down.

```
SELECT user_name, client_environment:AUTHENTICATOR::string AS authn
FROM SNOWFLAKE.ACCOUNT_USAGE.LOGIN_HISTORY
WHERE event_timestamp >= DATEADD(day,-7,CURRENT_TIMESTAMP())
  AND user_name IN (SELECT name FROM SNOWFLAKE.ACCOUNT_USAGE.USERS WHERE
  disabled='FALSE')
GROUP BY 1,2;
```

Are Connections Coming from Approved Networks?

Correlate recent logins with IPs; tighten the allow-list if needed.

```
SELECT user_name, client_ip, COUNT(*) AS hits
FROM SNOWFLAKE.ACCOUNT_USAGE.LOGIN_HISTORY
WHERE event_timestamp >= DATEADD(day,-7,CURRENT_TIMESTAMP())
GROUP BY 1,2
ORDER BY hits DESC;
```

16.2.6 Common Pitfalls ➤ Smallest Safe Fix

Symptom	Likely cause	Fix that sticks
Users can still set local passwords	SSO enabled but password auth left on	Set `LOGIN_NAME` but require SSO at the role/process level; remove local passwords for SSO users
Long-lived tokens in apps	OAuth without rotation/short TTL	Enforce short token lifetimes; refresh via client credentials; cache per-request minimally
Vendors connecting from the Internet	No private endpoint; broad allow-list	Enable PrivateLink/Private Link/PSC; narrow `ALLOWED_IP_LIST` to egress IPs
Orphaned users with lingering access	Manual user lifecycle	Turn on SCIM; map IdP groups ➤ Snowflake roles; review quarterly

16.2.7 Secrets for External APIs and Integrations

When Snowflake code calls external services, do not embed API keys, OAuth refresh tokens, or cloud credentials directly in function DDL or application code. Store them as Snowflake `SECRET` objects and allow their use only through the specific integration or handler that needs them. Secrets are schema-level objects, protected by RBAC, and encrypted within Snowflake's key hierarchy.

```
CREATE OR REPLACE SECRET my_api_key
  TYPE = GENERIC_STRING
  SECRET_STRING = 'my-secret-value';
```

> After creation, the plain text is not exposed back through normal SQL inspection; the intended pattern is that supported components such as external functions, integrations, UDFs, and procedures consume the secret indirectly. When external network access is needed, bind the reachable destinations and the allowed secrets through an external access integration rather than leaving credentials scattered in code or config.

Bottom line: put humans behind **SSO + MFA**, put services on **OAuth or key pairs**, and make the **network** say "only from here." Because these are SQL-managed objects (security integrations, network policies), they live happily in version control and promotion pipelines—security by design, not by exception.

16.3 Data Encryption and Key Management

Snowflake encrypts data **end-to-end by default**—at rest with a hierarchical key scheme and in transit with TLS. You choose how much of the key lifecycle you want to own. Most teams rely on Snowflake-managed keys (simple, rotated automatically). Most teams rely on Snowflake-managed encryption. More regulated deployments may add a customer-managed key (CMEK) so the customer retains control over part of the key lifecycle. Tri-Secret Secure is a related but stronger dual-control model and should not be treated as identical to generic CMEK. Use the term that matches the specific Snowflake feature and operating model you have enabled. For lakes and external stages, encryption is governed by your cloud bucket settings but can be enforced from Snowflake DDL.

Posture	Fit	Strengths	Trade-offs
Snowflake-managed keys	Most analytics workloads	Lowest operational burden	No customer-controlled revocation point
Customer-managed key (CMEK)	Regulated workloads needing customer KMS control	Customer controls key lifecycle in cloud KMS	More KMS dependency and operational complexity
Tri-Secret Secure	Highest-control environments requiring dual control	Stronger separation of duties and revocation semantics	Strictest operating requirements

16.3.1 What's Encrypted, Where Keys Live, and Who Turns the Knobs

Layer	What's protected	Keys and rotation	Who controls it	Notes that matter
Transit	Client ↔ Snowflake, Snowflake ↔ cloud storage	TLS (modern ciphers)	Snowflake	No extra setup; use modern drivers to keep TLS current.
At rest (internal storage)	Micro-partitions, results, Time Travel/ Fail-safe	Hierarchical keys (account ➤ database/ schema/table/ partition); automatic rotation and periodic re-key	Snowflake (default)	Per-object keys limit the blast radius; re-keying is online.
At rest with customer key	Same as above, but "dual control"	Your CMK in KMS/Key Vault + Snowflake key (Tri-Secret/CMEK)	You + Snowflake	Disable/rotate your CMK to instantly revoke access (crypto-shred).
External stages/ lakes	Files in S3/ADLS/ GCS	SSE-KMS/platform encryption; optional client-side encryption	You (bucket policy, KMS)	Enforce KMS on the stage; compact files to keep costs/perf sane.

16.3.2 Choosing the Posture

Posture	Fit	Strengths	Trade-offs
Snowflake-managed keys (default)	Most analytics estates	Zero ops; automatic rotation; online re-key	You don't hold a "kill switch."
Customer-managed key (CMEK/Tri-Secret Secure)	Regulated data, explicit separation of duties	Dual-key control; crypto-shred via CMK disable; provable ownership	You manage the KMS key lifecycle and availability.
Strict lake encryption	Shared lakehouse, external tables	Uniform KMS policy across engines	You must enforce bucket/KMS policy and rotate keys.

16.3.3 Compact DDL Patterns

1) **Customer-Managed Key Via a Security/Storage Integration**
(Names vary slightly by cloud; the shape is similar.)

```
-- App registration differs per cloud; this binds Snowflake to
your KMS and storage
CREATE OR REPLACE STORAGE INTEGRATION si_lake
  TYPE = EXTERNAL_STAGE
  ENABLED = TRUE
  STORAGE_PROVIDER = 'S3'                    -- or 'AZURE' / 'GCS'
  STORAGE_AWS_ROLE_ARN = 'arn:aws:iam::123456789012:role/
  snowflake-stage'
  STORAGE_ALLOWED_LOCATIONS = ('s3://my-bucket/data/')
  -- enforce bucket-side KMS
  COMMENT = 'Access via role; bucket enforces SSE-KMS';

-- Stage that *requires* KMS on S3 objects
CREATE OR REPLACE STAGE lake_sales
  URL = 's3://my-bucket/data/sales/'
  STORAGE_INTEGRATION = si_lake
```

```
  ENCRYPTION = (TYPE = 'AWS_SSE_KMS' KMS_KEY_ID = 'arn:aws:kms:us-
  east-1:123456789012:key/abcd-...');

-- GCS/ADLS equivalents:
-- ENCRYPTION = (TYPE='GCS_SSE_KMS', KMS_KEY='projects/.../
cryptoKeys/...')
-- ENCRYPTION = (TYPE='AZURE_CEK', ... ) when applicable.
```

2) **Optional Client-Side Encryption for Files You PUT to a Stage**
(Useful when compliance requires you to encrypt *before* upload.)

```
-- For S3 external stages only (illustrative)
CREATE OR REPLACE STAGE ext_raw
  URL='s3://my-bucket/raw/'
  ENCRYPTION = (TYPE='AWS_CSE' MASTER_KEY='base64-encoded-key-
  material');
```

3) **Account-Level Customer Key (Tri-Secret/CMEK)**
Provision the CMK in your cloud KMS, create the security integration in Snowflake, then enable for the account. Rotation/disable are performed in KMS; Snowflake respects it automatically.

Administrative steps vary by cloud; the operating model is consistent: **rotate** the CMK on a schedule; **disable** to revoke access; **reenable** to restore. All re-keys are online.

16.3.4 Rotation, Revocation, and Recovery: What's Actually Happening

Action	What Snowflake does	What you do	Impact on workloads
Snowflake key rotation	Periodically re-wraps data keys; cascading re-key in background	Nothing	Transparent; no downtime.
CMK rotation (CMEK)	Uses new CMK to wrap Snowflake's root	Rotate CMK in KMS; confirm integration health	Transparent; plan during low-risk windows.
CMK disable (crypto-shred)	Fails decryption attempts	Disable or revoke CMK in KMS	Immediate access loss to encrypted data until reenabled.
Stage/ bucket KMS rotation	Reads new KMS-encrypted files fine	Update `ENCRYPTION`/ bucket policy as needed	Transparent to queries; writer jobs must have KMS perms.

16.3.5 External Tables and Iceberg: Don't Forget the Lake

Storage	Enforce encryption	Snowflake control point	Extra notes
S3	Bucket policy `s3:x-amz-server-side-encryption = aws:kms`	Stage `ENCRYPTION = (TYPE='AWS_SSE_KMS' ...)`	Align object ACLs/IAM; block public access.
ADLS Gen2	Account-level CMK (Azure SSE with CMK)	Storage account and container policy	Private Endpoints pair well with network policies.
GCS	CMEK on bucket	Stage `ENCRYPTION = (TYPE='GCS_SSE_KMS' ...)`	Service account needs `cloudkms.cryptoKeyEncrypterDecrypter`.

16.3.6 Auditing and "Trust But Verify"

Minimal queries you can keep in an ops runbook:

```
-- Recent stage activity (confirms KMS-encrypted writes/reads by user/file)
SELECT *
FROM SNOWFLAKE.ACCOUNT_USAGE.COPY_HISTORY
WHERE START_TIME >= DATEADD(day,-7,CURRENT_TIMESTAMP());

-- Who accessed sensitive tables (tie to masking/row policies separately)
SELECT *
FROM SNOWFLAKE.ACCOUNT_USAGE.ACCESS_HISTORY
WHERE QUERY_START_TIME >= DATEADD(day,-7,CURRENT_TIMESTAMP())
  AND OBJECT_NAME IN ('PII.CUSTOMERS','PII.PAYMENTS');
```

For governance audits, pair entitlement review with actual usage review. ACCOUNT_USAGE.ACCESS_HISTORY shows which tables and columns were touched, and Snowflake uses the same lineage foundations to trace data flow across write operations. That makes it useful for "who read this sensitive column?", "what downstream objects depend on it?", and "did masking coverage follow the data?" reviews.

16.3.7 Practical Guardrails

Risk	Smallest effective control	Why it works
Accidental unencrypted lake files	Bucket policy + stage ENCRYPTION=...	Enforces SSE-KMS at both ends
Lost keys or operator error	Dual control (CMEK) with documented rotation	Clear roles; reversible when reenabled
Over-broad decryption rights	Least-privilege IAM/KMS grants to Snowflake's role	Readers can't decrypt outside intended prefixes
Silent drift	Weekly COPY/ACCESS history review; alert on policy changes	You discover gaps before audits do

Bottom line: encryption is always on; your decision is whether Snowflake manages **all** keys or you add a **customer root key** for dual control. Enforce KMS on external stages, keep rotations boring and scripted, and verify with simple history queries. That gets you strong cryptography without turning key management into a source of outages.

16.4 Governance Automation: Classification, Lineage, and Posture

Snowflake security is no longer just roles and policies. In practice, enterprise governance also depends on automated sensitive data discovery, lineage, and posture review. Horizon Catalog brings these together by combining access controls, masking and row access policies, sensitive data classification, and metadata/lineage views in a shared governance layer.

At a minimum, sensitive data classification should be part of the baseline because it directly informs where masking, tagging, and policy review should be applied. Snowflake supports automated classification with system semantic and privacy categories, and classification results can be surfaced in the Trust Center or managed with SQL classification profiles.

Access History complements RBAC by showing which objects and columns were touched and by supporting lineage analysis for governance and audit work. That makes it the operational bridge between "who is allowed" and "what was actually accessed."

Trust Center should be called out as the account-level posture view. It now provides high-level security findings and, on the Data Security side, setup and review for sensitive data classification, including whether sensitive data is already protected by masking.

```
-- Classify a table and auto-apply recommended system tags
CALL SYSTEM$CLASSIFY('FINANCE.MART.DIM_CUSTOMER', {'auto_tag': true});
-- Optional: map detected semantic categories to your own governance tags
-- through a classification profile
```

Use classification to discover candidate PII and sensitive fields before writing masking policies. `SYSTEM$CLASSIFY` can classify an object and auto-apply recommended system tags, while classification profiles can map detected semantic categories to user-defined tags, so governance rules become repeatable instead of manual.

16.5 Minimum Security Baseline

Control area	Minimum baseline	Why this is the minimum
Authentication	Humans use SSO; password users are MFA-protected; service users use key-pair or OAuth, not passwords	Snowflake is rolling out stronger defaults around MFA for human password sign-ins and expects service users to use stronger non-interactive methods.
RBAC	Grant users domain roles, not admin roles; separate compute roles from data roles; use future grants carefully	Keeps least privilege durable and auditable.
Sensitive data	Run automated classification on key databases; review tagged columns; attach masking where needed	Classification feeds masking and governance coverage.
Row/column protection	Use masking policies and row access policies on the data edge; key RAPs to role-based mappings, not `CURRENT_USER()` for shared-service patterns	Keeps policy behavior stable across BI tools and service identities.
Network perimeter	Apply an account-level allow-list or private connectivity for server workloads; use narrower user overrides only when needed	Makes the source network part of the access decision.
Encryption	Accept Snowflake encryption defaults internally; enforce KMS-backed encryption on external stages and lakes; use CMEK/Tri-Secret only where policy requires it	Strong default cryptography without unnecessary operational risk.
Secrets	Store API keys and OAuth secrets in `SECRET` objects; reference them through integrations	Avoids plain text secrets in code and DDL.
Audit and posture	Review `SHOW GRANTS`, `ACCESS_HISTORY`, classification coverage, and Trust Center findings on a fixed cadence	Combines entitlement review, actual access review, and posture review.

A practical Snowflake baseline is simple: identities authenticate strongly, roles grant the minimum required access, sensitive data is discovered automatically, policies attach at the data edge, secrets stay in governed objects, network access is narrowed, and audit review combines grants, access history, and posture findings. The goal is not to maximize controls; it is to make secure defaults the easiest defaults to operate.

CHAPTER 17

Data Governance and Compliance

Governance in Snowflake works best when it lives **in SQL next to the data**. Tags, policies, and lineage are first-class objects you can create, review, and move through environments like any other DDL. The goal is a small, boring set of conventions: a few canonical tags, a repeatable way to discover sensitive fields, and lineage you can query for audits and impact analysis.

17.1 Object Tagging and Lineage

Tags answer "**what is this?**" and "**who owns it?**". Lineage answers "**where did it come from, who touched it, and what will break if we change it?**". Together, they let you prove controls (PII masking, residency), route costs, and predict change impact.

M. Hander, *Snowflake Data Warehouse Engineering*, https://doi.org/10.1007/979-8-8688-2628-3_17

17.1.1 A Compact Tagging Scheme (And Where It Goes)

Tag (key)	Typical values	Attach to	Why it matters
classification	pii.email, pii.ssn, restricted, public	Columns/Tables/Views	Drives masking and access reviews
owner	fin-data, mkt-analytics	Databases/Schemas/Tables	Clear responsibility and escalation
cost_center	fin, mkt, ops, eng	Warehouses/Databases	Chargeback/showback (see Chapter 15)
retention_policy	7d, 30d, indefinite	Schemas/Tables	Confirms Time Travel and purge windows
data_domain	sales, customer, product	Schemas/Tables	Search, catalog, and doc grouping

Create and use tags like any other object:

```
-- Define tags once (often in a governance schema)
CREATE TAG classification     ALLOWED_VALUES ('pii.email','pii.phone','pii.
ssn','restricted','public');
CREATE TAG owner ALLOWED_VALUES ('fin-data','mkt-analytics','ops-
data','eng-data');
CREATE TAG retention_policy ALLOWED_VALUES ('7d','14d','30d','indefinite');

-- Attach tags to objects
ALTER TABLE core.dim_customer
  SET TAG owner = 'fin-data', data_domain = 'customer';

-- Tag sensitive columns for enforcement & audits
ALTER TABLE core.dim_customer
  MODIFY COLUMN email SET TAG classification = 'pii.email';
```

Automatic discovery to seed tags (Snowflake semantic classification helpers):

```
-- Inspect likely categories for a table's columns
SELECT SYSTEM$EXTRACT_SEMANTIC_CATEGORIES('CORE','PUBLIC','DIM_CUSTOMER');
```

```
-- Apply recommended tags (review before running in prod)
CALL SYSTEM$ASSOCIATE_SEMANTIC_CATEGORY_TAGS('CORE','PUBLIC','DIM_
CUSTOMER');
```

Treat auto-classification as a starting point. Keep the final tags authoritative and code-reviewed.

17.1.2 Binding Policy to Tags (So Tags Do Work)

Tags become useful when policies **read them**. The stronger default pattern is tag-based masking: create masking policies by data type, assign them to a governance tag with `ALTER TAG ... SET MASKING POLICY`, and let Snowflake automatically protect tagged columns whose data type matches the policy signature. Use `SYSTEM$GET_TAG_ON_CURRENT_COLUMN()` only for advanced conditional logic inside a policy, not as the main governance pattern.

```
-- Define separate policies by data shape
CREATE OR REPLACE MASKING POLICY mask_email AS (val STRING) RETURNS
STRING ->
  CASE
    WHEN IS_ROLE_IN_SESSION('ROLE_PII_READER') OR IS_ROLE_IN_
    SESSION('SECURITYADMIN') THEN val
    ELSE REGEXP_REPLACE(val, '(^.).+(@.+$)', '\\1***\\2')
  END;
CREATE OR REPLACE MASKING POLICY mask_phone AS (val STRING) RETURNS
STRING ->
  CASE
    WHEN IS_ROLE_IN_SESSION('ROLE_PII_READER') OR IS_ROLE_IN_
SESSION('SECURITYADMIN') THEN val
    ELSE REGEXP_REPLACE(val, '(.{2}).+(.{2})', '\\1****\\2')
  END;
-- Bind policies to the classification tag
ALTER TAG classification
  SET MASKING POLICY mask_email, MASKING POLICY mask_phone;
-- Tag columns; Snowflake applies the matching policy automatically by
data type
```

```
ALTER TABLE core.dim_customer
  MODIFY COLUMN email SET TAG classification = 'pii.email';
ALTER TABLE core.dim_customer
  MODIFY COLUMN phone SET TAG classification = 'pii.phone';
```

Retention and residency checks can also be expressed with tags and a small control query (e.g., alert if `retention_policy='7d'` but `DATA_RETENTION_TIME_IN_DAYS > 7`).

17.1.3 Lineage You Can Query (And Trust)

Snowflake exposes several views for **structural lineage** (object dependencies) and **operational lineage** (who read/wrote what).

Question	Source to query	What you learn
"What does this view depend on?"	`INFORMATION_SCHEMA.OBJECT_DEPENDENCIES`	Upstream tables/views/UDFs; change impact
"Who used this table and when?"	`ACCOUNT_USAGE.ACCESS_HISTORY`	Consumers (queries, roles, apps); audit trail
"Which queries created this table?"	`ACCOUNT_USAGE.QUERY_HISTORY` (+ `BY_OBJECT`)	DDL/DML provenance, owners, timings
"Which tags touch which objects?"	`ACCOUNT_USAGE.TAG_REFERENCES`	Tag coverage and drift
"Which policies apply?"	`ACCOUNT_USAGE.POLICY_REFERENCES`	Masking/row access policy attachments

Compact examples:

```
-- Upstream/downstream structure for a view
SELECT REFRESHED_ON, REFERENCED_OBJECT_NAME, REFERENCED_OBJECT_DOMAIN,
REFERENCING_OBJECT_NAME
FROM <db>.<schema>.INFORMATION_SCHEMA.OBJECT_DEPENDENCIES
WHERE REFERENCING_OBJECT_NAME = 'V_SALES_MART';
```

```
-- Who read table X in the last 7 days

SELECT a.QUERY_ID, a.QUERY_START_TIME, a.ROLE_NAME, a.USER_NAME
FROM SNOWFLAKE.ACCOUNT_USAGE.ACCESS_HISTORY a,
     LATERAL FLATTEN(input => a.BASE_OBJECTS_ACCESSED) obj
WHERE a.QUERY_START_TIME >= DATEADD(day,-7,CURRENT_TIMESTAMP())
  AND obj.value:objectName::STRING = 'CORE.DIM_CUSTOMER'
  AND obj.value:objectDomain::STRING = 'Table';

-- Tag coverage report
SELECT OBJECT_DATABASE, OBJECT_SCHEMA, OBJECT_NAME, COLUMN_NAME,
       TAG_NAME, TAG_VALUE, MODIFIED
FROM SNOWFLAKE.ACCOUNT_USAGE.TAG_REFERENCES
WHERE TAG_NAME IN ('CLASSIFICATION','OWNER')
ORDER BY OBJECT_DATABASE, OBJECT_SCHEMA, OBJECT_NAME;
```

17.1.4 Making Lineage Actionable (Ops-Ready Views)

Publish a couple of stable views so engineers and auditors don't have to remember raw tables. For usage audits, flatten the ACCESS_HISTORY object arrays first; do not treat BASE_OBJECTS_ACCESSED or DIRECT_OBJECTS_ACCESSED as scalar arrays.

```
CREATE OR REPLACE VIEW gov.v_object_lineage AS
SELECT REFERENCING_OBJECT_DATABASE  AS db,
       REFERENCING_OBJECT_SCHEMA    AS sch,
       REFERENCING_OBJECT_NAME      AS obj,
       REFERENCED_OBJECT_DATABASE   AS src_db,
       REFERENCED_OBJECT_SCHEMA     AS src_sch,
       REFERENCED_OBJECT_NAME       AS src_obj,
       REFERENCED_OBJECT_DOMAIN     AS src_type
FROM <db>.<schema>.INFORMATION_SCHEMA.OBJECT_DEPENDENCIES;

CREATE OR REPLACE VIEW gov.v_pii_usage AS
SELECT a.QUERY_ID, a.QUERY_START_TIME, a.USER_NAME, a.ROLE_NAME,
       r.OBJECT_DATABASE, r.OBJECT_SCHEMA, r.OBJECT_NAME,
       r.COLUMN_NAME, r.TAG_VALUE AS classification
FROM SNOWFLAKE.ACCOUNT_USAGE.ACCESS_HISTORY a,
```

```
    LATERAL FLATTEN(input => a.BASE_OBJECTS_ACCESSED) obj
JOIN SNOWFLAKE.ACCOUNT_USAGE.TAG_REFERENCES r
  ON r.OBJECT_DATABASE = obj.value:objectName::STRING::VARCHAR SPLIT_
  PART('.', 1)
WHERE r.TAG_NAME = 'CLASSIFICATION';
```

17.1.5 Operational Checks You Can Schedule

Control goal	Quick query to schedule	What to alert on
PII columns always masked	Join TAG_REFERENCES to POLICY_REFERENCES	Any classification LIKE 'pii.%' column lacking a masking policy
Retention matches tag	Compare retention_policy tag to DATA_RETENTION_TIME_IN_DAYS	Mismatches (policy stronger than runtime, or vice versa)
Orphaned objects	Objects without owner tag in prod	Add tag or remove object
Impact before change	Read v_object_lineage for target	Downstream objects beyond a threshold (e.g., >10)

Example: "PII but no mask":

```
SELECT t.OBJECT_DATABASE, t.OBJECT_SCHEMA, t.OBJECT_NAME, t.COLUMN_NAME
FROM SNOWFLAKE.ACCOUNT_USAGE.TAG_REFERENCES t
LEFT JOIN SNOWFLAKE.ACCOUNT_USAGE.POLICY_REFERENCES p
  ON p.OBJECT_DATABASE = t.OBJECT_DATABASE
 AND p.OBJECT_SCHEMA   = t.OBJECT_SCHEMA
 AND p.OBJECT_NAME     = t.OBJECT_NAME
 AND COALESCE(p.COLUMN_NAME,'') = COALESCE(t.COLUMN_NAME,'')
WHERE t.TAG_NAME = 'CLASSIFICATION'
  AND t.TAG_VALUE LIKE 'pii.%'
  AND p.POLICY_NAME IS NULL;
```

17.1.6 Lifecycle and Promotion (Keep Tags with the Code)

Step	What to put in code	Why it works
DDL migration	Tag creation, tag sets on objects, policy attachments	Tags travel across environments like schema
CI check	Fails if new columns lack `classification` in PII schemas	Prevents drift at the PR gate
Release note	Link lineage diff for changed marts	Gives consumers an impact before deploy

Bottom line: use a handful of tags everywhere, prefer tag-based masking associations over repetitive per-column policy attachment, and publish a few flattened lineage and access views that auditors and engineers can query safely. With that in place, governance becomes a SQL-review habit instead of a separate, brittle process.

17.2 Row Access Policies and Column Masking

Snowflake enforces governance **at query time**. Two native controls do the heavy lifting:

- **Row Access Policies (RAP)** decide **which rows** a session may see, returning only rows where a Boolean expression is true.
- **Column Masking Policies (CMP)** decide **how a value appears** to a session, returning redacted or transformed values when needed.

Both are SQL objects you attach to tables (and views). They evaluate at query time, including through joins, views, Time Travel reads, and shares. For cloned or moved objects, review policy and tag inheritance explicitly, especially when using tag-based masking at the database or schema level, because protection behavior depends on where the tag, policy, and target object live.

17.2.1 When to Use Which (And Together)

Need	Use	Why
Per-tenant/project scoping	**Row Access Policy** on a table keyed by `tenant_id` (or similar)	Keeps unwanted rows out of result sets entirely
PII obfuscation by role	**Masking Policy** on sensitive columns (`email`, phone, `ssn`)	Same rows for all, but different **views** per consumer
Both multi-tenant **and** PII	Apply **RAP** for scoping **and CMP** for columns	Least data movement; single source of truth

17.2.2 Policy Building Blocks

- **Context:** `CURRENT_ROLE()`, `CURRENT_USER()`, `IS_ROLE_IN_SESSION('ROLE')`, session variables (`SELECT $var`), and lookup joins to mapping tables (e.g., user ➤ tenants).
- **Tag-aware logic**: Policies can branch on a column's tags with `SYSTEM$GET_TAG_ON_CURRENT_COLUMN('<tag>')`, letting one policy protect many columns.
- **Attachment**: One RAP per table (it can reference multiple columns). One masking policy per column.

17.2.3 Example: Tenant Scoping with a Row Access Policy

1) **A mapping table** (who can see which tenants)

```
CREATE OR REPLACE TABLE gov.role_tenants (
  role_name STRING,
  tenant_id STRING
) COMMENT='RBAC map for row access';
```

2) **The policy** (return TRUE only for permitted rows)

```
CREATE OR REPLACE ROW ACCESS POLICY rap_tenant_guard
AS (tenant_id STRING)
RETURNS BOOLEAN ->
  EXISTS (
    SELECT 1
    FROM gov.role_tenants m
    WHERE m.role_name = CURRENT_ROLE()
      AND m.tenant_id = tenant_id
  );
```

3) **Attach to fact tables** (one policy, many tables)

```
ALTER TABLE mart.fact_events
  ADD ROW ACCESS POLICY rap_tenant_guard ON (tenant_id);

ALTER TABLE mart.fact_orders
  ADD ROW ACCESS POLICY rap_tenant_guard ON (tenant_id);
```

Now every query against those tables (directly or through views) returns only rows for tenants authorized to the active role. This is the safer production pattern because Snowflake's authorization model is role-based, and shared service users or BI tool identities often do not map one-to-one to business users.

17.2.4 Example: Reusable Masking with Tag-Based Policy Association

1) Tag columns once

```
CREATE TAG classification ALLOWED_VALUES ('pii.email','pii.
phone','pii.ssn','public');
ALTER TABLE dim_customer
  MODIFY COLUMN email SET TAG classification = 'pii.email';
ALTER TABLE dim_customer
  MODIFY COLUMN phone SET TAG classification = 'pii.phone';
```

2) Create masking policies by data type

```
CREATE OR REPLACE MASKING POLICY mask_email AS (val STRING)
RETURNS STRING ->
  CASE
    WHEN IS_ROLE_IN_SESSION('SECURITYADMIN') OR IS_ROLE_IN_
    SESSION('ROLE_PII_READER') THEN val
    ELSE REGEXP_REPLACE(val, '(^.).+(@.+$)', '\\1***\\2')
  END;
CREATE OR REPLACE MASKING POLICY mask_phone AS (val STRING)
RETURNS STRING ->
  CASE
    WHEN IS_ROLE_IN_SESSION('SECURITYADMIN') OR IS_ROLE_IN_
    SESSION('ROLE_PII_READER') THEN val
    ELSE REGEXP_REPLACE(val, '(.{2}).+(.{2})', '\\1****\\2')
  END;
```

3) Associate masking policies with the tag once; Snowflake applies protection automatically to tagged columns whose data type matches the policy signature. A tag supports one masking policy per data type.

```
ALTER TAG classification
  SET MASKING POLICY mask_email, MASKING POLICY mask_phone;
```

Result: analysts see redacted values; authorized roles see plain text—without changing queries or duplicating tables.

17.2.5 Combining RAP and Masking (Order and Behavior)

Aspect	Behavior
Evaluation	RAP filters **rows first**, then CMP renders the remaining column values
Joins	Policies apply to each protected table before join logic; rows you can't see don't participate in joins
Clones/Time Travel	Policies and attachments are preserved; historical reads still enforce RAP/CMP
Shares	Policies travel with shared objects; consumers see filtered/masked data

Tag-based masking has a few important limits: a tag supports only one masking policy per data type, a column cannot simultaneously be protected by both a row access policy signature and a masking policy signature, and a table containing a column protected by a tag-based masking policy cannot be used as a mapping table. These constraints are worth calling out because they affect how you design governance schemas and policy lookup tables.

17.2.6 Operational Views and Quick Checks

Which objects have policies attached?

```
SELECT *
FROM SNOWFLAKE.ACCOUNT_USAGE.POLICY_REFERENCES
ORDER BY OBJECT_DATABASE, OBJECT_SCHEMA, OBJECT_NAME;
```

Which PII columns remain unmasked (gap check)?

```
SELECT t.OBJECT_DATABASE, t.OBJECT_SCHEMA, t.OBJECT_NAME, t.COLUMN_NAME
FROM SNOWFLAKE.ACCOUNT_USAGE.TAG_REFERENCES t
LEFT JOIN SNOWFLAKE.ACCOUNT_USAGE.POLICY_REFERENCES p
  ON p.OBJECT_DATABASE = t.OBJECT_DATABASE
 AND p.OBJECT_SCHEMA   = t.OBJECT_SCHEMA
 AND p.OBJECT_NAME     = t.OBJECT_NAME
 AND COALESCE(p.COLUMN_NAME,'') = COALESCE(t.COLUMN_NAME,'')
WHERE t.TAG_NAME = 'CLASSIFICATION'
  AND t.TAG_VALUE LIKE 'pii.%'
  AND p.POLICY_NAME IS NULL;
```

Who queried a protected table (audit trail)?

```
SELECT QUERY_ID, QUERY_START_TIME, USER_NAME, ROLE_NAME
FROM SNOWFLAKE.ACCOUNT_USAGE.ACCESS_HISTORY
WHERE QUERY_START_TIME >= DATEADD(day,-7,CURRENT_TIMESTAMP())

AND EXISTS (
  SELECT 1
  FROM LATERAL FLATTEN(input => a.BASE_OBJECTS_ACCESSED) obj
```

```
  WHERE obj.value:objectName::STRING = 'MART.FACT_EVENTS'
    AND obj.value:objectDomain::STRING = 'Table'
);
```

17.2.7 Design Tips That Hold Up in Audits

Situation	Guidance	Why
Multi-tenant marts	Put a single RAP on the **fact** tables; avoid duplicating schemas per tenant	One definition to maintain; consistent logic across models
Many PII columns	Use tag-aware masking (one policy, many columns)	Prevents drift and copy-paste errors
App/service access	Prefer role checks (`IS_ROLE_IN_SESSION`) over user lists	Roles are easier to rotate and audit
Testing policies	Switch roles and `SELECT` the same row; verify redaction and row presence	Proves enforcement without separate data
Performance	Keep RAP predicates sargable and mapping tables small and indexed (clustered on join keys if needed)	Minimizes policy evaluation overhead

17.2.8 Promotion and Change Control

- Treat RAP/CMP DDL like schema: version in migrations, peer-review changes, promote through environments.
- Attach policies in the same migration that creates new tables/columns.
- Store small unit tests: a seed row + assertions for masked/unmasked views under different roles.

Bottom line: Row Access Policies keep the wrong rows out; Column Masking keeps sensitive values obscure. Define both once, key row access to roles rather than usernames, and let tags and roles drive behavior. For masking, use tag-based policy association as the default so new tagged columns inherit protection automatically. That yields consistent protection across queries, copies, and time—without multiplying datasets.

17.3 PII Handling, Auditing, and Compliance Patterns

The safest way to handle PII in Snowflake is to **treat protection as data-first logic**: classify columns, bind policies to those tags, keep lineage/audit trails queryable, and make retention and subject-rights workflows repeatable in SQL. The aim is boring compliance—clear rules, few exceptions, and evidence you can re-run.

17.3.1 What "Good" Looks Like in Practice

Area	Practical stance	Why it holds up
Classification	Tag sensitive columns (`classification=pii.*`) and keep owners on tables/schemas	Policies can key off tags; responsibility is obvious
Access	Enforce row scope with RAP; obfuscate values with masking	One copy of data, different lawful views
Environments	No real PII in dev/test; use masked **secure views** or synthesized data	Eliminates accidental exposure outside prod
Retention	Short, explicit windows on mutable data; transient for intermediates	Less to protect; easier DSAR and purge logic
Evidence	Lineage + access history + policy references published as views	Audits use SQL, not screenshots

17.3.2 Classify Once, Let Policies Do the Rest

```
-- Tags
CREATE TAG classification ALLOWED_VALUES ('pii.email','pii.phone','pii.
name','restricted','public');
```

```
-- Mark columns
ALTER TABLE core.dim_customer
  MODIFY COLUMN email SET TAG classification='pii.email';
ALTER TABLE core.dim_customer
  MODIFY COLUMN phone SET TAG classification='pii.phone';

-- Masking policy keyed on the tag
CREATE OR REPLACE MASKING POLICY mask_pii AS (val STRING) RETURNS STRING ->
  CASE
    WHEN IS_ROLE_IN_SESSION('ROLE_PII_READER') THEN val
    WHEN SYSTEM$GET_TAG_ON_CURRENT_COLUMN('classification')='pii.email'
      THEN REGEXP_REPLACE(val,'(^.).+(@.+$)','\\1***\\2')
    WHEN SYSTEM$GET_TAG_ON_CURRENT_COLUMN('classification')='pii.phone'
      THEN REGEXP_REPLACE(val,'(.{2}).+(.{2})','\\1****\\2')
    ELSE val
  END;

ALTER TABLE core.dim_customer MODIFY COLUMN email SET MASKING POLICY
mask_pii;
ALTER TABLE core.dim_customer MODIFY COLUMN phone SET MASKING POLICY
mask_pii;
```

If you're multi-tenant or partner-scoped, pair masking with a single **Row Access Policy** on facts keyed by `tenant_id`.

17.3.3 Pseudonymization and Irreversible Transforms

Goal	Pattern	Example
Stable join keys without plain text	Salted, deterministic hash in a separate column	`SHA2(LOWER(email) || ':SALT_v1', 256)`
Reversible token (strict custody)	Format-preserving tokenization via UDF backed by a key in KMS	Keep key material outside SQL; restrict UDF to secure role
One-way anonymization	Null/aggregate at source or policy-driven substitute	Use for exports/analytics where identity is not needed

```
-- Salted hash for joins
ALTER TABLE core.dim_customer ADD COLUMN email_hash STRING;
UPDATE core.dim_customer
SET email_hash = SHA2(LOWER(email) || ':SALT_v1', 256);
```

Keep salts/keys versioned. When you rotate, create *_v2 columns and deprecate old hashes.

17.3.4 DSAR (Access) and Erasure Workflows

Request	Minimal, auditable flow	Notes
Access (export what you hold)	Assemble via secure view that joins all subject keys; COPY INTO a private stage	Query runs under a controlled role; emit a manifest
Erasure (delete/obfuscate)	Soft-delete in sources ➤ propagate to marts; mask residual PII; purge in next retention window	For facts needed for finance, keep business keys but remove direct identifiers

Subject Export (PII Bundle View)

```
CREATE OR REPLACE SECURE VIEW gov.v_subject_bundle AS
SELECT c.customer_id, c.email, c.phone, a.*
FROM core.dim_customer c
LEFT JOIN mart.activity a USING (customer_id);
```

Erase by Key (Soft Delete + Masking)

```
-- Mark subject as erased
UPDATE core.dim_customer SET erased_at = CURRENT_TIMESTAMP()
WHERE customer_id = :subject_id;

-- Optional: blank direct identifiers
UPDATE core.dim_customer
SET email = NULL, phone = NULL
WHERE customer_id = :subject_id;
```

Downstream, ensure your pipelines respect erased_at (exclude or mask on read) and that retention purges old snapshots per policy.

17.3.5 Erasure Caveat: Time Travel, Fail-Safe, and Downstream Copies

Erasure in Snowflake is not only a row-level update problem. Historical values can remain recoverable through Time Travel for the object's retention period, and permanent objects then move into Snowflake Fail-safe for a non-configurable seven-day period. For transient tables, there is no Fail-safe, and Time Travel can be reduced to 0 or 1 day, which makes them a better fit for short-lived intermediate copies of sensitive data. A compliant erasure workflow therefore needs four checks: remove or mask current values, minimize Time Travel where policy allows, avoid permanent downstream copies of direct identifiers, and track all cloned or exported copies that fall outside the source table's retention settings.

```
-- Example: shorten retention where policy allows
ALTER TABLE core.dim_customer SET DATA_RETENTION_TIME_IN_DAYS = 0;
-- Better for short-lived sensitive intermediates
CREATE OR REPLACE TRANSIENT TABLE stage.subject_export AS
SELECT ...
```

> Setting `DATA_RETENTION_TIME_IN_DAYS = 0` disables Time Travel for the object, but on permanent tables, the deleted or modified data still moves into Fail-safe, and Snowflake notes that this background move may take a short time to complete. Do not describe this as "instant erasure."

17.3.6 Retention, Legal Hold, and Proof

Concern	Control	Example
Routine retention	`DATA_RETENTION_TIME_IN_DAYS` by DB/schema; transient for intermediates	`ALTER DATABASE prod SET DATA_RETENTION_TIME_IN_DAYS=7;`
Legal hold	Tag and suspend drops; snapshot with **zero-copy clone**	`CREATE DATABASE hold_2025_10_01 CLONE prod;` keep hold DB isolated
Proof of enforcement	Scheduled checks that compare `retention_policy` tag vs. runtime	Alert on mismatches; store results in `gov.audit_findings`

17.3.7 Auditing You Can Re-run

Who Accessed PII This Week

```
SELECT a.QUERY_ID, a.QUERY_START_TIME, a.USER_NAME, a.ROLE_NAME,
       obj.value:objectName::STRING AS object_name,
       obj.value:objectDomain::STRING AS object_domain
FROM SNOWFLAKE.ACCOUNT_USAGE.ACCESS_HISTORY a,
     LATERAL FLATTEN(input => a.BASE_OBJECTS_ACCESSED) obj
WHERE a.QUERY_START_TIME >= DATEADD(day,-7,CURRENT_TIMESTAMP())
  AND obj.value:objectDomain::STRING IN ('Table', 'View')
ORDER BY a.QUERY_START_TIME DESC;
```

For exact PII-column evidence, build a separate audit view that flattens the object and column structures in ACCESS_HISTORY before joining to TAG_REFERENCES; a direct scalar join on DIRECT_OBJECTS_ACCESSED is not reliable because these fields are arrays of objects, not scalar values.

Gaps: PII Columns Without Masking

```
SELECT t.OBJECT_DATABASE, t.OBJECT_SCHEMA, t.OBJECT_NAME, t.COLUMN_NAME
FROM SNOWFLAKE.ACCOUNT_USAGE.TAG_REFERENCES t
LEFT JOIN SNOWFLAKE.ACCOUNT_USAGE.POLICY_REFERENCES p
  ON p.OBJECT_DATABASE=t.OBJECT_DATABASE
 AND p.OBJECT_SCHEMA=t.OBJECT_SCHEMA
 AND p.OBJECT_NAME=t.OBJECT_NAME
 AND COALESCE(p.COLUMN_NAME,'')=COALESCE(t.COLUMN_NAME,'')
WHERE t.TAG_NAME='CLASSIFICATION' AND t.TAG_VALUE LIKE 'pii.%'
  AND p.POLICY_NAME IS NULL;
```

Lineage Impact Before Changing a PII Column

```
SELECT REFERENCING_OBJECT_DATABASE, REFERENCING_OBJECT_SCHEMA, REFERENCING_
OBJECT_NAME
FROM <db>.<schema>.INFORMATION_SCHEMA.OBJECT_DEPENDENCIES
WHERE REFERENCED_OBJECT_NAME='DIM_CUSTOMER';
```

17.3.8 Sharing and Clean Rooms (Keep PII Safe Across Boundaries)

Scenario	Safer pattern	Why
Share with partner	Share **secure views** with masking and RAP attached	Policies travel; no raw tables exposed
Joint analytics	Use clean room/restricted UDFs; exchange only aggregates	Reduces re-identification risk

17.3.9 Non-prod Data Strategy

Option	How to populate dev/test	Risk posture
Synthetic	Generate statistically similar but fake data	Ideal—no real PII leaves prod
Masked secure views	Developers see masked values via `SECURE VIEW`	Good—policies enforced everywhere
Tokenized sample	Deterministic tokens with no plain text	OK—keep tokens/salts out of dev

17.3.10 Operational Checklist in One View

Signal	Smallest effective change
New PII column added but unmasked	Tag `classification`; attach `mask_pii`
Access spike on PII by unexpected role	Review `ACCESS_HISTORY`; tighten RBAC or RAP
DSAR backlog	Publish `gov.v_subject_bundle`; script `COPY INTO` + ticket link
Retention drift	Align `DATA_RETENTION_TIME_IN_DAYS` with `retention_policy` tag; shorten staging

Bottom line: classify once, enforce at read, minimize copies, and make audits a `SELECT` built on flattened access-history structures. For erasure, be explicit that Time Travel, Fail-safe, and downstream copies are part of the compliance design, not edge cases. With tags driving policies, RAP scoping rows, and retention encoded in DDL, PII handling becomes a set of small, reviewable SQL changes—not emergency procedures.

CHAPTER 18

Data Sharing and Collaboration

Snowflake lets you publish data **without copying it**. Providers expose objects from their account; consumers query them as if local. The platform handles security, billing, and freshness, so collaboration looks like ordinary SQL. What makes this work in practice is a small, explicit contract: *what you share, to whom, under which policies, and how you version it.*

18.1 Secure Data Sharing and Reader Accounts

Secure Data Sharing is the default way to collaborate when your partner already has a Snowflake account. **Reader Accounts** extend the model to partners who don't: you host a managed, read-only Snowflake account for them. Both approaches avoid file exports, duplication, and sync jobs; both preserve governance you attach to the shared objects.

18.1.1 How the Sharing Models Compare

Topic	Secure Data Sharing (consumer has Snowflake)	Reader Account (consumer does not)
Setup	Provider creates a **share**; consumer creates a **database from share**	Provider creates a **managed reader account** and a share to it
Data copies	None; consumer queries provider's storage	None; consumer queries provider's storage
Governance travel	**Secure views**, masking and row access policies travel with objects	Same—policies remain effective in the reader

(*continued*)

M. Hander, *Snowflake Data Warehouse Engineering*, https://doi.org/10.1007/979-8-8688-2628-3_18

Topic	Secure Data Sharing (consumer has Snowflake)	Reader Account (consumer does not)
Billing	Consumer pays for **compute** in their account	Provider pays for **compute** used by the reader
Identity and SSO	Consumer manages their users/roles	Provider provisions users/roles in the reader
Cross-region/ cloud	Use **replication** to a per-region copy, then share locally	Same—replicate first, then attach the reader in that region
When to choose	Partner already on Snowflake; clear boundaries	Lightweight external analytics, short-term access, or small partners

18.1.2 Contract of a Good Share

Element	Guidance	Why it holds up
Surface	Share **secure views** that project a stable schema	You can evolve internals without breaking consumers
Scope	One domain per share (e.g., "sales_mart_public")	Principle of least surprise; clearer SLAs
Governance	Attach masking/row policies to base objects; prefer secure views	Protection travels; audits stay simple
Versioning	Publish a versioned schema or view name (`v1`, `v2`)	Consumers can migrate on their schedule
Catalog	Include tags/descriptions on objects in the share	Discoverability and compliance context

18.1.3 Minimal DDL: Provider ➤ Consumer

Provider (Create and Grant a Share)

```
-- Create a share
CREATE OR REPLACE SHARE sales_public COMMENT='Sales mart public v1';
```

```
-- Add objects (schemas/tables/views must be in databases you own)
GRANT USAGE ON DATABASE SALES TO SHARE sales_public;
GRANT USAGE ON SCHEMA   SALES.MART TO SHARE sales_public;

-- Share secure views or tables
GRANT SELECT ON VIEW SALES.MART.V_ORDERS_V1 TO SHARE sales_public;
GRANT SELECT ON VIEW SALES.MART.V_CUSTOMERS_V1 TO SHARE sales_public;

-- Allow a consumer account to see it (same region)
ALTER SHARE sales_public ADD ACCOUNTS = ('CONSUMER_ACCT_NAME');
```

Consumer (Create a Database from the Share)

```
-- In consumer account
CREATE DATABASE SALES_FROM_PROVIDER FROM SHARE PROVIDER_ACCT_NAME.
sales_public;

SELECT * FROM SALES_FROM_PROVIDER.MART.V_ORDERS_V1;
```

Cross-Region or Cross-Cloud

Replicate the provider database to the target region account, run the initial REFRESH before exposing it, and set a replication schedule that matches the freshness promise you publish to consumers. Replication targets must be specified with fully qualified organization_name.account_name identifiers.

```
-- Provider: enable replication on the source database

-- Step 1: enable replication from the primary account using org.
account format
ALTER DATABASE SALES
  ENABLE REPLICATION TO ACCOUNTS my_org.provider_acct_eu;
-- Step 2: in the target region account, create the replica
CREATE DATABASE SALES_REPLICA
  AS REPLICA OF my_org.provider_acct_primary.SALES;
-- Step 3: initial refresh is required before the replica contains data
ALTER DATABASE SALES_REPLICA REFRESH;
-- Step 4: keep it current on a schedule that matches the published
freshness promise
ALTER DATABASE SALES_REPLICA
```

```
  ENABLE REPLICATION
  REPLICATION_SCHEDULE = '60 MINUTE';
-- Step 5: create the share in the target region account
CREATE SHARE sales_public_eu;
GRANT USAGE ON DATABASE SALES_REPLICA TO SHARE sales_public_eu;
GRANT USAGE ON SCHEMA SALES_REPLICA.MART TO SHARE sales_public_eu;
GRANT SELECT ON VIEW SALES_REPLICA.MART.V_ORDERS_V1 TO SHARE sales_
public_eu;

-- Create share in the target region account, grant from SALES_REPLICA
CREATE SHARE sales_public_eu;
GRANT USAGE ON DATABASE SALES_REPLICA TO SHARE sales_public_eu;
...
```

18.1.4 Reader Accounts When Partners Don't Have Snowflake

You host a managed, read-only Snowflake account; you share points at the reader, and you control its users/roles. Compute charges accrue to you, so keep warehouses small and elastic.

Provider (Create Reader Account and Share)

```
-- Create a managed reader account
CREATE MANAGED ACCOUNT reader_acme
  ADMIN_NAME = 'acme_admin'
  ADMIN_PASSWORD = 'StrongPass!23'
  TYPE = READER;

-- Connect to the reader account as the admin user you created,
-- then create the warehouse inside the reader account
CREATE WAREHOUSE wh_reader
  WAREHOUSE_SIZE = 'XSMALL'
  AUTO_SUSPEND = 60
  AUTO_RESUME = TRUE;
-- Grant warehouse usage to a role in the reader account
GRANT USAGE ON WAREHOUSE wh_reader TO ROLE PUBLIC;
```

Reader account warehouses are created inside the reader account, not in the provider account. The provider manages the reader account, but compute access is still granted through reader-account roles such as PUBLIC or custom roles. Resource monitors remain part of the provider-side operational guardrails for controlling spend.

```
-- Share data to the reader
CREATE OR REPLACE SHARE sales_public_reader;
GRANT USAGE ON DATABASE SALES TO SHARE sales_public_reader;
GRANT USAGE ON SCHEMA   SALES.MART TO SHARE sales_public_reader;
GRANT SELECT ON VIEW SALES.MART.V_ORDERS_V1 TO SHARE sales_public_reader;

ALTER SHARE sales_public_reader ADD ACCOUNTS = ('reader_acme');
```

Reader (First Login)

The partner uses the credentials you set to log in to the managed account. They see the shared database, and they can query it using the warehouse you created and granted inside the reader account.

Keep reader warehouses tiny with short auto-suspend, and manage spend with a resource monitor as part of the provider's operating model for the reader account.

18.1.5 Governance That Travels with the Share

Control	Behavior in shares/readers	Notes
Secure views	Definition hidden; only results exposed	Prefer secure views as the shared surface
Masking/Row Access Policies	Enforced for consumers, including readers	Test with a consumer role before go-live
Tags	Values visible in consumer metadata (read-only)	Carry classification, owner, retention_policy
Access History	Provider sees reads in ACCESS_HISTORY	Useful for audit and sticky deprecation warnings

18.1.6 Versioning and Deprecation Without Breaking Consumers

Situation	Provider move	Consumer experience
Add columns	Add to V_ORDERS_V1; avoid breaking semantics	Non-breaking; documented in share notes
Breaking change	Publish V_ORDERS_V2; keep V1 for N weeks	Consumers migrate at their pace
Sunset	Remove V1 from share after window	Access errors signal an upgrade is required

A small **"release notes"** view (table of changes by date) inside the shared schema keeps consumers informed:

```
CREATE OR REPLACE VIEW MART.V_RELEASE_NOTES AS
SELECT '2025-09-12' AS dt, 'V_ORDERS_V2 adds column order_channel' AS note
UNION ALL
SELECT '2025-08-01', 'Initial V1 release';
```

18.1.7 Cost and Performance Hygiene

Risk	Posture that works	Why
Provider pays for reader compute	Tiny reader warehouse, AUTO_ SUSPEND=60, resource monitor	Keeps surprises small
Hot partitions thrashed by many consumers	Encourage result caching and projections via **secure views**	Less scanned data per query
Wide, "SELECT *" queries	Publish narrow, purpose-built views	Better pruning and stable schema
Cross-region latency	Replicate database to consumer region before sharing	Queries stay local; no egress surprises

18.1.8 Observability and Audit (Provider's Side)

Who Queried Shared Objects Recently

```
SELECT QUERY_ID, QUERY_START_TIME, USER_NAME, ROLE_NAME, DIRECT_OBJECTS_
ACCESSED
FROM SNOWFLAKE.ACCOUNT_USAGE.ACCESS_HISTORY
WHERE QUERY_START_TIME >= DATEADD(day, -7, CURRENT_TIMESTAMP())
  AND obj.value:objectName::STRING = 'SALES.MART.V_ORDERS_V1'
  AND obj.value:objectDomain::STRING = 'View';
```

In ACCESS_HISTORY, DIRECT_OBJECTS_ACCESSED and BASE_OBJECTS_ACCESSED are arrays of objects, not arrays of strings. Flatten them first; do not use ARRAY_CONTAINS with object names.

Observability note When querying ACCOUNT_USAGE.ACCESS_HISTORY, object-access fields such as DIRECT_OBJECTS_ACCESSED and BASE_OBJECTS_ACCESSED are semi-structured arrays of objects. Use LATERAL FLATTEN and filter on keys like objectName and objectDomain; do not treat them as scalar arrays and do not use lambda-style array functions from other SQL dialects.

Which Accounts Consume a Share

```
SHOW SHARES;          -- provider: see consumers and objects
DESCRIBE SHARE sales_public;
```

Reader Spend Guardrail

```
CREATE OR REPLACE RESOURCE MONITOR rm_reader_acme
  WITH CREDIT_QUOTA = 100
  FREQUENCY = MONTHLY
  TRIGGERS ON 80 PERCENT DO NOTIFY
           ON 100 PERCENT DO SUSPEND;

ALTER WAREHOUSE wh_reader_acme SET RESOURCE_MONITOR = rm_reader_acme;
```

18.1.9 Security Posture That Keeps Legal Happy

Topic	Practical stance	Effect
PII	Share **secure views** with masking; keep RAP on base tables	Consumers see only lawful values/rows
Residency	Share from a **regional replica** that matches consumer residency	Avoid cross-region transfer
Lifecycle	Treat shares as code: DDL in migrations, peer-reviewed	Predictable changes and rollbacks
Offboarding	`ALTER SHARE … REMOVE ACCOUNTS = (…)` and rotate reader creds	Immediate revocation

Closing note Share **views, not raw tables**, replicate to where consumers are, and keep a slim, versioned contract. With reader accounts for non-Snowflake partners and secure data sharing for everyone else, collaboration stops being a pipeline and becomes a permission—fast to grant, easy to audit, and simple to take back.

18.2 Marketplace Publishing/Consumption

Snowflake Marketplace turns data sharing into a **product**: a discoverable listing with docs, versioning, and optional monetization. Technically, it's still Secure Data Sharing under the hood, but with catalog, entitlement, and (if you choose) billing handled for you. Treat listings like APIs—stable contracts, release notes, and observable usage.

18.2.1 What Changes vs. Ad Hoc Shares

Aspect	Private share	Marketplace listing
Discovery	Direct invite	Searchable catalog entry with docs and tags
Entitlement	Add a consumer account to share	Consumer "requests" ➤ you approve (or auto-approve)
Versioning	Your naming convention	Listing versions, deprecation notes
Monetization	N/A	Optional paid plans, trials, usage terms
Support	Email/Slack	Listing contact + changelog within the product

18.2.2 Designing a "Product" Surface

Publish **secure views** with a stable schema and keep raw tables private. Add a small "about" view for quick profiling.

```
-- Provider account
CREATE OR REPLACE SECURE VIEW DATASET.V_TRIPS_V1 AS
SELECT trip_id, pickup_ts, dropoff_ts, pickup_zone, dropoff_zone,
fare_amount
FROM RAW.TRIPS
WHERE trip_ts >= '2023-01-01';

CREATE OR REPLACE VIEW DATASET.V_RELEASE_NOTES AS
SELECT '2025-09-10' AS dt, 'V_TRIPS_V1 adds fare_amount' AS note
UNION ALL
SELECT '2025-07-01', 'Initial public release';
```

Tag objects so catalog filters work:

```
CREATE TAG data_domain ALLOWED_VALUES ('mobility','retail','finance');
ALTER VIEW DATASET.V_TRIPS_V1 SET TAG data_domain='mobility';
```

18.2.3 Provider Workflow (Compact)

Step	What you do	Why it helps
Prepare	Build **secure views**, attach masking/row policies to bases	Governance travels; schema stays stable
Replicate (if needed)	Create a regional **replica** near target buyers	Low latency; no cross-region egress
Create listing	Title, categories, terms, contact, regions; choose free or paid	Makes it discoverable and governed
Entitlements	Auto-approve or manual approval; optional trials	Control who gets it and for how long
Observe	Use Access History to see consumption	Plan deprecations and capacity

DDL for shares is created behind the scenes by Marketplace; you still control which views enter the listing.

18.2.4 Consumer Workflow (Compact)

Step	What you do	Outcome
Discover	Search Marketplace; read docs and sample schema	Decide fit without back-and-forth
Request	Click "Get" (free) or start trial/purchase	A database appears in your account once approved
Use	`CREATE DATABASE <name> FROM SHARE <provider>.<listing>`	Query like local objects; pay only compute
Track	Tag queries with `QUERY_TAG`	Attribute spend by team or project

18.2.5 Versioning and Deprecation

Keep versions side-by-side; publish a retire window.

Move	Provider action	Consumer experience
Non-breaking (columns added)	Update V1 docs; bump listing minor version	Queries keep working
Breaking	Publish V2 views; mark V1 deprecates in N days	Time to migrate without data copy
Sunset	Remove V1 from the listing after window	Failing queries surface "not found" with clear notes

18.2.6 Observability You Can Rely On (Provider Side)

Which shared views from this listing were queried this week. Flatten the access-history object arrays first; Snowflake SQL does not support the lambda-style ARRAY_EXISTS **syntax used in some other systems.**

```
SELECT a.QUERY_START_TIME, a.USER_NAME, a.ROLE_NAME
FROM SNOWFLAKE.ACCOUNT_USAGE.ACCESS_HISTORY a,
     LATERAL FLATTEN(input => a.DIRECT_OBJECTS_ACCESSED) obj
WHERE a.QUERY_START_TIME >= DATEADD(day,-7,CURRENT_TIMESTAMP())
  AND obj.value:objectName::STRING LIKE 'DATASET.V_TRIPS%'
  AND obj.value:objectDomain::STRING = 'View';
```

Top Consumers by Volume (Approx.)

```
SELECT VALUE:clientApplication::string AS client_app,
       COUNT(*) AS queries
FROM SNOWFLAKE.ACCOUNT_USAGE.QUERY_HISTORY,
     LATERAL FLATTEN(input => CLIENT_ENVIRONMENT)
WHERE START_TIME >= DATEADD(day,-30,CURRENT_TIMESTAMP())
  AND DATABASE_NAME='DATASET'  -- your shared DB/schema
GROUP BY 1
ORDER BY queries DESC;
```

(Granularity varies by account type; use Marketplace analytics where available for commercial metrics.)

18.2.7 Security and Compliance Posture

Topic	Provider stance	Effect
PII	Share **secure views** with masking; keep RAP on base tables	Buyers see only lawful values/rows
Residency	Publish from the **consumer's region** (via replica)	No cross-region transfer
Terms and contact	Fill listing legal terms + support email	Sets expectations and support channel
Audits	Keep tags (`classification`, `owner`, `retention_policy`) visible	Buyers' GRC can evaluate quickly

18.2.8 Cost and Performance Hygiene

Risk	Small fix	Why
Consumer scans too much	Publish narrow views; include date/tenant envelopes	Strong pruning on their side
Hot objects for many buyers	Encourage result caching (stable SQL text) in docs	Lower provider storage I/O and buyer compute
Region mismatch	Replicate first; list per region	Removes latency/egress surprises

18.2.9 Sample Consumer Onboarding Snippet

```
-- In consumer account after approval
CREATE DATABASE MOBILITY_TRIPS FROM SHARE PROVIDER_ACCT.trips_marketplace;

-- Optional: local view to pin a stable contract internally
CREATE OR REPLACE VIEW shared.trips_v1 AS
SELECT * FROM MOBILITY_TRIPS.DATASET.V_TRIPS_V1;
```

```
-- Tag queries for chargeback
ALTER SESSION SET QUERY_TAG='team=product;purpose=mobility_feature';
```

18.2.10 Operating Rhythm

- Treat a listing like software: docs, semantic versioning, release notes.
- Validate every change against a **consumer test** account before publishing.
- Watch weekly usage; communicate deprecations in-product and via email.
- Keep governance in SQL (policies, tags), so portability and audits are easy.

Marketplace makes collaboration a **click, not a pipeline**. With secure views as the interface, regional replicas for reach, and a small set of product habits—docs, versions, and observability—you can publish real data products without operational drag.

18.3 Cross-Account Contracts and SLAs

A data share is a **product with a promise**. The consumer is betting their dashboards, models, and downstream SLAs on your tables or views; you're committing to keep a stable interface, predictable freshness, and clear communications when something changes. Put those promises in a **data contract** that lives next to the objects—versioned, queryable, and easy to check during CI/CD.

A practical contract has four pillars:

1) **Interface stability.** Treat the shared surface as an API. Publish **secure views** that project a stable schema, keep raw tables private, and version explicitly (e.g., `V_ORDERS_V1`, later `V_ORDERS_V2`). Non-breaking additions land in place; breaking changes ship as a new view with a deprecation window for the old one. A tiny "release notes" view in the shared schema helps consumers validate their expectations without chasing emails.

2) **Freshness and availability.** Consumers don't need internal details; they need a simple *"how fresh is this and when should I expect updates?"* Define a cadence ("every 15 minutes," "hourly at :05," "daily by 02:00 UTC"), and surface it as data: a **freshness heartbeat** row in each share, plus a view that shows the newest partition per object. When cross-region sharing is required, replicate to the consumer's region, and state the expected replication lag window; availability is the share being queryable in that region.

3) **Security posture.** The governance you apply at home must travel: masking on PII, any row access logic, residency guarantees via regional replicas, and the list of tags you expose (`classification`, `owner`, `retention_policy`). Put that posture in the contract and keep it testable—consumers should be able to verify that PII remains masked from their roles.

4) **Change and incident process.** Changes are announced in-product, with dates: publish the target date for a version cutover, the deprecation date for older views, and a contact. Incidents are time-boxed and observable: you record start/stop, impact, and corrective actions in a small table the consumer can read. The point is to make your promises **machine-checkable** and your communication **repeatable**.

18.3.1 What to Publish in the Schema

Keep two small artifacts in the shared schema alongside the data views:

- `V_CONTRACT`: A view with one row per shared object stating its SLOs and version.
- `V_RELEASE_NOTES`: A changelog table or view with date, object, change note, and deprecation/retire dates.

Example shapes:

```
-- Consumer-friendly contract facts (you maintain the rows during releases)

CREATE OR REPLACE VIEW MART.V_CONTRACT AS
SELECT
  'V_ORDERS_V1'        AS object_name,
  'daily 02:00 UTC'    AS update_cadence,
  120                  AS max_expected_lag_minutes,
  '2025-12-31'::DATE   AS support_until,
  'v1.6.0'             AS version
UNION ALL
SELECT
  'V_CUSTOMERS_V1',
  'hourly at :05',
  30,
  '2025-12-31'::DATE,
  'v1.4.2';

CREATE OR REPLACE VIEW MART.V_RELEASE_NOTES AS
SELECT '2025-09-10'::DATE AS dt, 'V_ORDERS_V1', 'Add order_channel (non-
breaking)', NULL::DATE AS deprecates_on
UNION ALL
SELECT '2025-08-01','V_CUSTOMERS_V1','Initial release', NULL;
```

For freshness, expose a heartbeat:

```
CREATE OR REPLACE VIEW MART.V_FRESHNESS AS
SELECT 'V_ORDERS_V1' AS object_name,
       (SELECT MAX(order_dt) FROM MART.V_ORDERS_V1) AS newest_partition,
       CURRENT_TIMESTAMP() AS observed_at;
```

Consumers can alert when `DATEDIFF('minute', newest_partition, CURRENT_TIMESTAMP())` exceeds `max_expected_lag_minutes`. Using integer minutes keeps the contract view portable and valid in Snowflake SQL.

18.3.2 SLOs That Are Easy to Agree on and Verify

Dimension	Typical promise	How the consumer checks
Freshness	95% of days, newest data ≤ stated lag after cadence time	Compare `DATEDIFF('minute', newest_partition, CURRENT_TIMESTAMP())` to `V_CONTRACT.max_expected_lag_minutes`; alert when the difference is greater.
Schema stability	No breaking changes without 30-day overlap window	Presence of `V1` and `V2` side-by-side; deprecate date in `V_RELEASE_NOTES`
Availability	Share queryable in the consumer region 99.9% monthly	Consumer monitors `SELECT 1 FROM V_CONTRACT` from their scheduler
Residency	Data shared from region <X> only	Provider lists replica region; consumer checks database location
Security	PII columns always masked for non-privileged roles	Consumer role test against example rows; tags visible on objects

18.3.3 Cost, Performance, and Responsibility

Be explicit about who pays for what. For **Secure Data Sharing**, compute is always on the **consumer**; for **Reader Accounts**, compute is on the **provider**—and should be capped with a small warehouse and a resource monitor. Performance expectations belong in the interface, not in the consumer's tuning: share **narrow, prunable views** with envelope columns (dates, tenants) so consumers don't need to guess how to filter. If you denormalize for convenience, say so in the notes and keep slowly changing keys stable.

18.3.4 Cross-Region Contracts

When consumers sit in other regions or clouds, the promise is **replication-backed freshness**. That promise is only real once the replica has been refreshed and a replication schedule is in place. Creating the replica object alone does not make data available to consumers. Align the refresh cadence with the `max_expected_lag_minutes` value you publish in `V_CONTRACT`. State the replication interval (e.g., hourly) and the

maximum replication lag. If a region is temporarily behind, the consumer's freshness alert should point to the published lag number; your incident notes should reflect the replica catch-up time. For regulated datasets, include the residency assertion explicitly: "Data originates and is served from `<region>`; no cross-region transfer occurs."

18.3.5 Incidents and Deprecations

Treat both as data. Keep a tiny table `MART.SERVICE_LOG` with columns `ts`, `severity`, `object_name`, `summary`, `status`. During a disruption, you insert a row ("Freshness exceeded lag due to upstream backlog; mitigation underway") and close it when resolved. For deprecations, add a note in `V_RELEASE_NOTES` with a deprecate date—your CI should refuse to drop the old view before that date.

```
CREATE OR REPLACE TABLE MART.SERVICE_LOG (
  ts TIMESTAMP_NTZ, severity STRING, object_name STRING, summary STRING,
  status STRING
);
```

18.3.6 Testing and Promotion

Before you change what consumers see, test in a **consumer sandbox**. Create a separate consumer account (or a reader) wired to your pre-prod share, and run the same checks your customers will: schema diff against `V_CONTRACT`, freshness under load, and masking behavior under a least-privilege role. Promotion becomes a metadata step—granting the new view to the production share and adding a row to `V_RELEASE_NOTES`.

18.3.7 Closing Guidance

Write your promises down where queries live. A consumer should be able to answer, with SQL, what version they're on, how fresh their data is supposed to be, when a change is coming, and who to contact when something looks off. If the contract is **data**, both sides can automate checks and plan migrations without meetings—and your cross-account collaboration stays fast, predictable, and calm.

CHAPTER 19

Business Continuity

Business continuity in Snowflake is about **recovering state, preserving service, and proving it**. The platform already maintains highly durable storage and separates compute from data, so the craft is deciding which recovery points matter (minutes vs. days), where they must exist (same region vs. cross-region), and how to exercise them without drama. You get three native levers: **Time Travel** for short-horizon rewinds, **zero-copy cloning** for instant snapshots and safe changes, and **Fail-safe** as the last resort. Around those, you wrap promotion, replication, and a runbook that turns incidents into straightforward SQL.

19.1 Backups, Time Travel, and Fail-Safe

Snowflake does not ask you to ship nightly dumps. Instead, it records **immutable micro-partitions** and histories that let you query or restore a past state. Think in tiers: **operational oops** (minutes to days) ➤ Time Travel or a clone; **environment recovery** (schema/data at a point) ➤ clones and scripted replays; catastrophe or regional loss ➤ a replicated copy first; if all operator-controlled recovery windows are exhausted, Fail-safe is the last resort through Snowflake support. The shorter the horizon, the faster and cheaper the move.

M. Hander, *Snowflake Data Warehouse Engineering*, https://doi.org/10.1007/979-8-8688-2628-3_19

19.1.1 What Each Mechanism Is Really For

Capability	Best used for	Recovery feel	Notes that matter
Time Travel	Undo recent mistakes, inspect prior versions, rebuild a table "as of"	Query/restore to an exact timestamp or statement	Retention is configurable per object; costs scale with churn and window length
Zero-copy clone	Instant snapshot for dev/test, hotfixes, audits, backfills	Metadata-only; available in seconds	Writes after the clone diverge; drop when finished to release deltas
Fail-safe	Last-resort recovery beyond Time Travel	Provider-assisted restore window	Not user-initiated on demand; meant for disasters, not routine oops

19.1.2 Time Travel in Practice

Use Time Travel to answer *"what did this look like at 03:00?"* or to reverse an accidental merge. It works at database, schema, and table scope. Keep the window **short but useful** for production (7–14 days is common) and **tiny** for staging/transient areas.

Set retention where it belongs:

```
-- Safer default on the prod database
ALTER DATABASE prod SET DATA_RETENTION_TIME_IN_DAYS = 7;

-- Leaner on staging/intermediate layers
ALTER SCHEMA stage SET DATA_RETENTION_TIME_IN_DAYS = 1;
```

Query a past state without restoring anything:

```
SELECT *
FROM mart.fact_orders
  AT (TIMESTAMP => TO_TIMESTAMP_TZ('2025-09-12 03:00:00 +00:00'));
```

Recover a bad write into a new table first, validate it, and only then promote it with `SWAP`. Avoid restoring directly with `CREATE OR REPLACE` during an incident, because that removes your easiest fallback if the chosen timestamp is wrong.

```
CREATE OR REPLACE TABLE mart.fact_orders AS
SELECT * FROM mart.fact_orders
  AT (STATEMENT => '01b1a3a3-0000-1234-0000-abcdef123456');  -- the last
  good statement id
```

Or surgically insert missing rows:

```
INSERT INTO mart.fact_orders (SELECT * FROM mart.fact_orders AT (TIMESTAMP
=> DATEADD('minute', -15, CURRENT_TIMESTAMP()))
MINUS SELECT * FROM mart.fact_orders);
```

Cost sense: retention holds extra history only for **changes**; wide, high-churn tables are expensive to keep long windows on. Right-size by layer, not one blanket number.

19.1.3 Zero-Copy Clones: Snapshots You Can Work With

Clones are your instant backups, branches, and experiment sandboxes. They share storage with the source at creation; you only pay for new or changed partitions after the split. Use them to validate a hotfix, to backfill without touching prod, or to freeze an audit state.

Create and promote without copying data:

```
-- Freeze prod at a point (optional: specify AT <timestamp>)
CREATE DATABASE prod_candidate CLONE prod;

-- Run validation/backfill in the clone, then promote by swapping
ALTER DATABASE prod_candidate SWAP WITH prod;  -- atomic rename; sessions
continue with the new contents
```

Scope clones narrowly when possible:

```
CREATE SCHEMA mart_dev CLONE prod.mart;
CREATE TABLE  mart.fact_orders_dev CLONE prod.mart.fact_orders;
```

Drop clones when finished; the shared partitions are reference-counted, and space is reclaimed as deltas disappear.

19.1.4 Fail-Safe: What It Is and What It Is Not

Fail-safe is the safety net beyond your Time Travel window, but it is a Snowflake-managed, support-assisted recovery service provided on a best-effort basis. It is not a fast undo button, not available for transient or temporary tables, and Snowflake does not present it as something you should include in your published RTO commitments. Recovery can take hours to days and is intended only after all other recovery options have been exhausted. Treat Fail-safe as the last resort, not as an operational recovery plan.

19.1.5 A Calm Recovery Playbook

When something goes wrong, choose the smallest tool that gets truth back into place:

1. **Identify the good point.** Use `QUERY_HISTORY`, object timestamps, or a known checkpoint time.
2. **Prefer Time Travel** to observe and copy rows back into the live object or to rebuild the object wholesale with `CREATE OR REPLACE … AT (…)`.
   2a. If you plan to rebuild an object from Time Travel, restore into a new table, schema, or database first. Validate there, then promote with `SWAP` or controlled cutover. Do not overwrite the live object first during an incident.
3. **Prefer a clone** when you need to test fixes, compare diffs, or perform complex rewrites safely; promote by `SWAP` rather than editing prod live.
4. **Document the incident** in a small ops table (who, what, when, how recovered) so audits and postmortems use data, not lore.
5. **Escalate to Fail-safe** only if the window has elapsed or objects were dropped beyond Time Travel; coordinate with support.

A compact example—bad `MERGE` at 01:12:

```
-- Verify good state at 01:00 without touching prod
SELECT COUNT(*)
FROM mart.fact_orders AT (TIMESTAMP => '2025-09-12 01:00:00');
```

```
-- Restore into a new object first
CREATE OR REPLACE TABLE mart.fact_orders_restore AS
SELECT *
FROM mart.fact_orders AT (TIMESTAMP => '2025-09-12 01:00:00');
-- Compare live vs. restored before cutover
WITH live AS (SELECT * FROM mart.fact_orders),
     restored AS (SELECT * FROM mart.fact_orders_restore)
SELECT 'missing_from_live' AS kind, * FROM restored MINUS SELECT *
FROM live
UNION ALL
SELECT 'extra_in_live', * FROM live MINUS SELECT * FROM restored;
-- Promote after validation
ALTER TABLE mart.fact_orders_restore SWAP WITH mart.fact_orders;
```

Keep the swapped-out table until the incident is fully closed. That gives you a fast rollback path if validation misses something on the first pass. The same idea applies at schema or database scope: clone or restore to a separate object first, then promote.

19.1.6 Designing RPO/RTO with Native Pieces

Recovery isn't just "can we get it back"; it's **how much we can lose** (RPO) and **how long it takes** (RTO). Snowflake's primitives map neatly:

Target	How you meet it	Typical numbers
RPO (minutes)	Time Travel + frequent, idempotent ELT; optional Dynamic Tables	RPO near zero for most marts
RTO (minutes)	Clone + SWAP; runbooks instead of manual edits	Seconds to a few minutes
Regional outage tolerance	Database replication or, preferably, account/failover-group replication to a secondary region, with an initial refresh completed and an explicit ongoing refresh cadence	RPO ≈ refresh cadence; RTO minutes with a rehearsed promotion and client cutover

(Cross-region replication and failover/failback are covered in the next chapter; they extend this local continuity model to regional events.)

19.1.7 Guardrails That Keep Continuity Cheap and Reliable

- **Right-size retention.** Keep 7–14 days on shared prod databases; 1–2 days on staging/transient layers; temp for scratch.
- **Clone for change.** Any risky DDL/DML runs in a clone; promotion is an atomic `SWAP`.
- **Make ELT repeatable.** Build from source truth idempotently, so rewinds are replays, not hand edits.
- **Log what matters.** Store checkpoints and incident notes in a tiny `ops.event_log` table for audits and drills.
- **Practice.** Run a quarterly "oops drill": restore a table with Time Travel and promote a clone under a test role.

Bottom line: let Time Travel handle oops, clones handle change, and Fail-safe guard the unthinkable. With retention tuned by layer and a couple of practiced SQL moves, most recoveries are quick, reversible, and boring—which is exactly what continuity should feel like.

19.2 Database and Account Replication (Cross-Region/Cloud)

When continuity must survive a **regional** problem—or you need to serve data close to partners—you replicate. Snowflake offers two scopes: **database replication** for data products and **account replication** for the whole control plane (roles, users, warehouses, pipes, tasks, and selected account objects). Promotion is metadata—no bulk copy at switchover—and you can fail back once the source returns.

Replication works across regions and clouds. The contract you set is simple: which objects move, how often they refresh, who is allowed to promote a secondary, and what clients should do during a failover. Keep it boring: one primary, one or two secondaries, a named cadence, and a runbook you rehearse.

19.2.1 Choosing the Right Scope

Scope	What moves	Typical use
Database replication	All objects inside a database (schemas, tables, secure views, policies, tasks, streams, etc.)	Keep a product (or set of marts) queryable in another region/cloud; publish Marketplace/Share close to consumers
Account replication and failover	Databases **plus** account objects (users, roles, warehouses, resource monitors, integrations, shares, tags)	Region-level business continuity for an entire account; one switch promotes *everything* you've selected

A common pattern is: replicate the **core product DBs** frequently for consumer latency, and set up **account replication** on a slower cadence for disaster scenarios.

19.2.2 Database Replication in Practice

On the **primary** account, you authorize destinations; on the **secondary**, you create the replica. Refreshes copy only changes since the last cycle, but the first REFRESH is required to populate the secondary before it is usable. Your published RPO should match the refresh cadence you actually run.

```
-- Step 1: authorize the target using fully qualified org.account format
ALTER DATABASE SALES
  ENABLE REPLICATION TO ACCOUNTS my_org.prod_dr_eu;
-- Step 2: in the target account, create the replica
CREATE DATABASE SALES_REPL
  AS REPLICA OF my_org.prod_us.SALES;
-- Step 3: initial population is required before the replica is usable
ALTER DATABASE SALES_REPL REFRESH;
-- Step 4: keep it current on a cadence that matches the published RPO
-- for example, run this via a scheduled task or your orchestration layer:
-- ALTER DATABASE SALES_REPL REFRESH;
```

For broader DR coverage, Snowflake recommends account-level replication and failover groups rather than relying only on the limited database-replication feature. If this chapter keeps a database-level example for simplicity, say explicitly that it is the narrow product-level pattern, not the whole-estate default.

Consumer shares and Marketplace listings should point at the **replica** in that region, so queries never cross regions.

Promotion (regional incident): in the target account, promote the secondary and route workloads there. When the original region returns, repeat the process in reverse ("fail back") after reconciling changes.

19.2.3 Account Replication and Failover

Account replication adds the **control plane**: users, roles, warehouses, resource monitors, and other account objects replicate along with databases. Promotion switches the **entire** replicated estate to a new region or cloud with a single, audited action.

Operationally, you define

- A protected set of databases and account objects to replicate.
- A refresh cadence (e.g., every 15–60 minutes for hot estates).
- Who can initiate **failover** and **failback** (usually ORGADMIN/SECURITYADMIN via an elevated break-glass role).
- Client re-pointing behavior (preferred region/URL, DNS record, or driver connect list)

Keep the blast radius intentional: start with **one** well-documented secondary per account.

19.2.4 Cadence, RPO/RTO, and What "Fresh" Means

Replication is **near-real-time**, not synchronous. Your **RPO** equals the refresh interval plus in-flight work. Your **RTO** is the time to promote the secondary and bring clients back. State them explicitly:

Target	How to meet it	What to publish (internally)
RPO ≤ 15 min	Database refreshes every 10–12 min; idempotent ELT	A small "freshness" view in the secondary (max partition per table)
RTO ≤ 10 min	Pre-approved promotion runbook; client failover tested quarterly	Steps and contacts; a dry-run checklist and success criteria

Do not include Fail-safe in externally published recovery commitments. Your committed RPO/RTO should be based on Time Travel, clone-based promotion, and replication/failover mechanics that you can exercise and measure yourself. Fail-safe remains a last-resort provider service, not an operator-controlled recovery tier.

Make ELT **replayable** so any lost minutes are quickly rebuilt after the switch.

19.2.5 Client and Share Routing

Consumers should never have to cross regions. Two clean patterns:

- **Per-region share**: Replicate the database, create the share **from the replica**, and instruct regional consumers to bind to that share.
- **Connection preference**: For internal apps, publish a driver DSN list or DNS name that favors the primary but can swing to the secondary.

For Marketplace, publish a listing **in each region** you serve, fed by its local replica.

19.2.6 Governance and Security Travel with Replicas

Masking, row access policies, tags, and secure views replicate and enforce on the secondary. Roles/users (with account replication) come along as well. Keep secrets and integrations aligned: storage integrations, external access, and network rules should be defined in both regions with the same names so replicated objects bind cleanly after promotion.

19.2.7 Testing and Drills That Keep You Honest

Rehearse two paths:

1. **Read-only validation**: Point a staging client at the secondary; compare row counts, newest partitions, and critical joins against the primary at a checkpoint time.
2. **Promotion dry-run** (in non-prod): Promote the secondary, run a subset of jobs, then fail back. Document the exact operator steps and timings.

Capture results in a small `ops.dr_failover_runs` table with start/stop timestamps and pass/fail checks. Treat drills as change-controlled work.

19.2.8 Cost, Egress, and Change Discipline

Replication stores an extra copy of data in the secondary region and moves deltas between regions. The cost drivers are

- **Storage** in the secondary (comparable to primary, minus Time Travel deltas if you choose different retention).
- **Data transfer** for the change stream—keep refresh intervals reasonable and avoid thrashing high-churn staging DBs unless truly needed.
- **Compute** for ELT you choose to run in both regions.

Keep retention windows smaller in secondaries unless regulators require parity.

19.2.9 A Calm Runbook (Outline)

- **Detect and decide**: Incident meets your promotion criteria; declare the RTO window.
- **Quiesce risky writers**: Pause heavy ELT in the failing region to reduce divergence.
- **Promote**: Switch the secondary to primary for the selected scope (database or account).

- **Flip clients**: Update DNS/connection strings or let drivers use secondary preference.
- **Verify**: Run freshness and integrity probes; unpause critical workloads.
- **Communicate**: Internal status and expected catch-up time; add an entry to a service log table.
- **Fail back** when safe: Re-establish the original as secondary, refresh, promote back, and clean up.

The key is **idempotence**: if your jobs can replay a window deterministically, the RPO you can tolerate becomes easy to achieve.

Bottom line: replicate databases to be **near** your readers; replicate the account when you need a **regional escape hatch** for the whole estate. Keep a single secondary per scope, refresh on a named cadence, and drill until promotion and failback feel routine. Business continuity then becomes metadata and muscle memory—not a heroic copy job.

19.3 DR Strategies and Runbooks

Disaster recovery in Snowflake is less about rebuilding servers and more about **switching context to a good copy** and **replaying deterministically**. You already have the primitives: Time Travel and cloning for local rollbacks and database/account replication for regional incidents. A DR strategy decides which lever to use for each failure class, how fresh the secondary must be, and who executes which step. The runbook makes that decision repeatable under pressure.

19.3.1 Strategy at a Glance (Match Failure to Recovery Move)

Event class	Primary goal	First lever	Promotion path	When you escalate
Bad DDL/DML, corrupt load, missing partition	Restore the correct state locally	Time Travel or clone + SWAP	Stay in-region; promote clone into prod namespace	If the window exceeded or the object dropped long ago ➤ Fail-safe
Pipeline outage (upstream late)	Preserve integrity; catch up	Pause consumers; replay ELT idempotently	No promotion; resume when inputs good	N/A
Region-level impairment	Preserve service	Database or Account failover	Promote secondary in the target region/ cloud; flip clients	Fail back when primary healthy
Partner-facing share degraded	Maintain SLA to consumers	Replicate DB; route shares to replica	Shares point to a regional replica	N/A

The posture is simple: the **smallest repair that restores truth** and **promotion rather than editing** when risk is high.

19.3.2 Preparation That Makes Runbooks Work

Keep three ingredients current: an **RTO/RPO table** by product, a **who-does-what** roster, and **machine-checkable probes**. Store them in the warehouse so they travel with replication and can be queried during drills.

```
-- Minimal catalog of targets
CREATE OR REPLACE TABLE ops.dr_slo (
  product STRING,          -- e.g., 'sales_mart'
  scope   STRING,          -- 'database'|'account'
  rpo_min NUMBER,          -- e.g., 15
  rto_min NUMBER,          -- e.g., 10
  cadence STRING           -- 'hourly at :05', etc.
);
```

```
-- Assign clear owners
CREATE OR REPLACE TABLE ops.dr_contacts (
  role STRING, person STRING, contact STRING
);
```

Probes (packaged as views) verify "is it safe?" quickly:

```
CREATE OR REPLACE VIEW ops.v_freshness AS
SELECT 'fact_orders' AS object_name,
       (SELECT MAX(order_dt) FROM mart.fact_orders) AS newest_partition;

CREATE OR REPLACE VIEW ops.v_health AS
SELECT 'row_count_check' AS check_name,
       (SELECT COUNT(*) FROM mart.dim_date) > 0 AS ok
UNION ALL
SELECT 'fact_orders_today', (SELECT COUNT(*) FROM mart.fact_orders WHERE
order_dt=CURRENT_DATE()) > 0;
```

19.3.3 Runbook: Bad Write or Schema Change (Local)

Objective: rewind or replace the object without widening the blast radius.

1. **Freeze edits.** Pause dependent Tasks and ELT steps.
2. **Confirm good point.** Use QUERY_HISTORY, audit notes, or a known checkpoint time.
3. **Recover:**
   - If small and precise: **Time Travel** the table and CREATE OR REPLACE from the good timestamp or statement id.
   - If complex or many objects: **clone** the affected schema/database, fix inside the clone, then SWAP atomically.
4. **Validate.** Compare counts/keys vs. probes; run a diff if needed.
5. **Unfreeze** ELT/Tasks and backfill the gap.
6. **Record** incident in ops.event_log.

```
-- Rebuild from a known-good timestamp
CREATE OR REPLACE TABLE mart.fact_orders AS
SELECT * FROM mart.fact_orders AT (TIMESTAMP => '2025-09-12 01:00:00');

-- Or branch & promote
CREATE DATABASE prod_fix CLONE prod;
-- (run corrections in prod_fix...)
ALTER DATABASE prod_fix SWAP WITH prod;
```

19.3.4 Runbook: Upstream Pipeline Delay

Objective: keep consumers honest and catch up deterministically.

1. **Surface status.** Update a "late data" flag (view or table) that BI checks; optionally serve yesterday's snapshot.
2. **Pause write-heavy consumers.** Prevent partial dashboards.
3. **Repair window.** When inputs resume, run ELT for the missing interval only (idempotent MERGE/DT refresh).
4. **Clear flag** and resume schedules.

```
-- Example repair window merge
MERGE INTO mart.fact_orders t
USING (
  SELECT * FROM stage.orders_src
  WHERE ingested_at >= DATEADD('hour', -2, CURRENT_TIMESTAMP())
) s
ON t.order_id = s.order_id
WHEN MATCHED THEN UPDATE SET ...
WHEN NOT MATCHED THEN INSERT (...);
```

19.3.5 Runbook: Regional Failover (Database Scope)

Objective: keep reads and critical writes alive in another region.

1. **Criteria met.** Region health breach; leadership declares failover.
2. **Quiesce writers** in the impacted region to limit divergence.
3. **Promote the secondary** database in the target account/region.
4. **Flip clients.** Update DSNs/DNS or use driver secondary preference; point **shares/Marketplace** to the replica's share.
5. **Verify probes** (`ops.v_freshness`, key row counts).
6. **Run critical ELT** that can proceed with available sources.
7. **Communicate** RTO/RPO status and catch-up ETA.

Failback mirrors the steps: re-establish replication with the original as secondary, refresh to parity, promote back, resume writers, and retire the emergency routes.

19.3.6 Runbook: Account Failover (Whole Estate)

Objective: move users, roles, warehouses, and databases together.

1. **Elevate** into break-glass role (ORGADMIN/SECURITYADMIN per policy).
2. **Initiate account promotion** in the secondary region (per governance procedure).
3. **Flip DNS/entry points** for apps; ensure SSO/OAuth integrations in secondary are active.
4. **Verify** warehouse availability, critical roles, and probes.
5. **Throttle non-critical workloads** until catch-up completes.

Plan this scenario carefully in governance: integrations, private endpoints, and external access rules must exist in both regions with the same names.

19.3.7 Decision Table (Operators Love These)

Observation	Next move	Why
Wrong rows, right region	Time Travel or clone+SWAP	Smallest safe fix
Several objects drifted	Clone schema/DB and promote	Cohesive, atomic restoration
Queries OK, data late	Pause consumers, replay ELT	Integrity beats partial freshness
Region degraded	Promote replica; route locally	Keep service near readers
Secondary behind SLO	Accept RPO loss, then replay ELT	Deterministic rebuild after switch

19.3.8 Drills and Evidence

Run two drills per quarter: a **local rewind** (table rebuild from Time Travel) and a **regional promotion** in non-prod. Capture **start/stop**, steps, and probe outputs in a tiny table so you can prove RTO/RPO.

```
CREATE OR REPLACE TABLE ops.dr_runs (
  run_id STRING, started TIMESTAMP_NTZ, finished TIMESTAMP_NTZ,
  scenario STRING, rto_seconds NUMBER, rpo_minutes NUMBER, notes STRING
);
```

Populate from Tasks at the end of each drill.

19.3.9 Cost and Hygiene During DR

Keep secondaries lean: shorter Time Travel, only necessary warehouses, and resource monitors. In failover, favor **small multi-cluster** lanes for BI to absorb bursty catch-up and **single larger** lanes for ELT windows. After failback, **drop emergency clones** and retire temporary routes so costs return to baseline.

Bottom line: pick the lightest tool that restores truth, make promotions—not edits—your instinct for risky fixes, and rehearse until failover feels like a controlled deploy. With probes, owners, and cutover steps codified in SQL, DR stops being a heroic copy operation and becomes a routine metadata change.

PART VI

Operations and DevEx

CHAPTER 20

DevOps and DataOps for Snowflake

DevOps and DataOps in Snowflake are about **shipping small, reversible changes** to code *and* data. The platform's primitives—separation of storage/compute, zero-copy clones, Time Travel, role-based security—map neatly to modern delivery: you build and test in isolation, promote with metadata, and make every release auditable. A good practice keeps three threads braided: **software discipline** (version control, CI, artifacts), **data discipline** (tests, contracts, backfills), and **platform discipline** (roles, warehouses, monitors).

A simple rhythm works: developers branch and run fast local cycles (Snowpark notebooks, dbt dev runs, SQL migrations). CI validates style, security, and data tests against a **throwaway clone**. CD promotes by swapping clones or deploying versioned artifacts, then runs post-deploy probes and canary schedules. When something misbehaves, you **roll forward or swap back**—never hand-edit prod.

20.1 CI/CD for SQL, Snowpark, and dbt

20.1.1 One Repo, Clear Boundaries

Keep SQL DDL/DML, Snowpark code, and dbt models in a single mono-repo but with **separate build/deploy jobs**. This avoids tangled pipelines and lets each tool prove itself independently, while promotion steps remain coordinated.

```
repo/
  sql/
    migrations/             # 001_init.sql, 002_add_dim_customer.sql, ...
    policies/               # masking/row access DDL
    warehouses/             # IaC-like DDL for lanes/monitors
```


M. Hander, *Snowflake Data Warehouse Engineering*, https://doi.org/10.1007/979-8-8688-2628-3_20

```
snowpark/
  python/          # src/, pyproject.toml, requirements.txt
  scala_java/      # sbt/maven module if used
  tests/           # unit tests; fakes for UDFs
dbt/
  dbt_project.yml
  models/          # marts, staging, snapshots
  tests/           # generic + bespoke tests
  macros/
```

20.1.2 Promotion Model That Fits Snowflake

Developers work in **dev accounts/schemas**. CI stands up an **ephemeral clone** of the target database (or account objects in non-prod), applies migrations or packages, and runs tests. CD promotes by either

- **SWAP**: Build in `prod_candidate` cloned from `prod`, then `SWAP` atomically with `prod`.
- **Versioned deploy**: Publish a new UDF/JAR/wheel and flip a routing view or tag (`model_version`) for canary.

Both approaches leave a clean rollback: SWAP back, or flip the version tag.

20.1.3 Core Pipeline: End-to-End in Three Jobs

Stage	What it does	Success signal
Build and test	Lint SQL; build Snowpark wheels/JARs; dbt compile; unit tests	Green unit suite; artifacts published
Ephemeral integration	Clone DB; run SQL migrations; install Snowpark artifacts; run `dbt seed/run/test` with test vars	All dbt tests pass; contract checks green
Promote	Tag release; deploy to prod_candidate; smoke tests; SWAP or flip version; post-deploy probes	Freshness OK; BI canaries pass; rollback link recorded

Example: GitHub Actions Sketch

```
name: ci-cd-snowflake

on:
  pull_request:
  push:
    tags: ['v*.*.*']

jobs:
  build-test:
    runs-on: ubuntu-latest
    steps:
      - uses: actions/checkout@v4
      - name: Lint SQL
        run: sqlfluff lint sql/**/*.sql || (sqlfluff fix --force && exit 1)
      - name: Build Snowpark wheel
        run: |
          cd snowpark/python
          pip install -U build
          python -m build
      - name: dbt compile
        run: |
          cd dbt
          pip install dbt-snowflake
          dbt deps && dbt compile
      - name: Upload artifacts
        uses: actions/upload-artifact@v4
        with:
          name: build-artifacts
          path: |
            snowpark/python/dist/*.whl
            dbt/target
            sql/

  ephemeral-int:
    needs: build-test
```

```
    runs-on: ubuntu-latest
    steps:
      - uses: actions/checkout@v4
      - uses: actions/download-artifact@v4
        with: { name: build-artifacts, path: ./artifacts }
      - name: Prepare ephemeral clone & secrets
        env:
          SNOWSQL_PWD: ${{ secrets.SNOWSQL_PWD_CI }}
        run: |
          snowsql -q "CREATE DATABASE IF NOT EXISTS ci_base;
                      CREATE OR REPLACE DATABASE ci_run CLONE ci_base;"
      - name: Apply SQL migrations (idempotent)
        run: |
          for f in sql/migrations/*.sql; do snowsql -f "$f" -D TARGET_
          DB=ci_run; done
      - name: Install Snowpark package
        run: |
          snowsql -q "CREATE STAGE IF NOT EXISTS ops.pkg;
                      PUT file://artifacts/*.whl @ops.pkg AUTO_
                      COMPRESS=FALSE;"
      - name: dbt run + test
        run: |
          cd dbt
          dbt run --profiles-dir . --target ci --vars '{"target_
          db":"CI_RUN"}'
          dbt test --profiles-dir . --target ci

  promote:
    if: startsWith(github.ref, 'refs/tags/')
    needs: ephemeral-int
    runs-on: ubuntu-latest
    steps:
      - uses: actions/checkout@v4
      - uses: actions/download-artifact@v4
        with: { name: build-artifacts, path: ./artifacts }
      - name: Deploy to prod_candidate
```

```
    run: |
      snowsql -q "CREATE OR REPLACE DATABASE prod_candidate
      CLONE prod;"
      for f in sql/migrations/*.sql; do snowsql -f "$f" -D TARGET_
      DB=prod_candidate; done
      snowsql -q "PUT file://artifacts/*.whl @ops.pkg AUTO_
      COMPRESS=FALSE;"
      # smoke tests
      snowsql -q "SELECT COUNT(*) FROM prod_candidate.mart.fact_
      orders;"
  - name: Atomic swap
    run: snowsql -q "ALTER DATABASE prod_candidate SWAP WITH prod;"
```

(Use a scoped CI role with least privilege; never store secrets in code—use OIDC to mint short-lived keys or Snowflake OAuth.)

20.1.4 SQL Migrations That Are Safe to Re-run

Snowflake transactions are still useful for grouped DML, but they do not make DDL atomic across multiple statements. A migration runner should therefore treat each DDL change as independently committed and only write "applied" tracking records after the schema change has completed successfully.

Write migrations to be idempotent and small. In Snowflake, DDL auto-commits, so do the schema change first and record the migration only after the DDL succeeds. Do not rely on BEGIN/COMMIT to make DDL atomic.

```
-- 013_add_order_channel.sql
-- Note: DDL auto-commits in Snowflake. Record the migration only after DDL
succeeds.
CREATE TABLE IF NOT EXISTS ops.schema_migrations (
  id STRING PRIMARY KEY,
  applied_at TIMESTAMP_NTZ
);
-- Idempotent DDL first
ALTER TABLE IF EXISTS mart.fact_orders
  ADD COLUMN IF NOT EXISTS order_channel STRING;
```

```
-- Track completion only after the schema change succeeded
INSERT INTO ops.schema_migrations
SELECT '013', CURRENT_TIMESTAMP()
WHERE NOT EXISTS (
  SELECT 1
  FROM ops.schema_migrations
  WHERE id = '013'
);
```

For risky transformations, run them **in a clone** and promote with `SWAP`. That keeps RTO to seconds and rollback trivial.

20.1.5 Snowpark Delivery Without Surprises

Treat ML/compute code like any library: **build → scan → publish → deploy**.

- **Packaging**: Build a wheel/JAR with pinned deps; avoid giant binaries to keep UDF cold-starts low.
- **Staging**: Copy artifacts to a versioned internal stage (`@ops.pkg/snowpark_pkg/v1.2.3.whl`).
- **Create/replace UDFs** referencing the versioned artifact; route with a version param or a small view.

```
CREATE OR REPLACE FUNCTION ml.score_churn(v VARIANT, model_version STRING)
RETURNS FLOAT
LANGUAGE PYTHON
RUNTIME_VERSION = '3.10'
PACKAGES = ('joblib','numpy','scikit-learn==1.3.2')
IMPORTS = ('@ops.pkg/snowpark_pkg/v1_2_3.whl')
HANDLER = 'udf_entrypoint.score';
```

To canary, publish separate versioned functions or procedures, tag predictions with `model_version`, and route traffic in SQL through a view or explicit parameter until the new version is proven.

20.1.6 dbt CI That Catches Real Problems

Let dbt enforce model contracts and data quality before promotion.

- **Contracts**: Use `schema.yml` to declare columns, types, and accepted values.
- **Tests**: Built-ins (`unique`, `not_null`, `relationships`) plus custom macros for business logic.
- **Selections**: Run changed graph only on PR; full graph on main.

```
models:
  - name: fact_orders
    columns:
      - name: order_id
        tests: [unique, not_null]
      - name: order_channel
        tests:
          - accepted_values:
              values: ['web','store','partner']
```

In CI, fail the build if *any* data test fails. In CD, after SWAP, run a **post-deploy subset** (freshness, top joins) as smoke tests.

20.1.7 Observability and Rollout Safety Nets

Bake **probes** and **query tags** into deploys.

- Add `ALTER SESSION SET QUERY TAG='deploy=2025-10-19;team=platform';` to all pipeline steps.
- Publish views like `ops.v_pred_freshness`, `ops.v_contract`, `ops.v_release_notes` so operations and consumers can read the same truth.
- Use resource monitors on **deploy warehouses** to cap runaway jobs.

Tiny post-deploy checklist (all SQL):

```
-- Freshness and row counts compare
SELECT * FROM ops.v_freshness;
SELECT COUNT(*) FROM mart.fact_orders WHERE load_dt = CURRENT_DATE();

-- Access smoke
EXPLAIN USING TEXT SELECT * FROM mart.v_orders LIMIT 1;
```

20.1.8 Rollback That Isn't Scary

- **Schema/data**: `ALTER DATABASE prod_prev SWAP WITH prod;` (keep `prod_prev` for a day, then drop).
- **Snowpark**: Revert router to prior artifact; keep N-1 artifact staged.
- **dbt**: Pin to prior manifest (`dbt run --state ./manifests/v_prev/ --defer`), or route views back to prior models if you version views (`V_*`).

Because Snowflake tracks history, you can also **rebuild tables "AS OF"** with Time Travel if a bad MERGE slipped through.

20.1.9 Securing the Pipeline

- CI/CD connects with **External OAuth** or OIDC-minted tokens mapped to a **CI_ROLE** (no ACCOUNTADMIN in robots).
- Secrets live in the CI vault; **no passwords** in repo.
- Break-glass steps (e.g., account failover) are **manual** and gated.

20.1.10 Why This Works Long-Term

- SWAP and clones turn deployments into **metadata changes**, so RTO is tiny and rollback is clean.
- dbt contracts and data tests turn "it compiles" into **it's correct**.

- Snowpark wheels/JARs in stages make UDFs **versionable artifacts.**
- Tags, monitors, and ops views make releases **auditable** and **cost-aware**.

Ship small. Prove every step with SQL. Promote with a swap, not a prayer. That's CI/CD for Snowflake that stays fast on good days and calm on bad ones.

20.2 Testing Strategies (Unit/Integration/ Data Tests)

In Snowflake, testing succeeds when **logic, pipelines, and data quality** are validated separately but run together in CI. Unit tests keep pure code honest; integration tests verify that pipelines behave against a realistic schema; data tests act like constraints that run every build and every day. The goal is small, fast checks on branches and a deeper suite on main, all against **ephemeral clones** so prod stays untouched.

20.2.1 Philosophy in One Page

Layer	Question answered	Scope of truth	Where it runs best
Unit	"Does this function/SQL transform do the right math?"	Pure Python/Scala (Snowpark), SQL snippets, macros	Local/CI container—no database state needed
Integration	"Do models, tasks, and COPY steps work together end-to-end?"	Schemas, stages, permissions, small fixtures	Ephemeral **CLONE** of target DB; throw away after
Data tests	"Is the data valid today?"	Uniqueness, referential integrity, freshness, invariants	dbt tests + custom SQL in CI and on a nightly job

20.2.2 Unit Tests That Stay Fast

Keep pure code close to the language runtime and stub Snowflake I/O. For Snowpark, validate handlers and UDF logic with tiny, deterministic inputs.

```
# snowpark/python/tests/test_scoring.py
from mypkg.scoring import score_row

def test_score_row_simple():
    row = {"cnt_pos_7d": 3, "amt_30d": 120.0, "days_since_last": 5}
    assert 0.0 <= score_row(row) <= 1.0

# Example: validating a SQL macro with snapshots
SQL_IN  = "SELECT 1 AS x UNION ALL SELECT 2"
EXPECTED= [(1,), (2,)]
```

For SQL-heavy transforms, wrap logic in views or macros and sanity-check with **golden sets**: small inline tables that encode expected outputs. These can run as part of a dbt `unit` test or a lightweight SnowSQL job.

20.2.3 Integration Tests over an Ephemeral Clone

Create a throwaway environment per PR: clone the database (or schema), apply migrations, seed a **tiny fixture**, and run the pipeline exactly as prod would. The clone isolates permissions, Streams/Tasks wiring, and external stages.

```
-- Ephemeral environment
CREATE OR REPLACE DATABASE ci_run CLONE prod;

-- Seed minimal fixtures
CREATE OR REPLACE TABLE ci_run.stage.orders_src AS
SELECT * FROM prod.stage.orders_src WHERE load_dt >= CURRENT_DATE()-1
LIMIT 1000;

-- Execute the same steps as prod (dbt run, Streams & Tasks, COPY, etc.)
```

When the suite finishes, drop `ci_run`. If a step needs external I/O, point it to a **dev storage integration** with canned files to avoid side effects.

20.2.4 Data Tests As Living Constraints

dbt provides the backbone for repeatable data checks; extend it with custom SQL for domain rules.

Contracts and Generics (dbt)

```
models:
  - name: fact_orders
    columns:
      - name: order_id
        tests: [not_null, unique]
      - name: customer_id
        tests:
          - relationships:
              to: ref('dim_customer')
              field: customer_id
      - name: order_channel
        tests:
          - accepted_values: { values: ['web','store','partner'] }
```

Custom Invariants (SQL)

```
-- Fail if negative revenue appears
SELECT COUNT(*) AS bad_rows
FROM mart.fact_orders
WHERE revenue < 0;
```

Wire custom checks into CI so any non-zero result fails the build. In production, run the same tests nightly; store outcomes in `ops.data_test_runs` with the query tag `purpose=data_test` for auditing.

20.2.5 Test Data That Won't Bite You

Use three sources and keep each small.

Source	What it contains	Why it's safe
Synthetic fixtures	Hand-crafted rows that exercise edge cases	No PII; deterministic; great for unit/integration
Golden snapshots	Tiny CSV/Parquet with approved outputs	Catch regressions in joins/aggregations
Sampled prod	A few hundred recent rows, scrubbed or masked	Real shapes without real identities

Load fixtures via dbt seeds or `COPY INTO` from a test stage; never depend on "whatever is in prod today."

20.2.6 What to Automate in CI (Minimal But Meaningful)

1. **Unit**: Run Snowpark tests and macro checks; they must complete in seconds.
2. **Integration**: `CLONE` → apply migrations → `dbt run` → `dbt test` → destroy clone.
3. **Smoke**: After a successful main build, deploy to `prod_candidate` and run row-counts/freshness probes before a **SWAP**.

A tiny smoke probe that stays green only when the pipeline worked:

```
SELECT 'fact_orders_today' AS check,
       COUNT(*) > 0 AS ok
FROM prod_candidate.mart.fact_orders
WHERE load_dt = CURRENT_DATE();
```

20.2.7 Observability for Tests

Publish two simple views so engineers and auditors see the same truth.

```
CREATE OR REPLACE TABLE ops.test_results (
  run_id STRING, ts TIMESTAMP_NTZ, layer STRING, name STRING, ok BOOLEAN,
  details STRING
);

CREATE OR REPLACE VIEW ops.v_test_failures AS
SELECT * FROM ops.test_results WHERE ok = FALSE ORDER BY ts DESC;
```

Push CI outcomes here with `QUERY_TAG='purpose=test;layer=data|integration|unit'`. In prod, have a Task persist dbt test results nightly to the same table.

20.2.8 Handling Flaky Tests and Nondeterminism

Time-based windows and late-arriving data cause flakiness. Make tests deterministic by **fixing the evaluation window** and using **checkpoint times**.

```
-- Use a fixed checkpoint instead of CURRENT_DATE in CI
SET checkpoint_ts = TO_TIMESTAMP('2025-09-12 03:00:00');

SELECT COUNT(*) FROM mart.fact_orders
WHERE order_ts < $checkpoint_ts;
```

If the business rule tolerates lateness, encode it: e.g., a freshness test that allows "≤ 30 minutes late" instead of comparing to the wall clock.

20.2.9 Security and Governance in Tests

Run CI with a **scoped role** that mirrors production least privilege. Validate that sensitive views still enforce masking and RAP by switching roles mid-test:

```
USE ROLE ROLE_ANALYST;
SELECT EMAIL FROM mart.v_customer_sample LIMIT 1;    -- should be masked

USE ROLE ROLE_PII_READER;
SELECT EMAIL FROM mart.v_customer_sample LIMIT 1;    -- should be clear
```

Make these assertions part of integration tests so governance can't regress silently.

20.2.10 Bringing It Together

Keep **unit** tests fast and language-native, **integration** tests realistic in a **clone**, and **data** tests continuous and visible. Treat failures as deploy-blocking in CI and as alerts in production. With that split, you catch logic bugs before they reach Snowflake, wiring and permission issues before they reach prod, and data drift the day it appears—without turning every change into a fire drill.

20.3 Promotion Workflows and Environment Drift Control

Promotion in Snowflake should feel like switching a road sign, not moving a mountain. You build the new state in isolation, prove it, then **rename the door** that traffic uses. That keeps rollbacks trivial and makes drift something you *measure*, not debate.

20.3.1 A Calm Promotion Flow That Fits Snowflake

1. **Freeze a candidate**: Create `prod_candidate` as a **zero-copy clone** of `prod`. Apply SQL migrations, install Snowpark artifacts, and run dbt models into this clone.
2. **Prove it where it will live**: Execute smoke tests and business checks against `prod_candidate` (freshness, row counts, join integrity, masking/RAP behavior).
3. **Flip**: `ALTER DATABASE prod_candidate SWAP WITH prod;`—an atomic metadata move that keeps the name `prod` attached to the newly validated contents.
4. **Watch**: Post-deploy probes run with the same queries that green-lit the candidate; query tags tie them to the release.
5. **Rollback** (if needed): Swap back; the old `prod` is sitting there under the name `prod_candidate`.

```
-- Build the candidate
CREATE OR REPLACE DATABASE prod_candidate CLONE prod;
-- …apply migrations, load artifacts, run dbt…

-- One-time smoke probe
SELECT 'orders_today' AS check, COUNT(*) > 0 AS ok
FROM prod_candidate.mart.fact_orders
WHERE load_dt = CURRENT_DATE();

-- Flip
ALTER DATABASE prod_candidate SWAP WITH prod;
```

For surface changes (UDFs, services), publish **versioned artifacts** in a stage (`@ops.pkg/model/v1_3_0.whl`), and route through a stable function or view that reads a `model_version` setting. Promotion then becomes a single `ALTER … SET TAG model_version='v1_3_0'`.

20.3.2 Drift Control: Making "Prod = Code" Verifiable

Drift is anything in an environment that your repo and migrations didn't create. You keep it small by **code-owning** every object class, and you detect it with **comparative views** that diff metadata across environments.

What to Keep Under Code

Surface	Source of truth	Promotion shape
Warehouses, resource monitors, integrations	DDL files or IaC (Terraform)	`CREATE/ALTER` in a "platform" migration bundle
RBAC (roles, grants, future grants)	DDL grants in code	Run idempotent grant scripts on every deploy
Schemas, tables, views, policies, tags	SQL migrations (idempotent)	Apply to the candidate; validate with diffs
Snowpark artifacts/UDFs	Versioned wheels/JARs in stages	Recreate functions with pinned imports
dbt models	Compiled DAG + manifest	`dbt run/test` against the clone

A Small "Drift Report" You Can Run Anytime

1) **Structure drift (objects exist in prod but not in code/candidate). Snowflake does not provide a single** `INFORMATION_SCHEMA.OBJECTS` **view, so build the inventory by unioning the relevant metadata views for tables/views, procedures, and functions.**

```
-- Compare prod vs candidate object lists

WITH prod_objs AS (
  SELECT
    CASE
      WHEN table_type = 'BASE TABLE' THEN 'TABLE'
      WHEN table_type = 'VIEW' THEN 'VIEW'
    END AS object_type,
    table_schema AS object_schema,
    table_name   AS object_name
  FROM prod.information_schema.tables
  WHERE table_type IN ('BASE TABLE', 'VIEW')
  UNION ALL
  SELECT
    'PROCEDURE' AS object_type,
    procedure_schema AS object_schema,
    procedure_name   AS object_name
  FROM prod.information_schema.procedures
  UNION ALL
  SELECT
    'FUNCTION' AS object_type,
    function_schema AS object_schema,
    function_name   AS object_name
  FROM prod.information_schema.functions
),
cand_objs AS (
  SELECT
    CASE
```

```
      WHEN table_type = 'BASE TABLE' THEN 'TABLE'
      WHEN table_type = 'VIEW' THEN 'VIEW'
    END AS object_type,
    table_schema AS object_schema,
    table_name   AS object_name
  FROM prod_candidate.information_schema.tables
  WHERE table_type IN ('BASE TABLE', 'VIEW')
  UNION ALL
  SELECT
    'PROCEDURE' AS object_type,
    procedure_schema AS object_schema,
    procedure_name   AS object_name
  FROM prod_candidate.information_schema.procedures
  UNION ALL
  SELECT
    'FUNCTION' AS object_type,
    function_schema AS object_schema,
    function_name   AS object_name
  FROM prod_candidate.information_schema.functions
)
SELECT 'missing_in_candidate' AS kind, p.*
FROM prod_objs p
LEFT JOIN cand_objs c USING (object_type, object_schema,
object_name)
WHERE c.object_name IS NULL
UNION ALL
SELECT 'extra_in_candidate' AS kind, c.*
FROM cand_objs c
LEFT JOIN prod_objs p USING (object_type, object_schema,
object_name)
WHERE p.object_name IS NULL;
```

2) **Column-level drift:**

```
WITH prod_cols AS (
  SELECT table_schema, table_name, column_name, data_type,
  ordinal_position
  FROM prod.information_schema.columns
),
cand_cols AS (
  SELECT table_schema, table_name, column_name, data_type,
  ordinal_position
  FROM prod_candidate.information_schema.columns
)
SELECT 'diff' AS kind, p.table_schema, p.table_name,
p.column_name,
       p.data_type AS prod_type, c.data_type AS cand_type
FROM prod_cols p
FULL JOIN cand_cols c
  USING(table_schema, table_name, column_name)
WHERE (p.data_type IS DISTINCT FROM c.data_type)
   OR (p.table_name IS NULL OR c.table_name IS NULL);
```

3) **Policy/tag attachment drift:**

```
-- Policies
SELECT 'policy_missing_in_candidate' AS kind, p.*
FROM SNOWFLAKE.ACCOUNT_USAGE.POLICY_REFERENCES p
WHERE p.object_database = 'PROD'
  AND NOT EXISTS (
    SELECT 1 FROM SNOWFLAKE.ACCOUNT_USAGE.POLICY_REFERENCES c
    WHERE c.object_database = 'PROD_CANDIDATE'
      AND c.object_name = p.object_name
      AND COALESCE(c.column_name,'') = COALESCE(p.column_name,'')
      AND c.policy_name = p.policy_name
  );

-- Tags (classification/owner/etc.)
SELECT 'tag_missing_in_candidate' AS kind, t.*
```

```
FROM SNOWFLAKE.ACCOUNT_USAGE.TAG_REFERENCES t
WHERE t.object_database = 'PROD'
  AND NOT EXISTS (
    SELECT 1 FROM SNOWFLAKE.ACCOUNT_USAGE.TAG_REFERENCES c
    WHERE c.object_database = 'PROD_CANDIDATE'
      AND c.object_name = t.object_name
      AND COALESCE(c.column_name,'') = COALESCE(t.column_name,'')
      AND c.tag_name = t.tag_name
      AND c.tag_value = t.tag_value
  );
```

4) **Grants drift (who can do what changed outside code):**

```
-- Minimal: show grants on schema level that differ
SELECT 'grant_in_prod_only' AS kind, g.*
FROM prod.information_schema.applicable_roles g
WHERE NOT EXISTS (
  SELECT 1 FROM prod_candidate.information_schema.
  applicable_roles c
  WHERE c.grantee = g.grantee AND c.role_name = g.role_name
);
```

Fold these into a single view (`ops.v_env_drift`) and fail promotion if any rows appear.

20.3.3 Idempotent Grants and Future-Grants (Drift Prevention)

Make every deploy **re-assert** the intended privileges, so strays get overwritten.

```
-- Domain role can SELECT all current & future tables in MART
GRANT USAGE ON DATABASE prod TO ROLE ROLE_FIN_READER;
GRANT USAGE ON SCHEMA prod.mart TO ROLE ROLE_FIN_READER;
GRANT SELECT ON ALL TABLES IN SCHEMA prod.mart TO ROLE ROLE_FIN_READER;
GRANT SELECT ON FUTURE TABLES IN SCHEMA prod.mart TO ROLE ROLE_FIN_READER;
```

Do the same for masking/row policies and tag attachments in your migrations so a dropped-and-recreated table comes back protected.

20.3.4 Versioned Routes for Low-Risk Flips

For UDFs and model scoring, keep version routing in SQL or pass the version explicitly as a parameter. Do not make the Python handler depend on querying Snowflake objects at runtime, and do not import two versions of the same package into one UDF.

```
CREATE OR REPLACE FUNCTION ml.score_customer_v1(x VARIANT)
RETURNS FLOAT
LANGUAGE PYTHON
RUNTIME_VERSION = '3.10'
IMPORTS = ('@ops.pkg/model/v1_2_2.whl')
HANDLER = 'scorer.score';
CREATE OR REPLACE FUNCTION ml.score_customer_v2(x VARIANT)
RETURNS FLOAT
LANGUAGE PYTHON
RUNTIME_VERSION = '3.10'
IMPORTS = ('@ops.pkg/model/v1_3_0.whl')
HANDLER = 'scorer.score';
CREATE OR REPLACE SECURE VIEW ml.v_scored AS
SELECT order_id,
       CASE
         WHEN CURRENT_TIMESTAMP() < '2025-11-15'::TIMESTAMP
           THEN ml.score_customer_v1(features)
         ELSE ml.score_customer_v2(features)
       END AS score
FROM mart.fact_orders;
```

Promotion is a one-line view edit or a caller-side parameter change; rollback is the inverse. That keeps the version switch visible in SQL instead of hidden inside the UDF runtime.

20.3.5 Contract Checks Before the Flip

Two fast checks keep downstreams safe:

- **Schema contract**: Confirm required columns and types on public views.

```
SELECT column_name, data_type
FROM prod_candidate.information_schema.columns
WHERE table_schema='MART' AND table_name='V_ORDERS'
QUALIFY COUNT(*) OVER (PARTITION BY column_name) = 1;
```

- **Governance contract**: Confirm PII columns carry masking and RAP still binds.

```
-- PII must be masked
SELECT object_name, column_name
FROM SNOWFLAKE.ACCOUNT_USAGE.TAG_REFERENCES t
LEFT JOIN SNOWFLAKE.ACCOUNT_USAGE.POLICY_REFERENCES p
  ON p.object_name=t.object_name AND p.column_name=t.column_name
WHERE t.object_database='PROD_CANDIDATE'
  AND t.tag_name='CLASSIFICATION' AND t.tag_value LIKE 'pii.%'
  AND p.policy_name IS NULL;
```

If either returns rows, the candidate cannot be promoted.

20.3.6 Keeping Environments in Lockstep

- **Promote RBAC and platform first**, then data objects. A missing warehouse or integration will masquerade as flaky tests.
- **Pin driver/session settings** (timezone, timestamp types) in deploy jobs so CI and prod behave the same.
- **Query tags everywhere** (`deploy=...;team=...`), and store outcomes in `ops.release_log` to make audits readable.

```
CREATE OR REPLACE TABLE ops.release_log (
  release_id STRING, started_at TIMESTAMP_NTZ, finished_at TIMESTAMP_NTZ,
  commit_sha STRING, actor STRING, status STRING, notes STRING
);
```

Populate this from CI after SWAP, along with the drift report hash.

20.3.7 Why This Works

Clones and SWAP turn releases into **metadata flips**; drift reports make “state = code” a **query**, not a meeting. With grants and policies re-asserted every deploy and contracts checked on the candidate, environments stop diverging quietly. Promotion becomes a short, well-lit path—and undoing it is just as short.

CHAPTER 21

Observability and SRE

Operating Snowflake well means turning **platform behavior into queries** you can run every day. Instead of log files and agents, you get first-class system views that describe queries, warehouses, costs, storage, security events, replication, and data movement. The SRE posture is simple: collect a **small set of golden queries**, materialize them into an **ops schema**, and wire alerts from SQL, not shell scripts. When something goes wrong, investigations start with the same views you already use for weekly health reports.

A practical footprint has three layers:

- **Raw telemetry** in `SNOWFLAKE.ACCOUNT_USAGE` (per account) and `SNOWFLAKE.ORGANIZATION_USAGE` (across accounts), plus `INFORMATION_SCHEMA` for live metadata.
- **Ops marts**: Compact tables and views that summarize load, spend, freshness, and errors.
- **Runbooks** that answer "who, what, when, how much" with one or two repeatable SELECTs.

Note `SNOWFLAKE.ACCOUNT_USAGE` views are eventually consistent and can lag from about 45 minutes to several hours, depending on the view. Use them for trending, budgets, audits, and longer-horizon SLO reporting. For incident response, 5–15 minute alerting, and near-real-time operational checks, use the Information Schema table functions such as `QUERY_HISTORY()` and related live metadata surfaces.

M. Hander, *Snowflake Data Warehouse Engineering*, https://doi.org/10.1007/979-8-8688-2628-3_21

21.1 Account Usage and ORGANIZATION_USAGE

ACCOUNT_USAGE is your per-account black box: it records queries, credits, logins, storage, tasks, pipes, COPY history, access to sensitive objects, and more. ORGANIZATION_USAGE rolls up the same kinds of facts across **all accounts in your org**, so you can see total spend, region splits, and replication. Together, they are the backbone of Snowflake observability.

21.1.1 What Each Surface Is Best At

Question you ask	View family to start with	Why this surface
"What's slow, queued, or expensive right now? "	INFORMATION_SCHEMA.QUERY_HISTORY() plus live warehouse metadata	Near-real-time monitoring and alerting require low-latency surfaces. Use ACCOUNT_USAGE.QUERY_HISTORY later for trending and post-incident analysis.
"What was slow, queued, or expensive over the last day/week?"	ACCOUNT_USAGE.QUERY_HISTORY (optionally joined to warehouse metering/load history)	Longer retention and richer historical analysis, but not suitable for 5–15 minute alerts due to latency.
"Who used how many credits?"	ACCOUNT_USAGE.WAREHOUSE_METERING_HISTORY	Per-warehouse/day credits for chargeback and trend lines
"Are our warehouses saturated?"	INFORMATION_SCHEMA.WAREHOUSE_LOAD_HISTORY	Near-current concurrency and queued minutes
"Is data valid and fresh?"	Your dbt test tables + COPY_HISTORY, TASK_HISTORY, DYNAMIC_TABLE_REFRESH_HISTORY	Ties freshness to pipeline steps and errors
"Who touched PII?"	ACCOUNT_USAGE.ACCESS_HISTORY + TAG_REFERENCES	Row-level audit with object and column context

(*continued*)

Question you ask	View family to start with	Why this surface
"Why did storage grow?"	DATABASE_STORAGE_USAGE_HISTORY, STAGE_STORAGE_USAGE_HISTORY	Active vs. Time Travel vs. Fail-safe; stage bloat
"Is SSO/network policy working?"	LOGIN_HISTORY, USERS, SESSIONS	Auth method, client IP, failure reasons
"How healthy is cross-region sharing/replication?"	REPLICATION_USAGE_HISTORY, REPLICATION_GROUP_USAGE_HISTORY	Volume moved, lag, and target regions
"Org-wide spend and regions?"	ORGANIZATION_USAGE.METERING_DAILY_HISTORY	One rollup across all accounts and regions

21.1.2 First-Class SRE Queries (Drop-in and Adapt)

Hot or Queued Workload (Last Two Hours)

```
SELECT
  WAREHOUSE_NAME,
  AVG((QUEUED_OVERLOAD_TIME + QUEUED_PROVISIONING_TIME)/1000) AS avg_
queued_s,
  AVG(EXECUTION_TIME/1000)                                    AS avg_exec_s,
  COUNT(*)                                                    AS q_count
FROM SNOWFLAKE.ACCOUNT_USAGE.QUERY_HISTORY
WHERE START_TIME >= DATEADD(hour, -2, CURRENT_TIMESTAMP())
GROUP BY 1
ORDER BY avg_queued_s DESC;
```

Top Queries by Bytes Scanned (Find Pruning Issues)

```
SELECT QUERY_ID, USER_NAME, WAREHOUSE_NAME,
       BYTES_SCANNED/POWER(1024,3) AS gb_scanned,
       TOTAL_ELAPSED_TIME/1000     AS elapsed_s,
       QUERY_TEXT
FROM SNOWFLAKE.ACCOUNT_USAGE.QUERY_HISTORY
WHERE START_TIME >= DATEADD(day, -1, CURRENT_TIMESTAMP())
```

```
ORDER BY BYTES_SCANNED DESC
LIMIT 20;
```

Credits by Warehouse and Day (FinOps Baseline)

```
SELECT
  WAREHOUSE_NAME,
  TO_DATE(START_TIME) AS day,
  SUM(CREDITS_USED)   AS credits
FROM SNOWFLAKE.ACCOUNT_USAGE.WAREHOUSE_METERING_HISTORY
WHERE START_TIME >= DATEADD(day, -30, CURRENT_TIMESTAMP())
GROUP BY 1,2
ORDER BY 2,1;
```

Storage Composition by Database (Current Snapshot)

```
WITH latest AS (
  SELECT *,
         ROW_NUMBER() OVER (PARTITION BY DATABASE_NAME ORDER BY USAGE_DATE
DESC) rn
  FROM SNOWFLAKE.ACCOUNT_USAGE.DATABASE_STORAGE_USAGE_HISTORY
)
SELECT DATABASE_NAME,
       ACTIVE_BYTES/POWER(1024,3)      AS active_gb,
       TIME_TRAVEL_BYTES/POWER(1024,3) AS tt_gb,
       FAILSAFE_BYTES/POWER(1024,3)    AS fs_gb
FROM latest
WHERE rn = 1
ORDER BY (active_gb + tt_gb + fs_gb) DESC;
```

Who Accessed PII This Week (Joins Tags to Access)

```
SELECT a.QUERY_START_TIME, a.USER_NAME, a.ROLE_NAME,
       obj.value:objectName::STRING   AS object_name,
       obj.value:objectDomain::STRING AS object_domain,
       r.COLUMN_NAME,
       r.TAG_VALUE AS classification
FROM SNOWFLAKE.ACCOUNT_USAGE.ACCESS_HISTORY a,
```

```
    LATERAL FLATTEN(input => a.DIRECT_OBJECTS_ACCESSED) obj
JOIN SNOWFLAKE.ACCOUNT_USAGE.TAG_REFERENCES r
  ON r.OBJECT_NAME = obj.value:objectName::STRING
WHERE a.QUERY_START_TIME >= DATEADD(day, -7, CURRENT_TIMESTAMP())
  AND r.TAG_NAME = 'CLASSIFICATION'
  AND r.TAG_VALUE LIKE 'pii.%'
ORDER BY a.QUERY_START_TIME DESC;
```

> `DIRECT_OBJECTS_ACCESSED` and `BASE_OBJECTS_ACCESSED` are semi-structured arrays of objects in `ACCESS_HISTORY`. For audit joins, flatten them first and then filter or join on keys such as objectName and objectDomain; do not use scalar path extraction like `DIRECT_OBJECTS_ACCESSED:objectName`.

Org-Wide Credits by Account and Day

```
SELECT ACCOUNT_NAME,
       USAGE_DATE AS day,
       SUM(CREDITS_USED) AS credits
FROM SNOWFLAKE.ORGANIZATION_USAGE.METERING_DAILY_HISTORY
WHERE USAGE_DATE >= DATEADD(day, -30, CURRENT_DATE())
GROUP BY 1,2
ORDER BY 2,1;
```

21.1.3 An Ops Mart You Can Keep Forever

Create a small schema to **materialize** the above into durable, query-friendly tables. Refresh them with Tasks on a cadence that matches view latency, and always tag job runs for traceability.

```
CREATE SCHEMA IF NOT EXISTS ops;

CREATE TABLE IF NOT EXISTS ops.t_credits_daily AS
SELECT * FROM SNOWFLAKE.ACCOUNT_USAGE.WAREHOUSE_METERING_HISTORY WHERE 1=0;

-- Incremental load pattern (idempotent)
MERGE INTO ops.t_credits_daily t
USING (
```

```
  SELECT * FROM SNOWFLAKE.ACCOUNT_USAGE.WAREHOUSE_METERING_HISTORY
  WHERE START_TIME >= DATEADD(day, -2, CURRENT_TIMESTAMP())
) s
ON t.START_TIME = s.START_TIME
AND t.WAREHOUSE_ID = s.WAREHOUSE_ID
WHEN MATCHED THEN UPDATE SET CREDITS_USED = s.CREDITS_USED
WHEN NOT MATCHED THEN INSERT VALUES (s.START_TIME, s.WAREHOUSE_ID,
s.WAREHOUSE_NAME, s.CREDITS_USED);

-- A simple view SREs can memorize
CREATE OR REPLACE VIEW ops.v_bi_load AS
SELECT TO_DATE(START_TIME) AS day, WAREHOUSE_NAME,
       SUM(CREDITS_USED) AS credits
FROM ops.t_credits_daily
GROUP BY 1,2;
```

Pair these with alerting Tasks that raise notifications for low-latency operational signals such as queued-time spikes or task failures, and use the ops mart plus ACCOUNT_USAGE for slower-moving controls such as weekly cost trends, freshness history, and security audit review.

21.1.4 SLOs You Can Check with SQL

SLO	How to measure it	What to alert on
BI concurrency	Avg queued time per warehouse (last 15–30 min) from QUERY_HISTORY	Threshold per lane (e.g., >5 s avg queued)
Freshness	Max partition date in key marts vs. expected cadence	Late by more than your stated lag
Cost	Credits/day vs. budget from WAREHOUSE_METERING_HISTORY	Seven-day trend above target
Security	Access to PII by roles outside the allow-list	Any event matches the rule
Replication	Last refresh time vs. target cadence	Lag exceeds RPO

Example freshness probe:

```
SELECT 'fact_orders' AS object_name,
       MAX(load_dt)   AS newest_dt,
       CURRENT_DATE() - MAX(load_dt) AS days_late
FROM prod.mart.fact_orders;
```

21.1.5 Tying It Together with Tags and Query Tags

- **Tag** warehouses, databases, and views with `cost_center`, `owner`, `data_domain`, so usage is attributable without extra joins.
- **Set** `QUERY_TAG` in all jobs (`deploy=…;team=…;purpose=…`), so investigations filter quickly and chargeback reports stay clean.
- **Record releases** in an `ops.release_log` table, and join it to `QUERY_HISTORY` for "what changed when" clarity.

```
ALTER SESSION SET QUERY_TAG = 'team=platform;purpose=ops_refresh';
```

21.1.6 Operating Notes That Keep You Sane

- Build dashboards from **your ops mart**, not directly on usage views; you control retention and can add business context.
- During active incidents, favor `INFORMATION_SCHEMA.*` for near-real-time signals; reconcile with `ACCOUNT_USAGE` afterward.
- When you have multiple accounts, make **org-level dashboards** from `ORGANIZATION_USAGE`, so leadership sees one ledger.

Observability in Snowflake is just SQL. Learn the handful of usage views that matter, materialize them into an ops schema, and let small, reliable queries power your SLOs, alerts, and postmortems. With `ACCOUNT_USAGE` for depth and `ORGANIZATION_USAGE` for breadth, SRE becomes a habit you can version, review, and promote like any other part of the platform.

21.2 Query/Task Monitoring and Alerting

Snowflake doesn't ship an agent. You watch the platform by using two time horizons on purpose: low-latency Information Schema table functions for live monitoring and alerting and ACCOUNT_USAGE views for durable history, trending, and postmortems. Mixing those roles creates blind spots, because ACCOUNT_USAGE is not intended for 5–15 minute alert windows. A good setup has three pieces: (1) a handful of probe queries over QUERY_HISTORY, WAREHOUSE_LOAD_HISTORY, and task views; (2) a tiny **alerts** table so you can deduplicate and suppress noise; (3) one or two **Tasks** that evaluate probes on a cadence and notify through a notification integration. Everything is SQL, so you can version it and promote it like any other change.

Design rule: use Information Schema table functions for probe windows under about one hour, because they have no documented latency but shorter retention. Use ACCOUNT_USAGE for durable history, trending, audit evidence, and organization-level reporting. Treat them as complementary layers, not interchangeable ones.

21.2.1 The Minimal Footprint (One-Time Objects)

```
-- Where alerts live
CREATE SCHEMA IF NOT EXISTS ops;
CREATE OR REPLACE TABLE ops.alerts (
  id STRING,                    -- stable hash of the finding
  first_seen TIMESTAMP_NTZ,
  last_seen  TIMESTAMP_NTZ,
  severity   STRING,            -- info|warn|crit
  kind       STRING,            -- 'queued_spike' | 'task_fail' | ...
  subject    STRING,            -- warehouse, task, etc.
  details    STRING,
  is_open    BOOLEAN
);

-- Optional: email notifications (use your account name/region specifics)
CREATE OR REPLACE NOTIFICATION INTEGRATION ops_email
  TYPE = EMAIL
  ENABLED = TRUE
  ALLOWED_RECIPIENTS = ('dataops@example.com');
```

```
-- Helper view for operators
CREATE OR REPLACE VIEW ops.v_alerts_open AS
SELECT * FROM ops.alerts WHERE is_open ORDER BY last_seen DESC;
```

21.2.2 Probes That Matter (Compact and Actionable)

1) **Querying pressure and latency (last ten minutes, near-real-time). Use the Information Schema** `QUERY_HISTORY()` **table function here;** `ACCOUNT_USAGE.QUERY_HISTORY` **is for history and trend reporting, not live alerting.**

   If users are *waiting*, alert on **queued time**, not just "slow."

```
-- Avg queued vs. exec per warehouse (10-min window)

WITH q AS (
  SELECT WAREHOUSE_NAME,
         AVG((QUEUED_OVERLOAD_TIME + QUEUED_PROVISIONING_TIME) /
         1000.0) AS avg_queued_s,
         AVG(EXECUTION_TIME / 1000.0)
                                             AS avg_exec_s,
         COUNT(*)
                                                                AS q_count
  FROM TABLE(INFORMATION_SCHEMA.QUERY_HISTORY(
    END_TIME_RANGE_START => DATEADD(minute, -10, CURRENT_
    TIMESTAMP()),
    END_TIME_RANGE_END   => CURRENT_TIMESTAMP()
  ))
  GROUP BY 1
)
SELECT *
FROM q
WHERE q_count >= 20
  AND avg_queued_s > 5;
```

2) **Runaway/expensive queries (find pruning misses).**

```
SELECT QUERY_ID, USER_NAME, WAREHOUSE_NAME,
       BYTES_SCANNED / POWER(1024, 3) AS gb_scanned,
       TOTAL_ELAPSED_TIME / 1000.0    AS elapsed_s
FROM TABLE(INFORMATION_SCHEMA.QUERY_HISTORY(
  END_TIME_RANGE_START => DATEADD(minute, -15, CURRENT_TIMESTAMP()),
  END_TIME_RANGE_END   => CURRENT_TIMESTAMP()
))
WHERE BYTES_SCANNED > 200 * POWER(1024, 3)
ORDER BY BYTES_SCANNED DESC
LIMIT 20;
```

3) **Task health (failures/suspensions).**

```
-- Tasks that failed in the last run

SELECT NAME, DATABASE_NAME, SCHEMA_NAME, STATE, RETURN_VALUE,
SCHEDULED_TIME
FROM TABLE(INFORMATION_SCHEMA.TASK_HISTORY(
  SCHEDULED_TIME_RANGE_START => DATEADD(minute, -30, CURRENT_
  TIMESTAMP()),
  RESULT_LIMIT => 1000
))
QUALIFY ROW_NUMBER() OVER (
  PARTITION BY DATABASE_NAME, SCHEMA_NAME, NAME
  ORDER BY SCHEDULED_TIME DESC
) = 1
AND STATE IN ('FAILED', 'CANCELLED', 'SUSPENDED');
```

For alerting windows under an hour, prefer Information Schema task-history table functions for the same reason as query monitoring: ACCOUNT_USAGE.TASK_HISTORY can lag by up to 45 minutes.

```
-- Optional: dynamic table refresh issues
SELECT NAME, TARGET_ROWS_INSERTED, TARGET_ROWS_DELETED, ERROR_MESSAGE,
LAST_REFRESH_TIME
```

```
FROM SNOWFLAKE.ACCOUNT_USAGE.DYNAMIC_TABLE_REFRESH_HISTORY
WHERE LAST_REFRESH_TIME >= DATEADD(minute, -30, CURRENT_TIMESTAMP())
  AND ERROR_MESSAGE IS NOT NULL;
```

21.2.3 Turning Probe Rows into Alerts (Stateful, Low-Noise)

Wrap probes in a small procedure that **upserts** alerts by a stable id (hash of kind+subject) and **only notifies** on *newly opened* alerts. You can write the upsert in SQL; here's the core pattern:

```
-- Example: record queued-spike per warehouse
MERGE INTO ops.alerts t
USING (
  SELECT
    MD5('queued_spike:'||WAREHOUSE_NAME) AS id,
    CURRENT_TIMESTAMP()                  AS now_ts,
    'warn'                               AS severity,
    'queued_spike'                       AS kind,
    WAREHOUSE_NAME                       AS subject,
    TO_JSON(OBJECT_CONSTRUCT('avg_queued_s', avg_queued_s, 'avg_exec_s',
    avg_exec_s)) AS details
  FROM (
    /* paste the queued-time probe here */
  ) q
) s
ON t.id = s.id AND t.is_open = TRUE
WHEN MATCHED THEN
  UPDATE SET last_seen = s.now_ts, details = s.details
WHEN NOT MATCHED THEN
  INSERT (id, first_seen, last_seen, severity, kind, subject, details,
  is_open)
  VALUES (s.id, s.now_ts, s.now_ts, s.severity, s.kind, s.subject,
  s.details, TRUE);
```

Close alerts automatically when the condition clears:

```
-- Any open queued_spike that is no longer observed becomes closed
UPDATE ops.alerts
SET is_open = FALSE, last_seen = CURRENT_TIMESTAMP()
WHERE is_open = TRUE
  AND kind = 'queued_spike'
  AND subject NOT IN (SELECT WAREHOUSE_NAME FROM /* current queued probe */);
```

21.2.4 Notifying Humans (Without Spam)

Use the email notification integration (or your cloud queue integration) to message on **new** alerts only:

```
-- NEW alerts in the last run

SET alert_count = (
  SELECT COUNT(*)
  FROM ops.alerts
  WHERE is_open = TRUE
    AND DATEDIFF(minute, first_seen, last_seen) < 2
);
SET alert_body = (
  SELECT LISTAGG(
           subject || ' -> ' || kind || ' @ ' || TO_VARCHAR(last_seen),
           '\n'
         )
  FROM ops.alerts
  WHERE is_open = TRUE
    AND DATEDIFF(minute, first_seen, last_seen) < 2
);
CALL SYSTEM$SEND_EMAIL(
  'ops_email',
  'dataops@example.com',
  CONCAT('Snowflake alert(s): ', $alert_count::STRING),
  $alert_body
);
```

The recipient list passed to SYSTEM$SEND_EMAIL must be a comma-separated string, and if the notification integration defines ALLOWED_RECIPIENTS, every address in that string must already be on the allow-list or no email is sent. Validate the integration separately before wiring it into a monitoring task.

```
-- Optional guard to avoid empty notifications
-- Only call SYSTEM$SEND_EMAIL when $alert_count > 0
```

Tasks can point at this block on a cadence, but the probe queries inside the task should use Information Schema table functions for recent activity and reserve ACCOUNT_USAGE for later summarization into the ops mart.

```
CREATE OR REPLACE TASK ops.t_monitor_every_5m
  WAREHOUSE = wh_ops
  SCHEDULE  = 'USING CRON 0/5 * * * * UTC'
AS
BEGIN
  -- Upsert queued spikes
  /* MERGE block above */

  -- Upsert task failures (similar MERGE keyed by 'task_
  fail:<db>.<sch>.<name>')
  /* MERGE for TASK_HISTORY */

  -- Send one summary email for new alerts
  /* SYSTEM$SEND_EMAIL block */
END;
ALTER TASK ops.t_monitor_every_5m RESUME;
```

(If you prefer paging, swap email for a cloud **NOTIFICATION INTEGRATION** *to SNS/Event Grid/Pub/Sub, and have your incident tool subscribe.)*

21.2.5 Runbooks that Pair with Alerts

- **Queued spike (BI lane):** Check WAREHOUSE_LOAD_HISTORY to confirm concurrency pressure. If active clusters stay at 1, raise MAX_CLUSTER_COUNT or stagger schedulers; if execution dominates, size up ELT or fix pruning.

- **Runaway query:** Open the plan in Query Profile; if it's accidental SELECT * on a wide table, cancel (CALL SYSTEM$CANCEL_QUERY('<id>')), and follow up with pruning guidance or a view.
- **Task failed/suspended:** Read RETURN_VALUE/ERROR_MESSAGE, fix the source, then ALTER TASK … RESUME;. Set SUSPEND_TASK_AFTER_NUM_FAILURES = 3, and use ERROR_INTEGRATION = ops_email on critical tasks, so you get one crisp email when it breaks.

21.2.6 Keeping False Alarms Low

- Evaluate over **windows** (10–15 minutes), and require a **minimum sample** (e.g., ≥20 queries) before alerting.
- Deduplicate by (kind, subject), and keep alerts **open** until the condition clears; only notify on *open → true*.
- Tag probes (ALTER SESSION SET QUERY_TAG='team=platform;purpose=monitoring'), so they never show up as "top spenders."

```
ALTER SESSION SET QUERY_TAG='team=platform;purpose=monitoring';
```

21.2.7 A Single Page Operators Can Memorize

Signal	Probe	First move
Average queued time > SLO	QUERY_HISTORY window avg	Add lanes (MAX_CLUSTER_COUNT) or stagger starts
High bytes scanned outliers	QUERY_HISTORY bytes scanned	Review plan; add envelope filters; guide authors
Task failing/ suspended	TASK_HISTORY last run	Fix source; resume; consider SUSPEND_TASK_AFTER_NUM_FAILURES
Dynamic table errors	DYNAMIC_TABLE_REFRESH_HISTORY	Inspect error, upstream availability; re-run refresh

Treat observability as code. A couple of history queries become probes; a tiny ops.alerts table keeps memory; one Task ties it together and pings you only when something *new* breaks. From there, your SRE runbooks are just SQL and a calm checklist.

21.3 Incident Response and Performance SLOs

Incidents in Snowflake are rarely about servers; they are about **waiting**, **freshness**, and **governance**. A good response plan treats the warehouse like a product: clear SLOs, a small set of SLIs measured from system views, and a playbook that moves work to a safe state, restores truth, and records what happened. Everything lives in SQL, so it can be rehearsed and audited.

21.3.1 The Contract You Actually Operate To

SLO dimension	Target	How it's measured (SLI)	Alert when
Interactive latency (BI)	P95 end-to-end ≤ 3 s during business hours	Windowed average **queued** and **execution** time from `QUERY_HISTORY` by BI warehouse	P95 queued > 1 s for 3 consecutive windows or avg queued > 5 s over 10 min
Throughput (ELT)	Critical job window finishes by 02:30 UTC	Last successful run in `TASK_HISTORY` and rows processed	Missed schedule once → warn; twice → page
Freshness	Key marts ≤ 15 min behind stated cadence	Max partition time vs. clock in `ops.v_freshness`	Lag > 15 min for 2 windows
Availability (shares/ reader)	Queryable 99.9% monthly	Ability to `SELECT 1` from shared views from a canary session	2+ consecutive failures or 5 within 10 min
Security	Zero unmasked PII reads by non-privileged roles	Flattened joins between `ACCESS_HISTORY` object arrays and `TAG_REFERENCES`	Any violation is critical. Because this is an audit-style control rather than a sub-minute operational signal, `ACCOUNT_USAGE.ACCESS_HISTORY` is appropriate here even though it is not real-time.

These are small numbers you can publish weekly. They fit on one page and all derive from Snowflake's own telemetry.

21.3.2 SLIs As Queries (Copy/Paste Friendly)

Interactive Latency (BI Lane, Last Ten Minutes, P95)

```
WITH q AS (
  SELECT (QUEUED_OVERLOAD_TIME + QUEUED_PROVISIONING_TIME) / 1000.0 AS
  queued_s,
         EXECUTION_TIME / 1000.0                               AS exec_s
  FROM TABLE(INFORMATION_SCHEMA.QUERY_HISTORY(
    END_TIME_RANGE_START => DATEADD(minute, -10, CURRENT_TIMESTAMP()),
    END_TIME_RANGE_END   => CURRENT_TIMESTAMP()
  ))
  WHERE WAREHOUSE_NAME = 'WH_BI'
    AND QUERY_TYPE IN ('SELECT')
),
p AS (
  SELECT
    APPROX_PERCENTILE(queued_s, 0.95) AS p95_queued,
    APPROX_PERCENTILE(exec_s,   0.95) AS p95_exec
  FROM q
)
SELECT * FROM p;
```

For live SLI checks over ten-minute windows, use the Information Schema table function. Use ACCOUNT_USAGE.QUERY_HISTORY later for weekly SLO reporting and postmortems.

Freshness (Per Mart View)

```
SELECT 'fact_orders' AS object_name,
       MAX(load_dt)  AS newest_dt,
       DATEDIFF(minute, MAX(load_dt), CURRENT_TIMESTAMP()) AS lag_min
FROM prod.mart.fact_orders;
```

Task Window Completion

```
SELECT NAME, STATE, SCHEDULED_TIME, COMPLETED_TIME, DATEDIFF(minute,
SCHEDULED_TIME, COMPLETED_TIME) AS mins
FROM SNOWFLAKE.ACCOUNT_USAGE.TASK_HISTORY
WHERE NAME = 'elt_refresh_fact_orders'
ORDER BY SCHEDULED_TIME DESC
LIMIT 1;
```

Security SLI

```
SELECT COUNT(*) AS unmasked_reads
FROM SNOWFLAKE.ACCOUNT_USAGE.ACCESS_HISTORY a,
     LATERAL FLATTEN(input => a.DIRECT_OBJECTS_ACCESSED) obj
JOIN SNOWFLAKE.ACCOUNT_USAGE.TAG_REFERENCES t
  ON t.OBJECT_NAME = obj.value:objectName::STRING
WHERE a.QUERY_START_TIME >= DATEADD(day, -1, CURRENT_TIMESTAMP())
  AND t.TAG_NAME = 'CLASSIFICATION'
  AND t.TAG_VALUE LIKE 'pii.%'
  AND a.ROLE_NAME NOT IN ('ROLE_PII_READER', 'SECURITYADMIN');
```

21.3.3 Incident Lifecycle That Fits Snowflake

Incidents move through four beats: **declare**, **stabilize**, **repair**, and **review**. You can run the whole arc without leaving SQL.

Stage	What you do	Why it works here
Declare	Tag the session (QUERY_TAG='incident=...'), write a row to ops.service_log (who/when/what), page on SLO breach	Gives a timestamp and actor for later joins to QUERY_HISTORY
Stabilize	Stop the bleeding: pause offending Tasks, add lanes or size up temporarily, or route readers to a regional replica	Separation of compute makes "add lanes" and "pause writers" instant

(continued)

Stage	What you do	Why it works here
Repair	Rebuild with **Time Travel**; or fix in a clone and **SWAP**; replay idempotent ELT for the lost window	Recovery is metadata and deterministic replays
Review	Capture root cause, impact, and specific prevention (pruning, grants, schedule change) as data	Next time you can search for similar patterns

A minimal service log table:

```
CREATE TABLE IF NOT EXISTS ops.service_log (
  ts TIMESTAMP_NTZ, severity STRING, incident_id STRING,
  area STRING, summary STRING, actor STRING, status STRING
);
```

21.3.4 Triage: Is It Waiting or Working?

Most "slow" reports are either **queued** time (not enough lanes) or **execution** time (too much work). Separate them quickly:

```
SELECT
  AVG((QUEUED_OVERLOAD_TIME + QUEUED_PROVISIONING_TIME) / 1000.0) AS avg_
  queued_s,
  AVG(EXECUTION_TIME / 1000.0)                                     AS avg_exec_s
FROM TABLE(INFORMATION_SCHEMA.QUERY_HISTORY(
  END_TIME_RANGE_START => DATEADD(minute, -10, CURRENT_TIMESTAMP()),
  END_TIME_RANGE_END   => CURRENT_TIMESTAMP()
))
WHERE WAREHOUSE_NAME = 'WH_BI';
```

If queued dominates and active clusters stay at 1, raise `MAX_CLUSTER_COUNT` or stagger schedules. If execution dominates, open Query Profile for top offenders; teach pruning (date/tenant envelopes, drop `SELECT *`), or size up **temporarily** and plan a prune fix.

21.3.5 Templates You'll Reuse Under Pressure

Pause and Protect Readers

```
-- Quiet write-heavy tasks
ALTER TASK elt_refresh_fact_orders SUSPEND;

-- BI elasticity (temporary)
ALTER WAREHOUSE wh_bi SET MAX_CLUSTER_COUNT = 6;
```

Rebuild to a Known-Good Point

```
CREATE OR REPLACE TABLE mart.fact_orders AS
SELECT * FROM mart.fact_orders
AT (TIMESTAMP => '2025-09-12 01:00:00'::timestamp_tz);
```

Hotfix in a Clone

```
CREATE DATABASE prod_fix CLONE prod;
/* apply corrections and reprocess */
ALTER DATABASE prod_fix SWAP WITH prod;  -- atomic promotion
```

Regional Read Continuity

```
-- Use replicated DB in consumer region during incident
-- (Promotion described in §19.2)
```

21.3.6 Error Budgets and Weekly Hygiene

SLOs become behavior when you attach a **budget** and spend it deliberately. For BI, define an error as "P95 end-to-end > 3 s during business hours." Allow, for example, 43 minutes of budget per week (~1%). For freshness, allow N minutes late per day. Compute budgets from the same SLIs you alert on and store them in a tiny table for trend lines.

```
CREATE TABLE IF NOT EXISTS ops.error_budget_bi_weekly (
  week_start DATE PRIMARY KEY,
  minutes_budget NUMBER, minutes_spent NUMBER, notes STRING
);
```

A Task calculates spent minutes by summing windows where P95 > target; another Task writes a short note when you burn >50% mid-week.

21.3.7 Communications That Don't Guess

Post status directly from the warehouse, so notes and metrics match:

```
INSERT INTO ops.service_log
SELECT CURRENT_TIMESTAMP(), 'info', 'INC-2025-0912-01', 'BI',
       'Queued P95 above 3s; added lanes to WH_BI; root-cause: top queries
       scanning >200GB',
       CURRENT_USER(), 'mitigating';
```

In the postmortem, include **two charts or tables**: P95 queued vs. execution over time, and top queries by bytes scanned. The numbers will tell the story without debate.

21.3.8 Keeping Noise Down, Signal Up

Window your SLIs (10–15 min), and require a minimum number of queries before declaring a breach. Conflate repeated alerts into one **open** record per warehouse or task, and notify only on open/close transitions. Tag monitoring queries (`purpose=monitoring`), so your observability never shows up as a top spender.

21.3.9 What "Good" Looks Like After Three Months

- BI lanes rarely page; when they do, you can show queued vs. execution and the exact top offenders within minutes.
- Freshness alerts correlate with upstream status and clear after one replay window.
- Post-incident reviews are SQL screenshots and one prevention item, not novels.
- Error budgets trend steady; when spend rises, you can point to a change in schedules, volume, or schema.

Define small SLOs you can measure with one query each, keep a calm playbook that flips metadata instead of editing prod, and make every step—declare, stabilize, repair, review—a row in your warehouse. When observability is a SELECT and recovery is a SWAP, SRE becomes a repeatable habit rather than an adrenaline sport.

CHAPTER 22

Metadata, Catalog, and Lineage

Snowflake's catalog is not a separate system to sync; it's **part of the warehouse**. Object definitions, dependencies, usage history, tags, policies, and even cross-account access can be queried like ordinary tables. That makes metadata operational: you can gate releases on schema contracts, route cost and ownership with tags, and generate lineage or docs from the same place you run analytics.

A healthy approach keeps two layers. First, **authoritative sources**: INFORMATION_SCHEMA for live object metadata and SNOWFLAKE.ACCOUNT_USAGE/ORGANIZATION_USAGE for historical activity. Second, a small **governance mart**—views you publish for engineers and auditors: object inventory, lineage, policy coverage, freshness, and access reports. Because everything is SQL, you can version these views, promote them through environments, and let downstream tools read them instead of scraping.

22.1 INFORMATION_SCHEMA and Access History

INFORMATION_SCHEMA is the **live catalog** for the current database (and each database has its own). It tells you what objects exist and how they're shaped—schemas, tables, views, columns, routines, and dependencies. ACCOUNT_USAGE.ACCESS_HISTORY is the **operational trace**: who read what, when, and through which object. Together, they answer change impact ("what breaks if we alter this view?"), governance ("who touched PII?"), and documentation ("what does this model expose?").

M. Hander, *Snowflake Data Warehouse Engineering*, https://doi.org/10.1007/979-8-8688-2628-3_22

22.1.1 What Each Surface Is Best At

Need	Prefer	Why
Discover current objects, columns, data types	INFORMATION_SCHEMA.TABLES, ... COLUMNS, ...VIEWS	Near-real-time; scoped to the database; light to query
Understand dependencies	INFORMATION_SCHEMA.OBJECT_DEPENDENCIES	Structural lineage (view → table, view → view)
Verify grants and policies	INFORMATION_SCHEMA.OBJECT_PRIVILEGES, ACCOUNT_USAGE.POLICY_REFERENCES	Who can do what and where policies are attached
Trace reads and writers over time	ACCOUNT_USAGE.ACCESS_HISTORY, QUERY_HISTORY	Time-windowed audit with role/user context
Join tags to access	ACCOUNT_USAGE.TAG_REFERENCES ↔ ACCESS_HISTORY	Turns raw reads into PII and domain-aware audits

> **Tip** Use INFORMATION_SCHEMA during incidents (it's current), and reconcile with ACCOUNT_USAGE later (it lags but is richer).

22.1.2 Compact Queries You'll Run Every Week

Inventory: What does this schema expose (and in what order)?

```
SELECT table_schema, table_name, column_name, data_type, ordinal_position
FROM <DB>.INFORMATION_SCHEMA.COLUMNS
WHERE table_schema = 'MART'
ORDER BY table_name, ordinal_position;
```

Impact: If we change V_ORDERS, what breaks?

```
SELECT REFERENCING_OBJECT_SCHEMA, REFERENCING_OBJECT_NAME, REFERENCED_
OBJECT_DOMAIN
FROM <DB>.<SCHEMA>.INFORMATION_SCHEMA.OBJECT_DEPENDENCIES
WHERE REFERENCED_OBJECT_NAME = 'V_ORDERS';
```

Privilege check: Who can SELECT **from** FACT_ORDERS**?**

```
SELECT GRANTEE, PRIVILEGE
FROM <DB>.<SCHEMA>.INFORMATION_SCHEMA.OBJECT_PRIVILEGES
WHERE OBJECT_NAME = 'FACT_ORDERS' AND PRIVILEGE = 'SELECT';
```

Access audit: Who read PII in the last seven days?

```
SELECT a.QUERY_START_TIME, a.USER_NAME, a.ROLE_NAME,
       obj.value:objectName::STRING   AS object_name,
       obj.value:objectDomain::STRING AS object_domain,
       r.COLUMN_NAME,
       r.TAG_VALUE AS classification
FROM SNOWFLAKE.ACCOUNT_USAGE.ACCESS_HISTORY a,
     LATERAL FLATTEN(input => a.DIRECT_OBJECTS_ACCESSED) obj
JOIN SNOWFLAKE.ACCOUNT_USAGE.TAG_REFERENCES r
  ON r.OBJECT_NAME = obj.value:objectName::STRING
WHERE a.QUERY_START_TIME >= DATEADD(day,-7,CURRENT_TIMESTAMP())
  AND r.TAG_NAME = 'CLASSIFICATION'
  AND r.TAG_VALUE LIKE 'pii.%'
ORDER BY a.QUERY_START_TIME DESC;
```

DIRECT_OBJECTS_ACCESSED and BASE_OBJECTS_ACCESSED in ACCESS_HISTORY are arrays of objects, not scalar strings. Flatten them first, then join or filter on keys such as objectName and objectDomain.

Consumer trail: Which views/tables are actually used?

```
SELECT obj.value:objectName::STRING AS object_name,
       COUNT(*) AS reads
FROM SNOWFLAKE.ACCOUNT_USAGE.ACCESS_HISTORY a,
     LATERAL FLATTEN(INPUT => a.BASE_OBJECTS_ACCESSED) AS obj
WHERE a.QUERY_START_TIME >= DATEADD(day,-14,CURRENT_TIMESTAMP())
GROUP BY 1
ORDER BY reads DESC;
```

22.1.3 Turn Metadata into a Governance Mart

Rather than point dashboards at system views directly, materialize a few durable views/tables in a gov or ops schema. They become the **source of truth** for docs, SLOs, and audits.

```
CREATE SCHEMA IF NOT EXISTS gov;

-- Object inventory (tables and views only)

CREATE OR REPLACE VIEW gov.v_objects AS
SELECT
  table_catalog AS db,
  table_schema  AS sch,
  table_name    AS name,
  table_type
FROM <DB>.INFORMATION_SCHEMA.TABLES
WHERE table_schema <> 'INFORMATION_SCHEMA';

-- Column catalog (with simple profiling hooks)
CREATE OR REPLACE VIEW gov.v_columns AS
SELECT table_schema, table_name, column_name, data_type
FROM <DB>.INFORMATION_SCHEMA.COLUMNS;

-- Structural lineage
CREATE OR REPLACE VIEW gov.v_lineage AS
SELECT REFERENCING_OBJECT_SCHEMA AS sch,
       REFERENCING_OBJECT_NAME   AS obj,
       REFERENCED_OBJECT_SCHEMA  AS src_sch,
       REFERENCED_OBJECT_NAME    AS src_obj,
       REFERENCED_OBJECT_DOMAIN  AS src_type
FROM <DB>.<SCHEMA>.INFORMATION_SCHEMA.OBJECT_DEPENDENCIES;
```

For **operational lineage**, snapshot ACCESS_HISTORY nightly into a narrow table, so you can trend usage without re-scanning the raw view:

```
CREATE TABLE IF NOT EXISTS gov.t_access_daily (
  day DATE, object_name STRING, role_name STRING, user_name STRING,
  reads NUMBER
);
```

```
MERGE INTO gov.t_access_daily t
USING (
  SELECT TO_DATE(a.QUERY_START_TIME) AS day,
         obj.value:objectName::STRING AS object_name,
         a.ROLE_NAME,
         a.USER_NAME,
         COUNT(*) AS reads
  FROM SNOWFLAKE.ACCOUNT_USAGE.ACCESS_HISTORY a,
       LATERAL FLATTEN(input => a.DIRECT_OBJECTS_ACCESSED) obj
  WHERE a.QUERY_START_TIME >= DATEADD(day,-2,CURRENT_TIMESTAMP())
  GROUP BY 1,2,3,4
) s
ON t.day = s.day
AND t.object_name = s.object_name
AND t.role_name = s.role_name
AND t.user_name = s.user_name
WHEN MATCHED THEN UPDATE SET reads = s.reads
WHEN NOT MATCHED THEN INSERT VALUES (s.day, s.object_name, s.role_name,
s.user_name, s.reads);
```

22.1.4 Make Contracts and Reviews Data-Driven

With INFORMATION_SCHEMA and access history in hand, you can enforce **schema contracts** and prove **least privilege**.

Schema contract gate (pre-promotion):

```
-- Required columns on the public view
SELECT column_name, data_type
FROM prod_candidate.information_schema.columns
WHERE table_schema='MART' AND table_name='V_ORDERS'
QUALIFY COUNT(*) OVER (PARTITION BY column_name)=1;  -- returns 1 row per
expected column
```

Least-privilege scan:

```
SELECT GRANTEE, PRIVILEGE
FROM prod.information_schema.object_privileges
WHERE object_name='V_ORDERS' AND privilege='SELECT'
  AND GRANTEE NOT IN ('ROLE_FIN_READER','ROLE_BI_APP');
```

Deprecation helper (who still depends on `V_ORDERS_V1`**?):**

```
SELECT REFERENCING_OBJECT_SCHEMA, REFERENCING_OBJECT_NAME
FROM prod.information_schema.object_dependencies
WHERE REFERENCED_OBJECT_NAME='V_ORDERS_V1';
```

22.1.5 Practical Notes That Keep This Reliable

- **Scope and latency.** `INFORMATION_SCHEMA` is per database and current; `ACCOUNT_USAGE` spans the account but can lag by hours. Use each where it fits.
- **Names matter.** Keep schemas and objects in consistent, searchable namespaces (`STAGE`, `STG_`, `DIM_`, `FACT_`, `V_*`). Your queries get simpler and audits clearer.
- **Tag everything important.** When you tag objects with `classification`, `owner`, and `data_domain`, those tags show up in `TAG_REFERENCES` and make access history meaningful.
- **Promote your views.** As your governance mart stabilizes, point docs, lineage graphs, and SLO dashboards to **your** `gov.*` views, so people stop inventing one-off catalog queries.

Treat the catalog and access log as first-class datasets. `INFORMATION_SCHEMA` tells you what it is; `ACCESS_HISTORY` tells you who touched it and when. Put a thin layer of curated views on top, and metadata becomes a tool you use daily—during design reviews, promotions, audits, and incident response—without introducing another system to run.

`INFORMATION_SCHEMA.TABLES` already returns one row for each table and view in the database. Use `VIEWS` only when you specifically need view-only metadata such as the definition; join to it, do not `UNION` it into a general object inventory.

22.2 Integrations with Catalogs (Purview/ Collibra/Atlan)

External catalogs work best when Snowflake stays the **source of truth for technical metadata and policy**, while the catalog curates **business context**—glossaries, ownership, quality rules, and data product docs. The integration becomes a loop: **harvest** what Snowflake knows, **enrich** it in the catalog, and **push back** only the parts Snowflake must enforce (tags, policies, sometimes descriptions). Keep the contract small, versioned, and testable with SQL.

22.2.1 What Each Side Should Own

Topic	Snowflake owns	Catalog owns	Hand-off
Technical inventory	Objects, columns, dependencies (`INFORMATION_SCHEMA`)	Indexing and search	Pull from `INFORMATION_SCHEMA`
Operational lineage	Access and read paths (`ACCESS_HISTORY`)	Stitching + viz across tools	Pull daily; catalog renders end-to-end
Business metadata	N/A	Glossary, definitions, stewards, SLAs	Write-only to the catalog
Policy surface	Masking, row access, object tags	Policy templates, approvals	Push **tags** and policy bindings to Snowflake
Classification	Semantic detection seed	Curated classification and mappings	Catalog → Snowflake `TAG` values

22.2.2 A Thin Integration Layer You Can Ship

Expose compact, stable views in a dedicated schema; point **any** catalog's connector at these views instead of raw system tables. It reduces vendor-specific SQL and gives you one place to evolve.

```
CREATE SCHEMA IF NOT EXISTS gov_api;

-- Inventory (tables + views)

CREATE OR REPLACE VIEW gov_api.v_objects AS
SELECT
  table_catalog AS db,
  table_schema  AS sch,
  table_name    AS name,
  table_type
FROM <DB>.INFORMATION_SCHEMA.TABLES
WHERE table_schema <> 'INFORMATION_SCHEMA';

-- Columns
CREATE OR REPLACE VIEW gov_api.v_columns AS
SELECT table_catalog AS db, table_schema AS sch, table_name AS obj, column_
name, data_type
FROM <DB>.INFORMATION_SCHEMA.COLUMNS;

-- Structural lineage (within the DB)
CREATE OR REPLACE VIEW gov_api.v_lineage AS
SELECT REFERENCING_OBJECT_DATABASE  AS db,
       REFERENCING_OBJECT_SCHEMA    AS sch,
       REFERENCING_OBJECT_NAME      AS obj,
       REFERENCED_OBJECT_DATABASE   AS src_db,
       REFERENCED_OBJECT_SCHEMA     AS src_sch,
       REFERENCED_OBJECT_NAME       AS src_obj,
       REFERENCED_OBJECT_DOMAIN     AS src_type
FROM <DB>.<SCHEMA>.INFORMATION_SCHEMA.OBJECT_DEPENDENCIES;
```

```
-- Tags & classifications
CREATE OR REPLACE VIEW gov_api.v_tags AS
SELECT OBJECT_DATABASE AS db, OBJECT_SCHEMA AS sch, OBJECT_NAME AS obj,
       COLUMN_NAME AS col, TAG_NAME AS tag, TAG_VALUE AS val, MODIFIED
FROM SNOWFLAKE.ACCOUNT_USAGE.TAG_REFERENCES;
```

Your catalog connector (Purview scan, Collibra Edge job, Atlan Snowflake connector) reads these four views and gets **most** of what it needs with stable names and consistent types.

22.2.3 Direction of Travel (How Sync Should Flow)

Metadata facet	Direction	Mechanism in practice
Inventory, columns, lineage	Snowflake → Catalog	Scheduled pull of `gov_api.*` (hourly/daily)
Access/usage signals	Snowflake → Catalog	Snapshot `ACCESS_HISTORY` to a narrow table (daily) to avoid lag variance
Glossary, owners, domains	Catalog only	Managed by stewards; visible in Snowflake as comments if desired
Classifications and tags	Bidirectional (with guardrails)	Catalog proposes → generate `ALTER ... SET TAG ...` SQL; Snowflake exports authoritative tag state back
Policies (masking/row access)	Catalog → Snowflake (templated)	Catalog creates a **change set**; CD pipeline applies in Snowflake, not live from the catalog

Keep catalog → Snowflake writes **code-reviewed**. Treat them as migrations, not live API pushes.

22.2.4 Mapping Specifics Without Vendor Lock-in

Capability	Purview	Collibra	Atlan	Snowflake anchor
Discovery	Scans with MI/ connector; Azure auth	Edge crawler to Snowflake	Managed connector with reverse sync	`gov_api.v_objects`, `.v_columns`
Lineage	Native stitching + OpenLineage	Technical lineage graphs	Lineage from query logs and ETL integrations	`.v_lineage` + ACCESS_HISTORY snapshot
Glossary ↔ tags	Term-to-classification mapping	Policy center + communities	Domains/owners and attributes	`gov_api.v_tags` round-trip
Write-back	Limited: classifications to tags	Workflows produce SQL/ REST changes	Reverse sync to Snowflake tags/ comments	ALTER … SET TAG, masking policy DDL

(OpenLineage events from Airflow/DBT/Spark can be ingested by most catalogs and stitched to Snowflake objects using fully qualified names—use consistent naming.)

22.2.5 Pushback Safely: Turning Catalog Decisions into Snowflake DDL

Have the catalog emit a **change set** you run via CI/CD:

```
-- Catalog-approved classification updates
ALTER TABLE core.dim_customer
  MODIFY COLUMN email SET TAG classification = 'pii.email';

ALTER TABLE core.dim_customer
  MODIFY COLUMN phone SET TAG classification = 'pii.phone';

-- Masking attachment from a template policy
ALTER TABLE core.dim_customer
  MODIFY COLUMN email SET MASKING POLICY mask_pii;
```

For bulk changes, generate DDL from a CSV the catalog exports:

```
-- Staging table populated by catalog export (db, sch, obj, col, tag, val)
MERGE INTO gov.pending_tags t
USING @stage/catalog_tags.csv FILE_FORMAT=(TYPE=CSV SKIP_HEADER=1) s
ON t.db=s.$1 AND t.sch=s.$2 AND t.obj=s.$3 AND COALESCE(t.col,'')=
COALESCE(s.$4,'')
WHEN NOT MATCHED THEN INSERT (db,sch,obj,col,tag,val) VALUES (s.$1,s.$2,
s.$3,s.$4,s.$5,s.$6);
```

Then a SQL generator view produces `ALTER ... SET TAG` statements for your deploy job to execute.

22.2.6 Stitching Lineage That Survives Audits

Combine structural lineage with **operational** access for end-to-end paths:

```
-- Operational touch points (daily snapshot recommended)

CREATE OR REPLACE VIEW gov_api.v_usage AS
SELECT TO_DATE(a.QUERY_START_TIME) AS day,
       obj.value:objectName::STRING AS obj,
       a.ROLE_NAME,
       a.USER_NAME,
       COUNT(*) AS reads
FROM SNOWFLAKE.ACCOUNT_USAGE.ACCESS_HISTORY a,
     LATERAL FLATTEN(input => a.DIRECT_OBJECTS_ACCESSED) obj
WHERE a.QUERY_START_TIME >= DATEADD(day,-7,CURRENT_TIMESTAMP())
GROUP BY 1,2,3,4;
```

Catalogs ingest `.v_lineage` for static edges (view→table) and `.v_usage` for "who actually reads what." That lets lineage UIs prioritize **hot paths** over theoretical ones.

22.2.7 Security and Auth That Won't Page You

Give the connector a **read-only role** with USAGE on databases and SELECT on gov_api.* and system views you decided to surface. For write-back, use a separate **governance role** with only APPLY MASKING POLICY, APPLY ROW ACCESS POLICY, and APPLY/SET TAG on the target schemas. Federate with SSO/OAuth or (on Azure) Managed Identity for Purview; avoid long-lived passwords.

Minimal grants pattern:

```
-- Reader for the catalog
CREATE ROLE ROLE_CATALOG_READER;
GRANT USAGE ON DATABASE <DB> TO ROLE ROLE_CATALOG_READER;
GRANT USAGE ON SCHEMA gov_api TO ROLE ROLE_CATALOG_READER;
GRANT SELECT ON ALL VIEWS IN SCHEMA gov_api TO ROLE ROLE_CATALOG_READER;

-- Writer (applies curated changes via CI)
CREATE ROLE ROLE_GOVERNANCE_APPLIER;
GRANT APPLY MASKING POLICY, APPLY ROW ACCESS POLICY ON ACCOUNT TO ROLE
ROLE_GOVERNANCE_APPLIER;
GRANT APPLY TAG ON ACCOUNT TO ROLE ROLE_GOVERNANCE_APPLIER;
```

22.2.8 Keeping the Loop Healthy

- **Schedule** harvesting during off-peak; catalog runs hourly/daily depending on change volume.
- **Detect drift** by diffing gov_api.v_tags against catalog classifications; resolve in the catalog, not ad hoc in Snowflake.
- **Promote** catalog-to-Snowflake changes through the same environments you use for code; test masking/RAP effects in a clone before prod.
- **Name things consistently** (domains, owners, data products). The catalog can only stitch what your names allow.

Build one thin, vendor-neutral interface in Snowflake, let the catalog harvest and enrich, and push only enforceable decisions back as reviewed SQL. That keeps policy close to the data, business context close to stewards, and lineage trusted by both.

22.3 Automating Documentation

Good documentation should **fall out of the warehouse you already run**. Snowflake stores enough truth—object shapes, dependencies, tags, access, and usage—to generate living docs without a separate manual step. The pattern is simple: assemble a few *curated views* of your metadata, render them into Markdown/HTML, and publish them beside the data (in a stage or share). When schemas change, the docs change because the queries change.

22.3.1 What to Generate and Where It Lives

Artifact	Source in Snowflake	Output	Audience
Dataset index	`INFORMATION_SCHEMA.TABLES/VIEWS`	Markdown page listing datasets by domain	Analysts, product teams
Column dictionary	`INFORMATION_SCHEMA.COLUMNS` + `TAG_REFERENCES` + `COMMENT`	Markdown tables per dataset (name, type, description, classification)	Developers, auditors
Lineage stubs	`INFORMATION_SCHEMA.OBJECT_DEPENDENCIES`	"Upstream/Downstream" sections per view	Anyone planning a change
Usage hints	`ACCESS_HISTORY` rolled up by object	"Most-used columns/ consumers" footnotes	Platform and SRE
Governance badges	`TAG_REFERENCES`, `POLICY_REFERENCES`	Badges (PII, retention, masking attached)	Security, compliance

Treat these as **queries that write prose**. You keep the SQL under version control, and ship it through the same CI/CD you use for models.

22.3.2 Curated "Doc API" Inside Snowflake

Create one schema that exposes documentation-friendly views with stable column names. They're your internal API; everything downstream reads these, not raw system views.

```
CREATE SCHEMA IF NOT EXISTS docs;

-- Objects with owner and comments (if you use COMMENT ON)
CREATE OR REPLACE VIEW docs.v_objects AS
SELECT
  t.table_catalog    AS db,
  t.table_schema     AS sch,
  t.table_name       AS name,
  t.table_type       AS kind,
  t.comment          AS description,
  COALESCE(o.tag_value, 'unknown') AS owner
FROM <DB>.INFORMATION_SCHEMA.TABLES t
LEFT JOIN SNOWFLAKE.ACCOUNT_USAGE.TAG_REFERENCES o
  ON o.object_database = t.table_catalog
 AND o.object_schema   = t.table_schema
 AND o.object_name     = t.table_name
 AND o.tag_name        = 'OWNER';

-- Columns with type, comment, and classification
CREATE OR REPLACE VIEW docs.v_columns AS
SELECT
  c.table_schema AS sch,
  c.table_name   AS obj,
  c.column_name  AS col,
  c.data_type    AS type,
  c.character_maximum_length AS len,
  c.numeric_precision        AS p,
  c.numeric_scale            AS s,
  c.comment      AS description,
  COALESCE(t.tag_value,'') AS classification
FROM <DB>.INFORMATION_SCHEMA.COLUMNS c
LEFT JOIN SNOWFLAKE.ACCOUNT_USAGE.TAG_REFERENCES t
  ON t.object_database = c.table_catalog
 AND t.object_schema   = c.table_schema
 AND t.object_name     = c.table_name
 AND COALESCE(t.column_name,'') = c.column_name
```

```
 AND t.tag_name = 'CLASSIFICATION';

-- Structural lineage (what feeds this object)
CREATE OR REPLACE VIEW docs.v_lineage AS
SELECT
  REFERENCING_OBJECT_SCHEMA AS sch,
  REFERENCING_OBJECT_NAME   AS obj,
  REFERENCED_OBJECT_SCHEMA  AS src_sch,
  REFERENCED_OBJECT_NAME    AS src_obj,
  REFERENCED_OBJECT_DOMAIN  AS src_type
FROM <DB>.<SCHEMA>.INFORMATION_SCHEMA.OBJECT_DEPENDENCIES;

-- Operational usage (snapshot this nightly for stability)

CREATE OR REPLACE VIEW docs.v_usage_7d AS
SELECT
  obj.value:objectName::STRING AS obj,
  COUNT(*) AS reads_7d
FROM SNOWFLAKE.ACCOUNT_USAGE.ACCESS_HISTORY a,
     LATERAL FLATTEN(input => a.DIRECT_OBJECTS_ACCESSED) obj
WHERE a.QUERY_START_TIME >= DATEADD(day,-7,CURRENT_TIMESTAMP())
GROUP BY 1;
```

Prefer fully qualified object names in usage snapshots and doc joins. Plain `OBJECT_NAME` joins are fragile when the same object name exists in multiple schemas.

With these in place, you have a consistent backbone for every rendering job, every catalog connector, and every audit.

22.3.3 Rendering Markdown from SQL

Snowflake won't emit HTML directly, but it can produce strings that are already Markdown. Write one view per page shape, then *COPY* that text to a stage for your docs site (or send to your catalog to display as rich text).

Dataset Index (One Markdown Table per Schema)

```
CREATE OR REPLACE VIEW docs.v_markdown_index AS
SELECT
  CONCAT(
    '# ', sch, '  -  Datasets', '\n\n',
    '| Object | Kind | Owner | Description | Reads (7d) |', '\n',
    '|---|---|---|---|---|', '\n',
    LISTAGG(
      CONCAT(
        '| `', sch, '.', name, '` | ', kind, ' | ',
        owner, ' | ',
        COALESCE(description,'') , ' | ',
        COALESCE(CAST(u.reads_7d AS STRING),'0'), ' |'
      ), '\n'
    ) WITHIN GROUP (ORDER BY name)
  ) AS md
FROM (
  SELECT o.sch, o.name, o.kind, o.description, o.owner, u.reads_7d
  FROM docs.v_objects o
  LEFT JOIN docs.v_usage_7d u
ON u.obj = CONCAT(o.sch, '.', o.name)   -- simple join when names are
unique within schema
) x
GROUP BY sch;
```

Column Dictionary (One Dataset per Page)

```
CREATE OR REPLACE VIEW docs.v_markdown_columns AS
SELECT
  obj,
  CONCAT(
    '## ', sch, '.', obj, '\n\n',
    '| Column | Type | Description | Classification |', '\n',
    '|---|---|---|---|', '\n',
    LISTAGG(
      CONCAT(
```

```
        '| `', col, '` | ', type,
        CASE WHEN len IS NOT NULL THEN CONCAT('(', len, ')') ELSE '' END,
        CASE WHEN p   IS NOT NULL THEN CONCAT('(', p, ',', s, ')')
        ELSE '' END,
        ' | ', COALESCE(description,''),
        ' | ', COALESCE(classification,''), ' |'
      ), '\n'
    ) WITHIN GROUP (ORDER BY col)
  ) AS md
FROM docs.v_columns
GROUP BY sch, obj;
```

Lineage Section (Upstream List)

```
CREATE OR REPLACE VIEW docs.v_markdown_lineage AS
SELECT
  obj,
  CONCAT(
    '### Upstream sources for ', sch, '.', obj, '\n\n',
    LISTAGG(CONCAT('- `', src_sch, '.', src_obj, '` (', src_type, ')'), '\n')
      WITHIN GROUP (ORDER BY src_sch, src_obj)
  ) AS md
FROM docs.v_lineage
GROUP BY sch, obj;
```

Now you can export the Markdown blobs to an internal stage and hand them to whatever hosts your docs (static site, wiki, catalog).

```
-- Write files: one index per schema

COPY INTO @docs_stage/index/
  FROM (SELECT md FROM docs.v_markdown_index)
  FILE_FORMAT = (
    TYPE = CSV
    FIELD_DELIMITER = '\t'
    FIELD_OPTIONALLY_ENCLOSED_BY = '"'
    ESCAPE_UNENCLOSED_FIELD = NONE
  )
```

```
  HEADER = FALSE
  OVERWRITE = TRUE
  SINGLE = TRUE;

-- Write dataset pages (one row per object becomes a tiny file)

COPY INTO @docs_stage/datasets/
  FROM (
    SELECT obj || '.md' AS name, md
    FROM docs.v_markdown_columns
  )
  FILE_FORMAT = (
    TYPE = CSV
    FIELD_DELIMITER = '\t'
    FIELD_OPTIONALLY_ENCLOSED_BY = '"'
    ESCAPE_UNENCLOSED_FIELD = NONE
  )
  HEADER = FALSE
  OVERWRITE = TRUE;
```

Never use | as the export delimiter for Markdown content. Pipe characters are part of Markdown table syntax, so they must remain inside the text payload rather than be treated as file separators.

(If you prefer JSON, emit JSON strings, and your site generator can template them into HTML.)

22.3.4 Keep Descriptions Close to the Data

Snowflake supports `COMMENT ON` for databases, schemas, tables, views, and columns. Populate comments from your repo (YAML/CSV → SQL), so business definitions render automatically.

```
COMMENT ON TABLE mart.fact_orders IS 'Transactional fact table with one row
per order; late-arriving updates patched hourly.';
COMMENT ON COLUMN mart.fact_orders.order_channel IS 'Channel taxonomy:
web|store|partner.';
```

If your catalog is the stewarded source for definitions, generate these COMMENT statements from the catalog, and apply them via your CI job. Either way, INFORMATION_SCHEMA will surface them for rendering.

22.3.5 Badges and Governance in the Docs

Policies become legible when they're badges, not paragraphs. Use tags and policy references to stamp each dataset with what matters:

```
CREATE OR REPLACE VIEW docs.v_badges AS
SELECT
  x.sch,
  x.obj,
  ARRAY_TO_STRING(ARRAY_AGG(DISTINCT x.pii_label), ', ') AS pii,
  IFF(EXISTS(
        SELECT 1
        FROM SNOWFLAKE.ACCOUNT_USAGE.POLICY_REFERENCES p
        WHERE p.OBJECT_SCHEMA = x.sch
          AND p.OBJECT_NAME   = x.obj
          AND p.POLICY_KIND   = 'MASKING_POLICY'
      ), 'MASKED', '-') AS masking
FROM (
  SELECT
    COALESCE(r.OBJECT_SCHEMA, t.OBJECT_SCHEMA) AS sch,
    COALESCE(r.OBJECT_NAME,   t.OBJECT_NAME)   AS obj,
    CASE
      WHEN t.TAG_NAME = 'CLASSIFICATION' AND t.TAG_VALUE LIKE 'pii.%'
      THEN 'PII'
      ELSE NULL
    END AS pii_label
  FROM SNOWFLAKE.ACCOUNT_USAGE.TAG_REFERENCES t
  FULL OUTER JOIN SNOWFLAKE.ACCOUNT_USAGE.POLICY_REFERENCES r
    ON r.OBJECT_SCHEMA = t.OBJECT_SCHEMA
   AND r.OBJECT_NAME   = t.OBJECT_NAME
) x
GROUP BY 1,2;
```

Join docs.v_badges into your Markdown views to display **PII/MASKED** chips next to each object.

22.3.6 Scheduling and Change Control

Documentation should update itself. A single Task can refresh usage snapshots (for stability) and re-export pages on a cadence.

```
CREATE OR REPLACE TASK docs.t_generate
  WAREHOUSE = wh_ops
  SCHEDULE  = 'USING CRON 15 * * * * UTC'
AS
BEGIN
  -- Optional: snapshot ACCESS_HISTORY to docs.t_usage_daily first
  -- Then regenerate files
  COPY INTO @docs_stage/index/   FROM (SELECT md FROM docs.v_markdown_
  index)   FILE_FORMAT=(TYPE=CSV FIELD_DELIMITER='|') OVERWRITE=TRUE
  SINGLE=TRUE;
  COPY INTO @docs_stage/datasets/ FROM (SELECT obj||'.md', md FROM
  docs.v_markdown_columns) FILE_FORMAT=(TYPE=CSV FIELD_DELIMITER='|')
  OVERWRITE=TRUE;
END;
```

```
ALTER TASK docs.t_generate RESUME;
```

Promote this Task and the docs.* views like code: PR, review, deploy to prod_candidate, and flip with a SWAP.

22.3.7 "Docs As Data" Pays Off Quickly

- New columns appear in the dictionary the moment they're in INFORMATION_SCHEMA.
- A masking policy attachment flips a **MASKED** badge without someone editing a wiki.

- Lineage updates itself whenever you refactor views.
- Usage sections reflect what teams actually read last week, not what they said they read last quarter.

Most importantly, no one is hand-curating pages in a separate system. The catalog is the warehouse; the docs are just **SELECTs with nicer formatting**.

CHAPTER 23

Operational Playbooks

An operational playbook turns stressful fixes into **repeatable moves**. In Snowflake, those moves are metadata-first: freeze reality with a clone, prove a correction away from production, promote atomically, and replay what needs to be rebuilt. Because storage is immutable and compute is elastic, recovery is mostly about **precise cut lines** (where to branch) and **deterministic replays** (how to catch up). A good playbook reads like a short checklist and runs as SQL you already trust.

23.1 Hotfixes with Cloning and Backfills

Hotfixes feel risky when they happen live. Cloning makes them boring. You take a **zero-copy snapshot** at the moment of truth, fix the problem in isolation, and then **swap the sign on the door**, so traffic sees the corrected state. Any data lost or delayed is rebuilt by a narrow, well-scoped backfill—minutes, not weeks.

23.1.1 The Flow at a Glance

Situation	Safe move	Why it works
Bad transform/merge landed	Clone the affected scope, repair inside, **SWAP**	Clone is instant; promotion is atomic and reversible
Stale or missing partition	Backfill only the gap window	Idempotent ELT keeps double-writes harmless
Wrong schema change applied	Recreate in clone, validate contracts, **SWAP**	Schema and data roll together; no hand edits in prod
Partial reload created duplicates	Rebuild from **Time Travel** into clone, dedupe, **SWAP**	Immutable history gives a clean base

M. Hander, *Snowflake Data Warehouse Engineering*, https://doi.org/10.1007/979-8-8688-2628-3_23

23.1.2 Where to Draw the Clone Line

Clone the **smallest scope** that still contains all objects you must repair together:

- A **table** when a single object is wrong and constraints are local
- A **schema** when multiple tables and views must agree (typical mart fix)
- A **database** when tasks/streams/policies and several schemas must march together

Clones are metadata-only at creation, so start wider if you're unsure; you pay only for **new deltas** you write while fixing.

```
-- Choose the scope
CREATE OR REPLACE SCHEMA mart_fix CLONE prod.mart;          -- schema-level
-- or
CREATE OR REPLACE DATABASE prod_fix CLONE prod;             -- database-level
```

23.1.3 Freezing Time (Optional But Powerful)

If you need "what prod looked like **before** the bad run," create the clone **as of** a timestamp or statement:

```
CREATE OR REPLACE DATABASE prod_fix CLONE prod
  AT (TIMESTAMP => TO_TIMESTAMP_TZ('2025-09-12 01:00:00 +00:00'));
```

That guarantees your hotfix starts from the last-known-good state, not the current, possibly corrupted one.

23.1.4 Repairing Inside the Clone

Do all corrections in the clone—no exceptions. Typical fixes include

- Rebuilding a table from Time Travel or curated sources into a new object, validating it, and only then promoting it with SWAP
- **Deduping** by stable keys when a partial load doubles rows

- **Re-applying schema** with guarded DDL (ADD COLUMN IF NOT EXISTS, etc.)
- **Recomputing** downstream views/materializations (dbt run against the clone)

Examples:

-- Restore a damaged table into a new object first, validate it, and only then promote it with SWAP. Avoid restoring directly with CREATE OR REPLACE during an incident because it removes your easiest fallback if the timestamp is wrong.

```
-- Step 1: restore into a new table
CREATE OR REPLACE TABLE prod_fix.mart.fact_orders_restored AS
SELECT *
FROM prod.mart.fact_orders
  AT (TIMESTAMP => TO_TIMESTAMP_TZ('2025-09-12 01:00:00 +00:00'));
-- Step 2: validate before promotion
SELECT load_dt, COUNT(*) AS rows
FROM prod_fix.mart.fact_orders_restored
GROUP BY load_dt
ORDER BY load_dt DESC
LIMIT 5;
-- Step 3: promote only after validation
ALTER TABLE prod_fix.mart.fact_orders_restored
  SWAP WITH prod_fix.mart.fact_orders;
```

```
-- Remove accidental duplicates (keep newest by load_ts)
CREATE OR REPLACE TABLE prod_fix.mart.fact_orders AS
SELECT *
FROM (
  SELECT *, ROW_NUMBER() OVER (PARTITION BY order_id ORDER BY load_ts
  DESC) AS rn
  FROM prod_fix.mart.fact_orders
) WHERE rn = 1;
```

23.1.5 Validating Before You Flip

Proof beats intuition. Run **the same checks you use for promotion**: row counts by day, required columns and types, masking and RAP still attached, and a small business query (e.g., revenue by channel for yesterday) that you can eyeball.

```
-- Row counts and freshness
SELECT load_dt, COUNT(*) AS rows
FROM prod_fix.mart.fact_orders
GROUP BY load_dt
ORDER BY load_dt DESC
LIMIT 3;

-- Schema contract (example)
SELECT column_name, data_type
FROM prod_fix.information_schema.columns
WHERE table_schema='MART' AND table_name='FACT_ORDERS';

-- Policy attachment spot check
SELECT *
FROM SNOWFLAKE.ACCOUNT_USAGE.POLICY_REFERENCES
WHERE OBJECT_DATABASE='PROD_FIX' AND OBJECT_NAME='FACT_ORDERS';
```

23.1.6 Promoting with an Atomic Swap

When checks pass, promotion is a metadata flip. Sessions that reference `prod` keep working; the contents change beneath the stable name.

```
ALTER DATABASE prod_fix SWAP WITH prod;  -- or ALTER SCHEMA mart_fix SWAP
WITH prod.mart;
```

Rollback is symmetrical: swap back. Keep the previous state around under `prod_fix` until you're comfortable, then drop.

23.1.7 Backfilling the Gap

Clones fix correctness; **backfills** fix **completeness**. Compute the exact window you missed and replay only that slice. Keep your ELT **idempotent** so re-runs are safe.

Minimal patterns:

```
-- Parameterize the gap (start/end)
SET bf_start = '2025-09-12 01:00:00'::timestamp;
SET bf_end   = '2025-09-12 03:00:00'::timestamp;

-- Idempotent merge from stage into fixed table
MERGE INTO prod.mart.fact_orders t
USING (
  SELECT * FROM prod.stage.orders_src
  WHERE ingested_at >= $bf_start AND ingested_at < $bf_end
) s
ON t.order_id = s.order_id
WHEN MATCHED THEN UPDATE SET  -- only mutable columns
  load_ts = s.load_ts,
  amount  = s.amount
WHEN NOT MATCHED THEN INSERT (order_id, load_ts, amount, ...)
VALUES (s.order_id, s.load_ts, s.amount, ...);
```

For derived views or Dynamic Tables, kick a manual refresh or run dbt for the affected subset. For Streams/Tasks, consider **pausing** during the hotfix and **resuming** after promotion to avoid writing into the broken state.

```
ALTER TASK elt_refresh_fact_orders SUSPEND;
-- ...fix & swap...
ALTER TASK elt_refresh_fact_orders RESUME;
```

23.1.8 Streams, Tasks, and Dynamic Tables: Keeping Wiring Sane

- **Streams** are pointers to change tables; when you rebuild the base, the stream position may need attention. Prefer **recreating** streams after a full rebuild or backfilling via MERGE without consuming the stream.
- **Tasks** should be **suspended** for the repair scope to prevent concurrent edits. Resume after promotion.

- **Dynamic Tables** can be nudged with ALTER DYNAMIC TABLE ... REFRESH to recompute from the corrected sources; ensure upstream windowing produces deterministic output for the backfill range.

23.1.9 Guardrails That Prevent Repeat Incidents

- Write migrations and ELT to be **re-runnable**. If every step is safe to execute twice, hotfixes become cheaper.
- Keep **envelope columns** (date/tenant) on facts so your backfill windows can prune naturally.
- Attach **masking and RAP** via tag-aware policies, so schema rewrites don't drop protections.
- Log the event in a small table, so you can query the history of hotfixes alongside your telemetry.

```
CREATE TABLE IF NOT EXISTS ops.event_log
  (ts TIMESTAMP_NTZ, incident_id STRING, scope STRING, action STRING, notes
  STRING);

INSERT INTO ops.event_log
VALUES (CURRENT_TIMESTAMP(), 'INC-2025-0912-01', 'prod.mart', 'swap',
'Rebuilt fact_orders from 01:00 snapshot; backfilled 01:00-03:00');
```

23.1.10 Cost and Cleanup

Clones share storage with the source; you only pay for the **delta** you write during the fix. After a few days, drop the emergency clone and any temporary lanes you opened. If you created ad hoc warehouses for the backfill, attach a **resource monitor** so future hotfixes cannot run away with the budget.

```
CREATE OR REPLACE RESOURCE MONITOR rm_hotfix
  WITH CREDIT_QUOTA = 50
  FREQUENCY = MONTHLY
  TRIGGERS ON 80 PERCENT DO NOTIFY ON 100 PERCENT DO SUSPEND;
ALTER WAREHOUSE wh_hotfix SET RESOURCE_MONITOR = rm_hotfix;
```

Clone to make a safe place, fix until the checks are boring, swap to promote, and backfill just the gap. Because each step is SQL, you can version and rehearse, a "hotfix" becomes a small, practiced maneuver instead of a late-night gamble.

23.2 Schema Evolution and Contract Testing

Schema change should feel like a **planned negotiation**, not a surprise. In Snowflake, you evolve safely by separating three concerns: how producers add or change fields, how contracts promise compatibility to consumers, and how CI proves those promises against a **candidate clone** before anything reaches production. The mechanics are simple: additive-first patterns, guarded DDL, and a few compact "contracts-as-queries" that must return green.

23.2.1 Additive-First Evolution

Most changes can be additive: new columns with defaults, new views that project new columns, or new semantic versions (V_*). Deletions and type changes become **version bumps** with an overlap window, so downstreams migrate deliberately. Use guarded DDL to keep migrations idempotent and safe to rerun.

```
-- Add columns safely (no-op if exists)
ALTER TABLE mart.fact_orders
  ADD COLUMN IF NOT EXISTS order_channel STRING,
  ADD COLUMN IF NOT EXISTS source_system STRING DEFAULT 'core';

-- Record the change for docs/release notes
COMMENT ON COLUMN mart.fact_orders.order_channel IS 'web|store|partner';
```

When a breaking change is unavoidable, publish a new surface and keep both until a deprecate date:

```
CREATE OR REPLACE SECURE VIEW mart.v_orders_v2 AS
SELECT order_id, order_dt, order_channel, amount
FROM mart.fact_orders;

-- Keep v1 until <date>; announce in release notes
```

23.2.2 Contracts As Queries (Prove Before You Flip)

Contracts are short SELECTs that encode what consumers rely on—column presence and types, nullability, and a handful of invariants. Run them against prod_candidate (your clone with migrations applied), and fail promotion if any query returns rows or wrong shapes.

Required Columns and Types

```
-- Columns that must exist with expected types
WITH expected(col, dtype) AS (
  SELECT 'ORDER_ID','NUMBER' UNION ALL
  SELECT 'ORDER_DT','DATE'  UNION ALL
  SELECT 'AMOUNT','NUMBER'  UNION ALL
  SELECT 'ORDER_CHANNEL','STRING'
)
SELECT e.col, e.dtype, c.data_type AS found_type
FROM expected e
LEFT JOIN prod_candidate.information_schema.columns c
  ON c.table_schema='MART' AND c.table_name='FACT_ORDERS' AND c.column_
  name=e.col
WHERE c.column_name IS NULL OR UPPER(c.data_type) <> e.dtype;
```

Non-breaking Promise (No Removals)

```
-- Anything present in prod must still be present in candidate
SELECT p.column_name
FROM prod.information_schema.columns p
LEFT JOIN prod_candidate.information_schema.columns c
  ON c.table_schema=p.table_schema AND c.table_name=p.table_name AND
  c.column_name=p.column_name
WHERE p.table_schema='MART' AND p.table_name='FACT_ORDERS' AND c.column_
name IS NULL;
```

Governance Stays Attached

```
-- PII columns must remain masked after rebuilds
SELECT t.object_schema, t.object_name, t.column_name
FROM SNOWFLAKE.ACCOUNT_USAGE.TAG_REFERENCES t
```

```
LEFT JOIN SNOWFLAKE.ACCOUNT_USAGE.POLICY_REFERENCES p
  ON p.object_database='PROD_CANDIDATE'
 AND p.object_schema=t.object_schema AND p.object_name=t.object_name AND
 COALESCE(p.column_name,'')=COALESCE(t.column_name,'')
WHERE t.object_database='PROD_CANDIDATE'
  AND t.tag_name='CLASSIFICATION' AND t.tag_value LIKE 'pii.%'
  AND p.policy_name IS NULL;
```

Semantic Invariants

```
-- Values must remain in the accepted set
SELECT DISTINCT order_channel
FROM prod_candidate.mart.fact_orders
WHERE order_channel IS NOT NULL
  AND order_channel NOT IN ('web','store','partner');
```

23.2.3 Evolution with Routing Views

Routing views let you **publish change without breaking** consumers by pinning a stable name to a version behind the scenes.

```
-- Router exposes a stable surface while you cut over
CREATE OR REPLACE VIEW mart.v_orders AS
SELECT * FROM mart.v_orders_v1;  -- later switch to v2 in one-line edit
```

You can dry-run the new view (v2) in downstream staging by explicitly referencing it, then switch the router when tests pass.

23.2.4 Deferred Enforcement via Policies and Tags

When a change introduces sensitive data, enforce it immediately through **tag-aware masking** rather than waiting for every downstream to update.

```
ALTER TABLE mart.fact_orders
  MODIFY COLUMN customer_email SET TAG classification='pii.email';
```

```
-- Existing generic masking policy will apply without new code
```

This keeps contract drift from becoming a security gap.

23.2.5 CI Path That Keeps Owners Honest

- **Branch**: Developer adds guarded DDL + dbt schema.yml updates.
- **Ephemeral clone**: CI creates `prod_candidate` via `CLONE prod`, applies migrations, runs dbt, and installs UDFs.
- **Contract suite**: Run the queries above; store results in `ops.test_results`. Zero rows = pass.
- **Promotion**: Only after contracts pass, `SWAP` candidate with prod (or flip router view).

Attach a small, machine-readable **release note** row for discoverability:

```
INSERT INTO ops.release_log
VALUES (CURRENT_TIMESTAMP(), 'v2025.10.19', 'mart.fact_orders: add ORDER_
CHANNEL', CURRENT_USER(), 'pending');
```

23.2.6 Handling the Hard Cases

Type Widening (e.g., NUMBER → STRING)

Publish a **new column** with the widened type, populate deterministically, and deprecate the old one after an overlap window. Downstreams migrate by reading the new column; router views can alias `old_name AS new_name` during the window.

```
ALTER TABLE mart.fact_orders ADD COLUMN order_id_str STRING;
UPDATE mart.fact_orders SET order_id_str = TO_VARCHAR(order_id);
-- Eventually drop old, after v2 in place
```

Column Rename

Avoid direct renames for widely used columns. Add a new column, copy the data, expose via view aliasing, update consumers, then drop the original.

```
CREATE OR REPLACE VIEW mart.v_orders_v2 AS
SELECT order_id AS id, order_dt, amount, order_channel FROM mart.
fact_orders;
```

Breaking Partition Keys

If partitioning semantics change (implicit pruning via envelope columns), introduce the envelope first (e.g., event_dt), and keep both filters working in views. After adoption, refactor base tables.

23.2.7 Detecting Accidental Contract Breaks (Drift)

Keep a tiny **contract snapshot**, so you can diff shapes automatically.

```
CREATE TABLE IF NOT EXISTS ops.contract_columns (
  env STRING, sch STRING, obj STRING, col STRING, dtype STRING, snap_ts
  TIMESTAMP_NTZ
);

-- Snapshot prod and candidate
INSERT INTO ops.contract_columns
SELECT 'prod', table_schema, table_name, column_name, data_type, CURRENT_
TIMESTAMP()
FROM prod.information_schema.columns
WHERE table_schema='MART';

INSERT INTO ops.contract_columns
SELECT 'candidate', table_schema, table_name, column_name, data_type,
CURRENT_TIMESTAMP()
FROM prod_candidate.information_schema.columns
WHERE table_schema='MART';

-- Diff
SELECT p.obj, p.col, p.dtype AS prod_type, c.dtype AS cand_type
FROM ops.contract_columns p
FULL JOIN ops.contract_columns c
  ON c.env='candidate' AND p.env='prod'
 AND p.sch=c.sch AND p.obj=c.obj AND p.col=c.col
WHERE p.env='prod' AND (c.col IS NULL OR p.dtype <> c.dtype);
```

Fail the deployment if any rows appear.

23.2.8 Rollout, Rollback, and Overlap

Rollouts are either **SWAPs** (database/schema) or **router flips** (views/functions). Rollback is symmetrical: swap back, or point the router to the previous version. Keep both versions live for a defined **overlap window**, and track usage via ACCESS_HISTORY to know when it's safe to retire the old surface.

```
-- Who still reads v1?

SELECT a.USER_NAME, a.ROLE_NAME, COUNT(*) AS reads
FROM SNOWFLAKE.ACCOUNT_USAGE.ACCESS_HISTORY a,
     LATERAL FLATTEN(input => a.DIRECT_OBJECTS_ACCESSED) obj
WHERE obj.value:objectName::STRING = 'MART.V_ORDERS_V1'
  AND a.QUERY_START_TIME >= DATEADD(day,-7,CURRENT_TIMESTAMP())
GROUP BY 1,2
ORDER BY reads DESC;
```

DIRECT_OBJECTS_ACCESSED is an array of objects in ACCESS_HISTORY, not a scalar value. For deprecation checks, flatten it first so you count all consumers instead of only the first element or a null result.

Evolve schemas by adding first, versioning when you must, and proving contracts with a few precise queries in a candidate clone. Keep governance tied to tags so protections persist through change. With that, schema evolution is routine, and contract testing is a check box that actually means something.

23.3 Cost Anomalies and Runaway Queries

Spend spikes usually come from **too much work** (bytes scanned, bad joins), **too many lanes** (parallel clusters or warehouses left warm), or **unexpected movement** (replication, stage growth, external calls). The playbook is: **detect quickly**, **contain without collateral**, **trace to actors and texts**, then **remove the root** and set a guardrail so it can't recur.

23.3.1 Detect (Is Today Off the Rails?)

Detection rule of thumb: use WAREHOUSE_METERING_HISTORY for warehouse compute credits, QUERY_HISTORY for scan-heavy or cloud-services-heavy query behavior, and Information Schema query-history table functions for near-real-time kill switches. ACCOUNT_USAGE is for historical analysis, not ten-minute operational control loops.

Credits vs. Baseline (Last 24 h vs. Trailing Seven-Day Median)

```
WITH d AS (
  SELECT TO_DATE(START_TIME) AS day, WAREHOUSE_NAME, SUM(CREDITS_USED)
  AS credits
  FROM SNOWFLAKE.ACCOUNT_USAGE.WAREHOUSE_METERING_HISTORY
  WHERE START_TIME >= DATEADD(day,-14,CURRENT_TIMESTAMP())
  GROUP BY 1,2
),
b AS (
  SELECT WAREHOUSE_NAME,
         MEDIAN(credits) AS med7
  FROM d
  WHERE day < CURRENT_DATE()
  GROUP BY 1
)
SELECT d.WAREHOUSE_NAME, d.credits AS today, b.med7,
       (d.credits - b.med7) AS delta
FROM d JOIN b USING (WAREHOUSE_NAME)
WHERE d.day = CURRENT_DATE()
ORDER BY delta DESC;
```

Top drivers (last 2h): split compute credits from warehouse metering and query-level cloud-services/scan signals from QUERY_HISTORY. Compute credits are metered at the warehouse level, not per query.

```
-- Compute credits by warehouse (correct source)
SELECT
  WAREHOUSE_NAME,
  TO_DATE(START_TIME) AS day,
  SUM(CREDITS_USED) AS compute_credits
```

```
FROM SNOWFLAKE.ACCOUNT_USAGE.WAREHOUSE_METERING_HISTORY
WHERE START_TIME >= DATEADD(hour, -2, CURRENT_TIMESTAMP())
GROUP BY 1,2
ORDER BY compute_credits DESC;
-- Cloud services credits + scan volume by team (valid QUERY_
HISTORY fields)
SELECT
  COALESCE(PARSE_JSON(QUERY_TAG):team::STRING, 'unknown') AS team,
  ROLE_NAME,
  WAREHOUSE_NAME,
  SUM(CREDITS_USED_CLOUD_SERVICES) AS cloud_svc_credits,
  SUM(BYTES_SCANNED) / POWER(1024,3) AS gb_scanned,
  COUNT(*) AS query_count
FROM SNOWFLAKE.ACCOUNT_USAGE.QUERY_HISTORY
WHERE START_TIME >= DATEADD(hour, -2, CURRENT_TIMESTAMP())
GROUP BY 1,2,3
ORDER BY gb_scanned DESC
LIMIT 20;
```

Runaway Queries Right Now

```
SELECT QUERY_ID, USER_NAME, ROLE_NAME, WAREHOUSE_NAME,
       BYTES_SCANNED/POWER(1024,3) AS gb_scanned,
       TOTAL_ELAPSED_TIME/1000 AS sec,
       LEFT(QUERY_TEXT, 300) AS txt
FROM TABLE(INFORMATION_SCHEMA.QUERY_HISTORY_BY_WAREHOUSE(RESULT_
LIMIT=>1000))
WHERE START_TIME >= DATEADD(minute,-15,CURRENT_TIMESTAMP())
  AND BYTES_SCANNED > 200*POWER(1024,3)          -- >200 GB
ORDER BY BYTES_SCANNED DESC;
```

23.3.2 Contain (Make It Stop, Safely)

Actions should be **reversible** and **scoped**.

```
-- Pause noisy background tasks, not user BI
ALTER TASK IF EXISTS etl.hourly_refresh SUSPEND;
```

```
-- Cap elastic spread for BI temporarily (keep 1-2 clusters)
ALTER WAREHOUSE wh_bi SET MAX_CLUSTER_COUNT = 2;

-- Put a temporary statements timeout on a noisy warehouse
ALTER WAREHOUSE wh_ad_hoc SET STATEMENT_TIMEOUT_IN_SECONDS = 600;

-- Cancel the single worst offender (capture QUERY_ID from detection)
CALL SYSTEM$CANCEL_QUERY('01a1b-...-000');
```

If the spend is from a **Reader Account**, throttle its warehouse:

```
ALTER WAREHOUSE wh_reader_partner SET WAREHOUSE_SIZE = 'XSMALL', AUTO_
SUSPEND=60;
```

If a replication job is the culprit, increase the interval, or pause non-critical replicated DBs until the window passes.

23.3.3 Trace (Who, What Text, Which Objects?)

By Object Family

```
SELECT obj.value:objectName::STRING AS obj,
       COUNT(*) AS q
FROM SNOWFLAKE.ACCOUNT_USAGE.ACCESS_HISTORY a,
     LATERAL FLATTEN(input => a.DIRECT_OBJECTS_ACCESSED) obj
WHERE a.QUERY_START_TIME >= DATEADD(hour,-2,CURRENT_TIMESTAMP())
GROUP BY 1
ORDER BY q DESC
LIMIT 30;
```

ACCESS_HISTORY is good for tracing which objects were involved, but BYTES_SCANNED belongs to QUERY_HISTORY, not ACCESS_HISTORY. Use ACCESS_HISTORY for object attribution and QUERY_HISTORY for scan-heavy query analysis.

By Text Patterns (Quick Smells)

```
SELECT COUNT(*) AS q,
       IFF(QUERY_TEXT ILIKE '%SELECT *%', 'select_star','ok') AS smell
FROM SNOWFLAKE.ACCOUNT_USAGE.QUERY_HISTORY
WHERE START_TIME >= DATEADD(hour,-2,CURRENT_TIMESTAMP())
GROUP BY smell;
```

Warehouse Behavior (Did Autosuspend Fail to Help?)

```
SELECT *
FROM INFORMATION_SCHEMA.WAREHOUSE_LOAD_HISTORY
WHERE START_TIME >= DATEADD(hour,-2,CURRENT_TIMESTAMP())
ORDER BY START_TIME DESC;
```

Look for long **ACTIVE** windows with low concurrency → likely dashboards staggering tiles too far apart or AUTO_SUSPEND too long.

23.3.4 Fix (Today) and Prevent (Tomorrow)

Symptom you observe	Immediate containment	Lasting fix
Credits jump on one warehouse, bytes scanned high	Cap MAX_CLUSTER_COUNT; set STATEMENT_TIMEOUT_IN_SECONDS	Add envelope filters in shared views (date/tenant), remove SELECT *, teach pruning with examples
Many clusters spin but small bytes per query	Lower AUTO_SUSPEND and MAX_CLUSTER_COUNT; stagger schedules	Consolidate BI refresh windows; use result cache-friendly SQL; consider one medium warehouse over many tiny
External functions explode costs	Throttle calling job; increase batching	Cache results in a lookup table; batch requests; lower concurrency on that task
Replication egress spikes	Pause non-critical replicas	Lengthen cadence; replicate only needed DBs; co-locate consumers
Stage/storage grows with tiny files	Pause downstream COPY	Compact upstream to ~200 MB files (Parquet); use manifests; ingest by partition

23.3.5 Codify Guardrails

Resource Monitors (Per-Risky Lane)

```
CREATE OR REPLACE RESOURCE MONITOR rm_ad_hoc
  WITH CREDIT_QUOTA = 100
  FREQUENCY = DAILY
```

```
  TRIGGERS ON 75 PERCENT DO NOTIFY
           ON 100 PERCENT DO SUSPEND;
ALTER WAREHOUSE wh_ad_hoc SET RESOURCE_MONITOR = rm_ad_hoc;
```

Statement Limits for Ad Hoc Lanes

```
ALTER WAREHOUSE wh_ad_hoc SET
  STATEMENT_TIMEOUT_IN_SECONDS = 900,
  STATEMENT_QUEUED_TIMEOUT_IN_SECONDS = 300;
```

Kill Switch for Patterns (Optional)

Create a simple “bad patterns” view and page when it lights up:

```
CREATE OR REPLACE VIEW ops.v_bad_patterns AS
SELECT QUERY_ID, USER_NAME, WAREHOUSE_NAME, BYTES_SCANNED
FROM TABLE(INFORMATION_SCHEMA.QUERY_HISTORY(
  END_TIME_RANGE_START => DATEADD(minute, -10, CURRENT_TIMESTAMP()),
  END_TIME_RANGE_END   => CURRENT_TIMESTAMP()
))
WHERE QUERY_TEXT ILIKE '%SELECT * FROM%FACT_%'
   OR BYTES_SCANNED > 500 * POWER(1024,3);
```

Wire a five-minute Task to check this view and notify only on new offenders. For sub-hour detection windows, use Information Schema query-history table functions rather than ACCOUNT_USAGE, which is intentionally delayed for historical reporting.

23.3.6 A Calm, Five-Step Runbook

1. **Confirm spike** with credits-vs.-baseline and top drivers.
2. **Contain** with targeted caps/timeouts and cancel the worst one or two queries.
3. **Trace** to objects and texts; collect examples.
4. **Patch** shared views (add envelope filters), fix schedules, or compact files; resume Tasks.
5. **Guardrail** with a monitor, tuned autosuspend/cluster counts, and a lint/check in CI for SELECT * or missing date filters in shared surfaces.

Treat cost like latency—measure, localize, and shave it where it begins. The combination of quick containment (timeouts, caps, cancels), precise attribution (tags, query text, objects), and small structural fixes (pruning, schedules, file sizing) turns a spike into a short story, not a saga.

PART VII

Case Studies and Patterns

CHAPTER 24

Enterprise Data Warehouse

An enterprise data warehouse (EDW) is less a destination than a **trust machine**: numbers must be traceable, repeatable, and defensible months after they were produced. Snowflake's strengths—immutable storage, metadata-first cloning, policy enforcement at read—line up well with those demands. The pattern is to keep a **single, governed source of truth**; shape it for analytics without duplicating it; and preserve every step (code, inputs, lineage, approvals) as data you can query.

24.1 Finance and Regulatory Reporting

Finance cares about **final answers with provenance**. Reporting must withstand audit: how a figure was calculated, which inputs were used, who approved it, and whether you can recreate it byte-for-byte. The warehouse becomes the **system of reporting** over operational systems of record, with controls that make closing the books boring.

24.1.1 What "Good" Looks Like

Numbers flow from raw feeds into standardized ledgers and subject-area marts. Each transformation is small, versioned, and provable. When month-end arrives, you **freeze** the state for that period, compute disclosures, and publish extracts as secure views or files. If late adjustments appear, you capture them as **postings**, not rewrites, and you can reproduce both the "as first closed" and "as restated" views on demand.

M. Hander, *Snowflake Data Warehouse Engineering*, https://doi.org/10.1007/979-8-8688-2628-3_24

24.1.2 Blueprint at a Glance

Concern	Concrete stance
Truth source	Land raw, immutable copies (files or CDC) in `raw` and model a **financial ledger** in `core` (double-entry or fact+dim) with envelope columns (`txn_dt`, `period`, `entity`).
Controls	Tag objects with `owner`, `classification`, `retention_policy`; enforce **masking** for PII; scope rows by **Row Access Policies** (entity/cost center).
Reproducibility	Use zero-copy clones pinned to an explicit timestamp or statement boundary to freeze period state, and keep Time Travel available for validation and investigation around that freeze point. Store run parameters and code hashes in `ops.run_log`.
Adjustments	Represent corrections as **journal entries** or delta facts with `posted_at`, `restatement_id`; never overwrite original rows.
Outputs	Publish **secure views** for CFO dashboards and regulator extracts; share cross-account via regional replicas when needed.

24.1.3 Modeling the Ledger

A finance mart benefits from a ledger backbone that makes postings explicit and balances provable. One practical shape:

```
-- Chart of accounts
CREATE TABLE core.dim_account (
  account_id STRING PRIMARY KEY,
  account_code STRING,
  account_name STRING,
  account_type STRING,          -- Asset, Liability, Equity,
                                Revenue, Expense
  natural_sign NUMBER           -- +1 or -1 for balance direction
);
```

Note On standard Snowflake tables, PRIMARY KEY is informational and not enforced. Keep it for metadata and optimizer hints only if you can also prove uniqueness with data-quality checks or a scheduled exception query. If you need enforced keys, use hybrid tables.

```
-- Data-quality check for duplicate accounts
SELECT account_id, COUNT(*) AS dup_count
FROM core.dim_account
GROUP BY 1
HAVING COUNT(*) > 1;
-- Journal header
CREATE TABLE core.f_journal (
  journal_id STRING,
  period DATE,
  entity_id STRING,
  source_system STRING,
  posted_at TIMESTAMP_NTZ,
  restatement_id STRING,          -- NULL for original close
  PRIMARY KEY (journal_id)
);

-- Journal lines (double-entry)
CREATE TABLE core.f_journal_line (
  journal_id STRING,
  line_no NUMBER,
  account_id STRING,
  amount NUMBER(20,4),            -- signed
  currency STRING,
  txn_dt DATE,
  cost_center STRING,
  project STRING
);
```

Balances and P&L rollups then become stable views that can be re-evaluated for any period or restatement:

```
-- Trial balance as-of a snapshot
CREATE OR REPLACE VIEW fin.v_trial_balance AS
SELECT jl.entity_id,
       j.period,
       a.account_code,
       SUM(jl.amount) * a.natural_sign AS balance
FROM core.f_journal_line jl
JOIN core.f_journal j USING (journal_id)
JOIN core.dim_account a USING (account_id)
GROUP BY 1,2,3;
```

24.1.4 Closing the Period Calmly

Month-end is a choreography of freeze → compute → attest → publish. In practice, freeze the close at an explicit timestamp or statement boundary so the clone represents a clean, approved cutoff rather than an arbitrary live moment.

```
-- Freeze books at T-0:01 (or the approved checkpoint)

CREATE DATABASE fin_close_2025M09 CLONE prod
AT (TIMESTAMP => '2025-09-30 23:59:59'::TIMESTAMP_NTZ);

-- Compute disclosures inside the clone
CREATE OR REPLACE TABLE fin_close_2025M09.report_ifrs_income AS
SELECT ... FROM fin_close_2025M09.fin.v_trial_balance ...;

-- Attest by recording code hash and inputs
CREATE TABLE IF NOT EXISTS fin_close_2025M09.ops.attestation (
  report_name STRING, run_ts TIMESTAMP_NTZ, git_sha STRING, params VARIANT,
  approver STRING
);
INSERT INTO fin_close_2025M09.ops.attestation
VALUES ('ifrs_income', CURRENT_TIMESTAMP(), 'git:abcd1234', OBJECT_
CONSTRUCT('period','2025-09'), CURRENT_USER());
```

```
-- Publish as secure views to consumers
CREATE OR REPLACE SECURE VIEW prod_finpub.v_ifrs_income_2025M09 AS
SELECT * FROM fin_close_2025M09.report_ifrs_income;
```

If an adjustment arrives, you **post** it in `prod.core.f_journal*`, regenerate the same view **in a new clone** (e.g., `fin_close_2025M09_r1`), and publish alongside the original. You can keep both for audit: "first close" and "restated".

24.1.5 Reconciliations and Completeness

Reconciliation should be data, not screenshots. Compute each side once in a small CTE or joinable summary and compare them explicitly; avoid correlated subqueries against large ledger tables in operational reconciliation views. Keep small, queryable checks: row counts per source, control totals, and tie-outs to bank or subledger statements.

```
-- Source to ledger control total

CREATE OR REPLACE VIEW ops.v_recon_orders AS
WITH src AS (
  SELECT SUM(amount) AS src_sum
  FROM raw.orders_daily
  WHERE load_dt BETWEEN :period_start AND :period_end
),
ledger AS (
  SELECT SUM(jl.amount) AS ledger_sum
  FROM core.f_journal_line jl
  JOIN core.dim_account a USING (account_id)
  WHERE a.account_code = '4000'
)
SELECT
  'orders' AS source,
  src.src_sum,
  ledger.ledger_sum,
  src.src_sum - ledger.ledger_sum AS delta
FROM src
CROSS JOIN ledger;
```

When deltas are non-zero, record an exception with a ticket link:

```
CREATE TABLE IF NOT EXISTS ops.recon_exceptions (
  created_at TIMESTAMP_NTZ, source STRING, period STRING, delta
  NUMBER(20,4), ticket STRING, status STRING
);
```

24.1.6 Regulatory Extracts with Provenance

Regulators ask for a **fixed layout**, **fixed period**, and **provable lineage**. Generate from the close clone and preserve the manifest.

```
-- Fixed-schema extract
CREATE OR REPLACE VIEW fin_close_2025M09.v_regulatory_template AS
SELECT entity_id, period, account_code, balance
FROM fin_close_2025M09.fin.v_trial_balance
WHERE account_code IN ('1000','1100','2000', ...);

-- Emit file
COPY INTO @fin_exports/2025-09/reg_template.csv
FROM (SELECT * FROM fin_close_2025M09.v_regulatory_template)
FILE_FORMAT = (TYPE=CSV FIELD_DELIMITER=',' HEADER=TRUE)
OVERWRITE = TRUE;
CREATE TABLE IF NOT EXISTS fin_exports.manifest (
  path STRING,
  rows NUMBER,
  bytes NUMBER,
  created_at TIMESTAMP_NTZ,
  source_clone STRING,
  git_sha STRING
);
-- Record the logical row count from the exported dataset
INSERT INTO fin_exports.manifest
SELECT
  '2025-09/reg_template.csv',
  COUNT(*) AS rows,
  NULL AS bytes,
```

```
  CURRENT_TIMESTAMP(),
  'fin_close_2025M09',
  'git:abcd1234'
FROM fin_close_2025M09.v_regulatory_template;
```

If you also need physical file metadata such as size, query the staged file metadata after export rather than trying to reference `METADATA$...` columns directly from the `COPY INTO` command body. `METADATA$FILE_ROW_COUNT` and related metadata columns are available when querying staged files.

24.1.7 Access, Segregation, and Attestations

Finance data is sensitive by default. Keep entity or legal-unit scoping as a role-based Row Access Policy. In production, keying entity access off `CURRENT_ROLE()` is more stable than `CURRENT_USER()` because applications and BI tools often run through shared service identities. And apply **tag-aware masking** to PII such as customer details tied to revenue subledgers. Attestation becomes a row in the warehouse, not a signature on a PDF:

```
-- Entity scoping

CREATE OR REPLACE ROW ACCESS POLICY rap_entity AS (entity_id STRING)
RETURNS BOOLEAN ->
  EXISTS (
    SELECT 1
    FROM gov.role_entities m
    WHERE m.role_name = CURRENT_ROLE()
      AND m.entity_id = entity_id
  );

ALTER TABLE core.f_journal ADD ROW ACCESS POLICY rap_entity ON (entity_id);
```

24.1.8 Change and Backfill Without Drama

Fixes to logic or schema flow through **candidate clones** and **SWAP**. Backfills are narrow and idempotent, always bounded by `txn_dt/period`. You never rewrite a closed clone; you re-close in a new clone and publish a new labeled surface.

```
-- Backfill a late subledger window safely

MERGE INTO prod.core.f_journal_line t
USING (
  SELECT *
  FROM stage.subledger
  WHERE txn_dt BETWEEN :s AND :e
) s
ON t.journal_id = s.journal_id
AND t.line_no   = s.line_no
WHEN MATCHED AND t.restatement_id IS NULL THEN
  UPDATE SET amount = s.amount
WHEN NOT MATCHED THEN
  INSERT (...);
```

Closed periods should not be mutated in place. Use `WHEN MATCHED AND ...` guards so only open or explicitly restated rows are updated; otherwise, the backfill should land as a new posting or restatement entry.

24.1.9 SRE Posture for Month-End

A handful of SLIs cover most risks:

- **Freshness** of subledgers vs. expected cadence.
- **Queued vs. execution** time for the close warehouse during the compute window.
- **Recon deltas** staying at zero.
- **Policy coverage** for PII columns in published outputs.

Drive alerts from `ACCOUNT_USAGE` and your reconciliation views; record incidents and remedial queries in `ops.service_log`.

24.1.10 Cross-Region and External Scrutiny

If filing requires **regional residency** or external review, replicate the close clone to the required region and share **secure views**. This keeps figures local and audited without copying files to desktops. Reader accounts work for partners without Snowflake: they see the period database and nothing else; you keep their warehouse tiny and metered by a resource monitor.

Finance and regulatory reporting thrive on **freezes, postings, and provenance**. Snowflake lets you freeze with a clone, post as deltas instead of edits, and keep every step—models, runs, manifests, approvals—as tables and views. That turns close and compliance from a brittle ritual into a repeatable, queryable process you can explain years later.

24.2 Sales Analytics and Planning

Sales teams need **fast answers that line up with finance** and a place to model "what if" without polluting the source of truth. In Snowflake, the pattern is straightforward: a **governed core** for pipeline and bookings, a **sales mart** with canonical metrics and slowly changing dimensions, and **planning sandboxes** fed by the same truth but isolated for scenario work. Queries stay quick because facts carry envelope columns for pruning (dates, territories, segments), and governance stays intact with masking and row access policies on PII and territories.

24.2.1 Shape of the Core and the Mart

Raw events arrive from CRM/CPQ/usage systems. You standardize them into a minimal core and then expose a mart that business tools can read without joins across unpredictable shapes.

```
-- Core opportunity (atomic grain: opportunity, day)
CREATE TABLE core.f_opportunity_daily (
  opp_id STRING,
  snapshot_dt DATE,
  account_id STRING,
  owner_user_id STRING,
  stage STRING,                -- e.g., 'Qualify','Commit'
  amount NUMBER(18,2),
  currency STRING,
  close_dt DATE,
  created_dt DATE,
```

```
  is_open BOOLEAN,
  is_won  BOOLEAN,
  is_lost BOOLEAN,
  source_system STRING
);

-- Slowly changing account dimension (type 2)
CREATE TABLE core.dim_account_scd (
  account_sk NUMBER AUTOINCREMENT,
  account_id STRING,
  name STRING,
  industry STRING,
  segment STRING,             -- Commercial/Enterprise
  territory STRING,
  effective_from DATE,
  effective_to   DATE,    -- '9999-12-31' when current
  is_current BOOLEAN
);
```

The mart should project stable definitions for pipeline and bookings using as-of joins to the SCD dimension. Historical snapshot facts must join on the business date that defines the snapshot, not on `is_current`, or past rows will inherit today's segment, territory, and ownership. It's better to compute these in SQL once than let every dashboard reinvent them.

```
-- Canonical pipeline snapshot (period end)
CREATE OR REPLACE VIEW sales_mart.v_pipeline AS
SELECT
  d.snapshot_dt AS as_of,
  a.segment,
  a.territory,
  d.stage,
  SUM(d.amount) AS pipeline_amt
FROM core.f_opportunity_daily d

JOIN core.dim_account_scd a
ON a.account_id = d.account_id
```

```
AND d.snapshot_dt BETWEEN a.effective_from AND a.effective_to
WHERE d.is_open
GROUP BY 1,2,3,4;

-- Bookings (won in period)
CREATE OR REPLACE VIEW sales_mart.v_bookings AS
SELECT
  d.close_dt AS booked_dt,
  a.segment, a.territory,
  SUM(d.amount) AS bookings_amt
FROM core.f_opportunity_daily d

JOIN core.dim_account_scd a
  ON a.account_id = d.account_id
 AND d.close_dt BETWEEN a.effective_from AND a.effective_to
WHERE d.is_won
GROUP BY 1,2,3;
```

24.2.2 Snapshots and Stage History

CRM data is mutable; to answer "what did we think on the last Friday of Q2?" you need **snapshots**. A lightweight approach is a daily materialization of opportunities that rewrites *only* the latest day and leaves prior days immutable. That gives you stage movement and forecast drift without heavy CDC logic.

```
-- Upsert the daily slice (idempotent)
MERGE INTO core.f_opportunity_daily t
USING stage.crm_opportunities s
ON t.opp_id = s.opp_id AND t.snapshot_dt = CURRENT_DATE()
WHEN MATCHED THEN UPDATE SET
  account_id = s.account_id, stage = s.stage, amount = s.amount,
  close_dt = s.close_dt, is_open = s.is_open, is_won = s.is_won, is_lost =
  s.is_lost
WHEN NOT MATCHED THEN INSERT (opp_id, snapshot_dt, account_id, stage,
amount, close_dt, created_dt, is_open, is_won, is_lost, source_system)
VALUES (s.opp_id, CURRENT_DATE(), s.account_id, s.stage, s.amount, s.close_
dt, s.created_dt, s.is_open, s.is_won, s.is_lost, 'sf');
```

24.2.3 Forecasts That Reconcile to Bookings

Planning workflows break down when forecasts don't reconcile with actuals. Keep **one forecast surface** that ties to period snapshots, and label each submission/version. Sales ops can then model upside/downside in a sandbox without touching the core.

```
CREATE TABLE plan.t_forecast (
  submitted_at TIMESTAMP_NTZ,
  horizon      STRING,            -- 'current_q','next_q'
  owner_user_id STRING,
  territory STRING,
  stage_floor STRING,    -- e.g., 'Commit' => include Commit+ stages
  forecast_amt NUMBER(18,2),
  version STRING    -- 'baseline','upside','downside'
);

-- Reconciliation: forecast vs. realized bookings for the same cohort
CREATE OR REPLACE VIEW plan.v_forecast_vs_actual AS
SELECT
  DATE_TRUNC('quarter', submitted_at)::DATE AS quarter,
  version,
  territory,
  SUM(forecast_amt) AS fcst,
  (SELECT COALESCE(SUM(bookings_amt),0)
     FROM sales_mart.v_bookings b
     WHERE DATE_TRUNC('quarter', b.booked_dt) = DATE_TRUNC('quarter',
     submitted_at)
       AND b.territory = p.territory) AS actual
FROM plan.t_forecast p
GROUP BY 1,2,3;
```

Planning **sandboxes** live in a separate schema or database (often a clone of prod). Users write what-if tables there; curated views union them back for comparison. When a plan is approved, you **snapshot** the inputs and treat them like any other dataset.

24.2.4 Territories, Quotas, and Attribution

Territory reassignments and split credit (e.g., overlay teams) need deterministic attribution. Encode the rules in a single mapping table and never sprinkle them across dashboards.

```
CREATE TABLE sales_rules.territory_map (
  as_of DATE,
  account_id STRING,
  owner_user_id STRING,
  territory STRING,
  credit_pct NUMBER(5,2) DEFAULT 1.00  -- supports split credit
);

-- Attribution view
CREATE OR REPLACE VIEW sales_mart.v_attributed_bookings AS
SELECT
  b.booked_dt, r.territory, r.owner_user_id,
  b.bookings_amt * r.credit_pct AS credited_amt
FROM sales_mart.v_bookings b
JOIN sales_rules.territory_map r
  ON r.account_id = b.account_id
 AND r.as_of <= b.booked_dt
QUALIFY ROW_NUMBER() OVER (
  PARTITION BY b.account_id, b.booked_dt
  ORDER BY r.as_of DESC
) = 1;  -- pick the latest mapping at booking time
```

Quota tracking becomes a simple join to monthly targets with the same territory keys.

24.2.5 Cohorts and Funnel Health

Sales leaders ask "Are we creating enough new pipeline?" and "What is the win rate by segment?" With snapshots, cohort math is clear and fast.

```
-- New pipeline created this month (first seen in month)
WITH first_seen AS (
  SELECT opp_id, MIN(snapshot_dt) AS first_dt
  FROM core.f_opportunity_daily
  GROUP BY opp_id
)
SELECT DATE_TRUNC('month', first_dt) AS cohort_month,
       a.segment,
       SUM(d.amount) AS new_pipeline
FROM first_seen f
JOIN core.f_opportunity_daily d USING (opp_id)

JOIN core.dim_account_scd a
   ON a.account_id = d.account_id
 AND d.snapshot_dt BETWEEN a.effective_from AND a.effective_to
WHERE DATE_TRUNC('month', first_dt) = DATE_TRUNC('month', CURRENT_DATE())
GROUP BY 1,2;
```

Win rate and cycle time are best computed off the **latest** snapshot per opportunity that shows a terminal state, not by counting stage changes.

24.2.6 Governance That Travels to Dashboards

Revenue data mixes **PII** and confidential commercial info. Keep emails and names masked for non-privileged roles and scope rows by **territory**, so regions only see their books. Because policies evaluate at read, the same mart can serve execs and reps safely.

```
-- Mask contact email; allow only specific roles to see plaintext
CREATE OR REPLACE MASKING POLICY mask_email AS (val STRING) RETURNS
STRING ->
  CASE WHEN IS_ROLE_IN_SESSION('ROLE_SALES_PII') THEN val
       ELSE REGEXP_REPLACE(val, '(^.).+(@.+$)', '\\1***\\2') END;

ALTER TABLE core.dim_account_scd
  MODIFY COLUMN owner_email SET MASKING POLICY mask_email;
```

```
-- Territory-based row access
CREATE OR REPLACE ROW ACCESS POLICY rap_territory AS (territory STRING)
RETURNS BOOLEAN ->
  EXISTS (SELECT 1 FROM gov.user_territories m
          WHERE m.user_name = CURRENT_USER()
            AND m.territory = territory);

ALTER VIEW sales_mart.v_pipeline ADD ROW ACCESS POLICY rap_territory ON
(territory);
```

24.2.7 Performance Notes

Sales workloads are **filter-heavy** by date, territory, and segment. Keep those as **scalar columns** on facts and cluster if—and only if—profiling shows broad scans with good predicates.

Small, durable practices:

- Store `snapshot_dt`, `close_dt`, `territory`, `segment` directly on facts for pruning.
- Avoid `SELECT *` in shared views; project just what BI needs.
- Compact CRM export files upstream; for Parquet aim ~200 MB per file and balanced row groups.

24.2.8 Sharing and Planning with Partners

When partners need access to bookings or pipeline for joint planning, publish **secure views** via **regional replicas**. For partners without Snowflake, use a **Reader Account** with a tiny warehouse and an explicit credit cap (resource monitor). Your governance policies travel with the share; partners see only what they should.

Build a small, honest core; project canonical pipeline and bookings once; snapshot relentlessly; and give planners a safe sandbox tied to the same truth. With policy at read and envelope columns for pruning, sales analytics stays fast, reconcilable, and ready for scenario work without breaking the books.

CHAPTER 25

Real-Time and Event-Driven Analytics

Event data arrives as a **stream of little truths**: clicks, sensor ticks, app beacons, payments, and detections. Snowflake's columnar engine and services layer let you land that firehose, **query it within seconds**, and fold it into marts without building a second "speed layer." The operating model is simple: ingest continuously, enrich lightly, publish **stable, prunable tables** for downstreams, and keep late/out-of-order behavior deterministic. Where milliseconds matter, push ingestion to **Snowpipe Streaming**; where minutes are fine, auto-ingest files. Either way, the same SQL powers both real-time views and historical rollups, so you don't fork logic.

25.1 Clickstream and IoT Pipelines

Think of the pipeline as three rings:

- **Edge → Landing:** Get events in fast, with producer identifiers and offsets that support exactly-once semantics at the pipeline level, and keep shape assumptions minimal. Downstream dedup by `event_id` remains part of the safety model.
- **Shaping → Enrichment:** Lift out a few envelope columns for pruning and join the smallest, most stable reference data.
- **Serving → Marts:** Expose sessionized click facts and IoT rollups with predictable keys and windowing that handles late data.

M. Hander, *Snowflake Data Warehouse Engineering*, https://doi.org/10.1007/979-8-8688-2628-3_25

25.1.1 Landing the Stream

For sub-minute latency, stream directly into a table with **Snowpipe Streaming** (from a client/connector). For near-real-time file drops, use COPY INTO + notifications. Both land in the same raw table, so downstream SQL doesn't care how it arrived.

```
CREATE OR REPLACE TABLE raw.events_json (
  src          STRING,
  received_ts  TIMESTAMP_TZ DEFAULT CURRENT_TIMESTAMP(),
  event_ts     TIMESTAMP_TZ, -- from payload (producer clock)
  event_date   DATE,  -- derived for pruning
  tenant_id    STRING,-- envelope for filters
  user_id      STRING,
  device_id    STRING,
  event_type   STRING,
  event_id     STRING,-- producer UUID for idempotence
  payload      VARIANT
);

-- Example: shaping stage for files (clickstream beacons in JSON)
CREATE OR REPLACE FILE FORMAT ff_json TYPE = JSON STRIP_OUTER_ARRAY=TRUE;

COPY INTO raw.events_json
FROM @landing/clicks/
FILE_FORMAT = (FORMAT_NAME=ff_json)
ON_ERROR = 'CONTINUE'
MATCH_BY_COLUMN_NAME = CASE_INSENSITIVE
-- Use SELECT to lift envelope fields (if files carry raw JSON rows)
-- COPY INTO supports column reordering; for complex transforms, stage →
SELECT → INSERT.
```

With **Snowpipe Streaming**, your producer (or Kafka Connector/Confluent) writes rows directly:

- Producer sets event_ts, event_id, and envelope fields; the pipe handles ordering and offset management.
- You still deduplicate by event_id downstream to guard against retries.

25.1.2 Idempotent Dedup and Late Data

A small staging table absorbs the latest batch or stream slice; a merge makes the raw table idempotent. Keep the natural key narrow (tenant_id, event_id).

```
MERGE INTO raw.events_json t
USING stage.events_ingest s
  ON t.tenant_id = s.tenant_id AND t.event_id = s.event_id
WHEN MATCHED THEN UPDATE SET
  payload = s.payload, event_ts = s.event_ts                    -- last-write-
                                                                wins if needed
WHEN NOT MATCHED THEN
  INSERT (src, received_ts, event_ts, event_date, tenant_id, user_id,
  device_id, event_type, event_id, payload)
  VALUES (s.src, s.received_ts, s.event_ts, TO_DATE(s.event_ts), s.tenant_
  id, s.user_id, s.device_id, s.event_type, s.event_id, s.payload);
```

Late events simply land with an older event_ts; your serving views use **watermarks**, so aggregates remain deterministic.

25.1.3 Light, Deterministic Enrichment

Lift scalar columns you'll filter on; keep the rest in payload. For clickstream, parse user-agent; for IoT, unpack metrics and normalize units. Snowpark UDFs help with small, reusable transforms.

```
-- Example: UA → device/os (Snowpark Python UDF skeleton)
CREATE OR REPLACE FUNCTION util.parse_ua(ua STRING)
RETURNS OBJECT
LANGUAGE PYTHON
RUNTIME_VERSION='3.10'
PACKAGES=('user-agents')
HANDLER='f'
AS
$$
from user_agents import parse
def f(ua):
```

```
    try:
        u = parse(ua or '')
        return {"device": u.device.family, "os": u.os.family, "browser":
        u.browser.family}
    except:
        return {"device": None, "os": None, "browser": None}
$$;

-- Shaped view for downstream SQL
CREATE OR REPLACE VIEW bronze.events_shaped AS
SELECT
  tenant_id, user_id, device_id, event_ts, event_date, event_type,
  event_id,
  util.parse_ua(payload:user_agent::string):device::string   AS
  device_family,
  util.parse_ua(payload:user_agent::string):os::string       AS os_family,
  payload                                                    AS v
FROM raw.events_json;
```

25.1.4 Sessionization and Funnels (Clickstream)

Session logic should be **pure SQL** so BI, ML features, and batch backfills share it. A classic approach uses gaps of inactivity (e.g., 30 minutes) per user/tenant.

```
WITH ordered AS (
  SELECT tenant_id, user_id, event_ts, event_date, event_type,
         LAG(event_ts) OVER (
           PARTITION BY tenant_id, user_id
           ORDER BY event_ts
         ) AS prev_ts
  FROM bronze.events_shaped
  WHERE event_type IN ('page_view','click','purchase')
    AND event_ts >= DATEADD('hour', -24, CURRENT_TIMESTAMP())
    -- watermark
    AND event_date >= CURRENT_DATE() - 1                        -- pruning
),
```

```
sessionized AS (
  SELECT *,
         CASE
           WHEN prev_ts IS NULL
             OR DATEDIFF('minute', prev_ts, event_ts) > 30
           THEN 1 ELSE 0
         END AS is_new,
         SUM(
           CASE
             WHEN prev_ts IS NULL
               OR DATEDIFF('minute', prev_ts, event_ts) > 30
             THEN 1 ELSE 0
           END
         ) OVER (
           PARTITION BY tenant_id, user_id
           ORDER BY event_ts
         ) AS session_seq
  FROM ordered
)
SELECT tenant_id, user_id,
       MIN(event_ts) AS session_start,
       MAX(event_ts) AS session_end,
       COUNT(*)      AS events,
       COUNT_IF(event_type = 'purchase') AS purchases
FROM sessionized
GROUP BY tenant_id, user_id, session_seq;
```

Enforce the watermark inside the SQL, not only in prose. In near-real-time serving, sessionization should only read the open horizon and rely on envelope columns such as event_date for pruning. If sessions can span the watermark boundary, keep a small overlap window, and treat only sessions older than the inactivity threshold plus overlap as closed. Dynamic Tables work best when the query boundary is explicit and incrementally maintainable.

Pin a **watermark** (e.g., "consider a session closed after 35 minutes of inactivity or when the event is > T-35 min from now"), and do not mutate closed sessions later.

25.1.5 IoT Rollups and Anomaly Features

IoT streams benefit from **tumbling windows** (e.g., one-minute) and device health KPIs. Keep unit conversions deterministic and derive features once.

```
-- Minute rollup per device
CREATE OR REPLACE VIEW silver.iot_minute AS
SELECT
  tenant_id, device_id,
  DATE_TRUNC('minute', event_ts) AS ts_min,
  AVG(TRY_TO_DOUBLE(v:temp_c))   AS avg_temp_c,
  MAX(TRY_TO_DOUBLE(v:rpm))      AS max_rpm,
  COUNT(*)                       AS msg_count
FROM bronze.events_shaped
WHERE event_type='telemetry'
GROUP BY 1,2,3;
```

```
-- Simple anomaly flag (example thresholds)
CREATE OR REPLACE VIEW silver.iot_anoms AS
SELECT *,
  (avg_temp_c > 80 OR max_rpm > 9000) AS is_anom
FROM silver.iot_minute;
```

25.1.6 Serving with Dynamic Tables (Near-Real-Time)

If you want outputs to **chase** the stream automatically, declare **Dynamic Tables** for sessionized or rolled-up layers and set a refresh lag you can afford (e.g., 1–5 minutes). They handle incremental recompute and dependencies; watermarks ensure deterministic outputs.

```
CREATE OR REPLACE DYNAMIC TABLE gold.click_sessions
TARGET_LAG = '5 minutes'
WAREHOUSE = wh_streaming
AS
SELECT * FROM /* sessionization SELECT above, wrapped as a view */;
```

Keep the Dynamic Table query bounded and incrementally friendly. Snowflake determines refresh strategy from the query definition at creation time; if the query is too broad or complex, refresh can become inefficient or fall back to full recomputation behavior. Watermarks and pruning columns should therefore be part of the SQL itself, not only an operational convention.

For scheduled control or complex joins, Streams+Tasks remain a great fit; Dynamic Tables remove boilerplate when dependencies are DAG-shaped and incremental.

25.1.7 Governance at Read, Even for Streams

Event payloads often contain PII such as IPs, emails, and device identifiers. Promote the sensitive fields you need to govern into typed top-level columns, then apply masking there. Policies attach cleanly at the column level; leaving everything buried inside a VARIANT payload makes precise governance harder to enforce and audit.

```
-- Mask email if present
CREATE OR REPLACE MASKING POLICY mask_email AS (val STRING) RETURNS
STRING ->
  IFF(IS_ROLE_IN_SESSION('ROLE_PII_READER'), val, REGEXP_REPLACE(val,
  '(^.).+(@.+$)', '\\1***\\2'));

-- Promote sensitive fields needed for governance to top level columns
ALTER TABLE raw.events_json
  ADD COLUMN IF NOT EXISTS email STRING;
-- Populate during shaping or backfill existing rows once
UPDATE raw.events_json
SET email = payload:email::string
WHERE email IS NULL;
-- Attach masking to the top-level column
ALTER TABLE raw.events_json
  MODIFY COLUMN email SET MASKING POLICY mask_email;

-- Row-level isolation by tenant

CREATE OR REPLACE ROW ACCESS POLICY rap_tenant AS (tenant_id STRING)
RETURNS BOOLEAN ->
```

```
  EXISTS (
    SELECT 1
    FROM gov.role_tenants m
    WHERE m.role_name = CURRENT_ROLE()
      AND m.tenant_id = tenant_id
  );

ALTER VIEW bronze.events_shaped ADD ROW ACCESS POLICY rap_tenant ON
(tenant_id);
```

25.1.8 Performance That Stays Cheap

Events read fast when you **prune**. Keep scalar envelopes (event_date, tenant_id, event_type) and filter on them first. Avoid SELECT * in shared views. Cluster only after profiling shows broad scans with good predicates (tenant hot spots are a common case). For Kafka/file ingest, prefer **balanced, fewer files** (e.g., ~200 MB Parquet) over tiny ones; Snowpipe Streaming sidesteps file sizing entirely.

25.1.9 Observability and SLOs

Two signals keep you honest:

- **Ingest freshness:** MAX(received_ts) - MAX(event_ts) over the last hour per source; alert when it exceeds your target (e.g., >2 minutes for streaming, >10 for files).
- **Serving lag:** For Dynamic Tables or Tasks, the **last refresh time** vs. now; page when beyond SLA.

```
-- Freshness
SELECT src, AVG(DATEDIFF('second', event_ts, received_ts)) AS avg_lag_s
FROM raw.events_json
WHERE received_ts >= DATEADD('minute', -10, CURRENT_TIMESTAMP())
GROUP BY src;

-- DT health (recent errors)
SELECT NAME, ERROR_MESSAGE, LAST_REFRESH_TIME
```

```
FROM SNOWFLAKE.ACCOUNT_USAGE.DYNAMIC_TABLE_REFRESH_HISTORY
WHERE LAST_REFRESH_TIME >= DATEADD('hour', -1, CURRENT_TIMESTAMP())
  AND ERROR_MESSAGE IS NOT NULL;
```

25.1.10 Folding Real-Time into the Warehouse

The trick is **no special path** for "real-time" logic. Your sessionization, rollups, and UDF enrichments are plain SQL/Snowpark that run for both streaming slices and historical backfills. When a rule changes, you rebuild in a **candidate clone** and **SWAP**, just like any other release. That keeps clickstream and IoT work from sprouting a shadow stack—and keeps your answers fast, consistent, and explainable.

25.2 Near-Real-Time Dashboards

Dashboards feel "real-time" when two things hold: **fresh data arrives predictably**, and **queries return predictably**. In Snowflake, you get there by letting ingestion run continuously (Snowpipe/Snowpipe Streaming or tight micro-batches), shaping a **serving layer** built for pruning, and giving BI a **stable, cache-friendly surface**. The art is choosing small, repeatable refresh moves rather than one heroic query that re-computes the world every minute.

25.2.1 Serving Layer Built for Speed (Not Just Truth)

Keep a thin set of **read-optimized views/tables** that the dashboard hits directly. Model them to prune by **time + tenant/segment**, avoid wide `SELECT *`, and compute expensive joins once.

```
-- "Live" fact: last 60 minutes, prunable first, cheap later

CREATE OR REPLACE VIEW live.v_events_60m AS
SELECT tenant_id, event_ts, event_type, metric_1, metric_2
FROM silver.events_flat
WHERE event_ts >= DATE_TRUNC('minute', CURRENT_TIMESTAMP()) - INTERVAL '60
minutes'
  AND tenant_id = CURRENT_ACCOUNT()::string;
```

This is still a freshness-oriented window, not a cache-maximizing one. Using CURRENT_TIMESTAMP() or other time-relative expressions is acceptable when the goal is a live surface, but it reduces persisted-result reuse because the query is no longer fully deterministic. For tiles where cache reuse matters more than second-by-second freshness, use discrete time buckets or refresh a serving table on a schedule and query that stable surface instead.

For continuously updated aggregates, prefer **Dynamic Tables** with a small TARGET_LAG (e.g., 1–5 minutes) over ad hoc refresh SQL in many places.

```
CREATE OR REPLACE DYNAMIC TABLE live.sessions_5m
TARGET_LAG = '3 minutes'
WAREHOUSE = wh_realtime
AS
SELECT tenant_id,
       DATE_TRUNC('minute', session_start) AS t_min,
       COUNT(*) AS sessions
FROM gold.click_sessions
WHERE session_start >= DATEADD(hour,-2,CURRENT_TIMESTAMP())
GROUP BY 1,2;
```

When you need second-level freshness but simple math (counts, sums), **materialized views** on the hot partitions can also work; keep their definition narrow and prunable.

25.2.2 A Refresh Rhythm the UI Can Trust

Dashboards don't need nanosecond truth; they need a **promise**. Publish a tiny "heartbeat" and have the UI display it.

```
CREATE OR REPLACE VIEW live.v_freshness AS
SELECT
  'events'  AS feed,
  MAX(event_ts) AS newest_event_ts,
  MAX(received_ts) AS newest_ingest_seen_ts,
  CURRENT_TIMESTAMP() AS observed_ts
FROM raw.events_json;
```

Most BI tools can read this alongside the main tiles, so users see "updated 90s ago," which is often all they need to trust the page.

25.2.3 Stable SQL for Cache Hits

Result cache works best when the query is deterministic and the effective query shape stays stable. Expose routing parameters as binds rather than splicing ad hoc literals. Keep shared views deterministic where you want cache reuse. Time-relative filters such as CURRENT_TIMESTAMP() are appropriate for freshness windows, but they trade away persisted-result reuse; for cache-friendly dashboard tiles, prefer discrete time buckets or pre-refreshed serving tables.

```
-- Parameterized via BI tool: :p_tenant_id, :p_win_min

-- Freshness-first version: acceptable when cache reuse is not the goal
SELECT event_type, COUNT(*) AS n
FROM live.v_events_60m
WHERE tenant_id = :p_tenant_id
  AND event_ts >= DATE_TRUNC('minute', CURRENT_TIMESTAMP()) - INTERVAL '60
  minutes'
GROUP BY event_type;

-- Cache-friendlier version: query a pre-bucketed serving surface
SELECT event_type, COUNT(*) AS n
FROM live.events_by_minute
WHERE tenant_id = :p_tenant_id
  AND t_min >= :p_window_start_minute
  AND t_min <  :p_window_end_minute
GROUP BY event_type;
```

25.2.4 Warehouse Posture for Spiky Pages

A small, elastic lane is better than a large, idle one. Keep **XSMALL/SMALL** with AUTO_RESUME=TRUE, short AUTO_SUSPEND (60–120s), and **multi-cluster** for fan-out (e.g., MIN_CLUSTER_COUNT=1, MAX_CLUSTER_COUNT=3). If your tiles fire in bursts, multi-cluster absorbs the spike while preserving per-query latency.

```
ALTER WAREHOUSE wh_realtime SET
  WAREHOUSE_SIZE='SMALL',
  AUTO_SUSPEND=120,
  AUTO_RESUME=TRUE,
  MIN_CLUSTER_COUNT=1,
  MAX_CLUSTER_COUNT=3,
  SCALING_POLICY='STANDARD';
```

Tag BI sessions so you can separate their behavior from ELT:

```
ALTER SESSION SET QUERY_TAG='app=bi;surface=realtime;page=overview';
```

25.2.5 Handling Late and Out-of-Order Events

Encode a **watermark** in serving logic: accept late arrivals up to *N* minutes into the past, and avoid rewriting beyond that horizon. For aggregates, recompute only the open windows; closed windows are treated as immutable.

```
-- Open window is last 10 minutes; older windows read from stable, pre-
aggregated table
WITH open AS (
  SELECT DATE_TRUNC('minute', event_ts) AS t_min, COUNT(*) AS n
  FROM bronze.events_shaped
  WHERE event_ts >= DATEADD(minute,-10,CURRENT_TIMESTAMP())
  GROUP BY 1
)
SELECT * FROM stable.events_by_minute
WHERE t_min < DATEADD(minute,-10,CURRENT_TIMESTAMP())
UNION ALL
SELECT * FROM open;
```

25.2.6 BI Model Hints That Matter

Keep the **semantic layer** thin and consistent with the SQL surface. Define canonical measures (e.g., "Active Users 5-minute," "Sessions 30-minute") once in a shared model, so every dashboard uses the same math. Pre-join small, stable dimensions (device/OS names, product taxonomy) into serving views to avoid N+1 joins from the BI tool.

25.2.7 Security at Read: So "One Surface Fits Many"

Row Access Policies (tenants, regions) and masking (emails, device IDs) let you publish **one set of views** to both internal and partner dashboards safely. Because policies evaluate at query time, you don't need separate "anonymized" tables for live use.

```
CREATE OR REPLACE ROW ACCESS POLICY rap_tenant AS (tenant_id STRING)
RETURNS BOOLEAN ->
  EXISTS (
    SELECT 1
    FROM gov.role_tenants t
    WHERE t.role_name = CURRENT_ROLE()
      AND t.tenant_id = tenant_id
  );

ALTER VIEW live.v_events_60m ADD ROW ACCESS POLICY rap_tenant ON
(tenant_id);
```

25.2.8 Observability and SLOs You Can Page On

Operate dashboards like a product: define **freshness** and **latency** SLOs, back them with one or two SLIs, and alert from SQL.

- **Freshness SLI:** `DATEDIFF(second, MAX(event_ts), CURRENT_TIMESTAMP())` over the last N minutes per feed.
- **Latency SLI:** P95 of `QUEUED_*` + `EXECUTION_TIME` for the BI warehouse over rolling 10-minute windows.

Keep a tiny Task that writes these to `ops.realtime_sli` and emails (or pushes to your incident tool) only on **new** breaches.

25.2.9 Common Pathologies and Small Fixes

- Tiles flicker between fast/slow → queries vary (no cache). Normalize parameters, remove `SELECT *`, and project only used columns.
- High queued time → fan-out spikes. Raise `MAX_CLUSTER_COUNT` or stagger tiles; ensure `AUTO_SUSPEND` isn't too aggressive.

- High execution time with low bytes scanned → expensive UDFs or over-windowing. Precompute heavy functions; reduce window sizes; switch to Dynamic Tables for rolling aggregates.
- Bytes scanned per tile huge → missing envelopes. Add `event_date`, `tenant_id`, and `event_type` filters first, then JSON fields.

Real-time dashboards don't need a separate stack. Land events continuously, shape a prunable serving layer, refresh with Dynamic Tables or tight micro-batches, and give BI stable SQL so result cache can help. With a small elastic warehouse and a visible freshness heartbeat, the experience feels live—and stays calm to run.

CHAPTER 26

Data Products and Sharing at Scale

At scale, your warehouse stops being a pile of tables and becomes a **portfolio of products**. Each product has a clear purpose, a stable interface, a small set of promises (freshness, availability, schema), and an owning team that operates it. Snowflake makes this practical because the "API surface" is just SQL: secure views, tags, policies, and a handful of small operational tables (contracts, release notes, incidents). You don't copy data to package it; you **name** and **govern** it so others can adopt it without ceremony. The rest is muscle memory—version, document, observe, and deprecate like software.

26.1 Internal Data Products

An internal data product is a **named dataset with a contract**. Consumers know what it is, how fresh it will be, which columns they can rely on, and who to ask when something looks off. Providers can change internals freely as long as the interface and promises hold. That separation keeps platform teams fast and consumers calm.

26.1.1 The Product Surface

Publish **secure views** that project a stable schema from private base tables. Put them in a product namespace and attach tags so governance and chargeback "see" them immediately.

```
-- Product namespace
CREATE SCHEMA IF NOT EXISTS prod_sales.dataset;

-- Stable interface (secure view)
CREATE OR REPLACE SECURE VIEW prod_sales.dataset.v_orders_v1 AS
```


M. Hander, *Snowflake Data Warehouse Engineering*, https://doi.org/10.1007/979-8-8688-2628-3_26

```
SELECT order_id, order_dt, customer_id, amount, channel
FROM prod_sales.core.fact_orders
WHERE order_dt >= '2023-01-01';

-- Ownership and classification
CREATE TAG IF NOT EXISTS owner ALLOWED_VALUES ('sales_ops','fin_
ops','platform');
CREATE TAG IF NOT EXISTS data_domain ALLOWED_VALUES ('sales','finance',
'marketing');
CREATE TAG IF NOT EXISTS classification;

ALTER VIEW prod_sales.dataset.v_orders_v1
  SET TAG owner='sales_ops', data_domain='sales';
```

For products that may later be shared outside the team or account, attach masking to the base-table columns rather than only to the interface view. That keeps the protection consistent when the product is published through secure views or shares.

```
CREATE OR REPLACE MASKING POLICY mask_email AS (val STRING) RETURNS
STRING ->
  IFF(IS_ROLE_IN_SESSION('ROLE_PII_READER'), val, REGEXP_REPLACE(
  val,'(^.).+(@.+$)','\\1***\\2'));

ALTER TABLE prod_sales.core.dim_customer
  MODIFY COLUMN email SET MASKING POLICY mask_email;
```

26.1.2 Contracts That Consumers Can Query

Contracts live as rows next to the product. They say **what** the interface is (columns/types), **how fresh** it should be, and **how long** the current version will be supported.

```
CREATE SCHEMA IF NOT EXISTS prod_sales.meta;

CREATE OR REPLACE TABLE prod_sales.meta.contract (
  object_name STRING,          -- 'V_ORDERS_V1'
  version     STRING,          -- 'v1.6.0'
  cadence     STRING,          -- 'hourly at :05'
  max_lag_min NUMBER,          -- 15
```

```
  support_until DATE,          -- '2026-06-30'
  owner_email STRING
);

INSERT INTO prod_sales.meta.contract
VALUES ('V_ORDERS_V1','v1.6.0','hourly at :05',15,'2026-06-30','salesops@
example.com');

CREATE OR REPLACE TABLE prod_sales.meta.release_notes (
  dt DATE, object_name STRING, note STRING, deprecates_on DATE
);

INSERT INTO prod_sales.meta.release_notes
VALUES (CURRENT_DATE(),'V_ORDERS_V1','Added column channel (non-
breaking)', NULL);
```

Consumers can self-serve status:

```
SELECT * FROM prod_sales.meta.contract WHERE object_name='V_ORDERS_V1';
SELECT * FROM prod_sales.meta.release_notes WHERE object_name='V_ORDERS_V1'
ORDER BY dt DESC;
```

26.1.3 Versioning Without Breakage

Additions land in-place on V1. Breaking changes become V2, with an overlap window and a router view for a smooth cutover.

```
CREATE OR REPLACE SECURE VIEW prod_sales.dataset.v_orders_v2 AS
SELECT order_id, order_dt, customer_id, amount, channel, currency
FROM prod_sales.core.fact_orders;

-- Router the BI layer uses
CREATE OR REPLACE VIEW prod_sales.dataset.v_orders AS
SELECT * FROM prod_sales.dataset.v_orders_v1;  -- later switch to v2

-- Announce deprecations
INSERT INTO prod_sales.meta.release_notes
VALUES (CURRENT_DATE(),'V_ORDERS_V1','Deprecates in 60 days; use V2',
DATEADD(day,60,CURRENT_DATE()));
```

26.1.4 Operating the Promises (SLOs As SQL)

Publish a tiny freshness heartbeat and a usage view; alert from these rather than guess. For freshness, measure an ingestion timestamp such as loaded_at or received_at on the base table, not a business date from the consumer-facing view.

Business dates such as order_dt can legitimately be old even when ingestion is healthy. Operational freshness should measure when data arrived or was loaded, not the semantic date carried by the row.

```
CREATE OR REPLACE VIEW prod_sales.meta.v_freshness AS
SELECT 'V_ORDERS_V1' AS object_name,
       MAX(loaded_at) AS newest_ingest_ts,
       DATEDIFF(minute, MAX(loaded_at), CURRENT_TIMESTAMP()) AS lag_min
FROM prod_sales.core.fact_orders;

CREATE OR REPLACE VIEW prod_sales.meta.v_usage_7d AS
SELECT f.value:objectName::STRING AS object_name,
       a.ROLE_NAME,
       COUNT(*) AS reads
FROM SNOWFLAKE.ACCOUNT_USAGE.ACCESS_HISTORY a,
     LATERAL FLATTEN(input => a.DIRECT_OBJECTS_ACCESSED) f
WHERE a.QUERY_START_TIME >= DATEADD(day,-7,CURRENT_TIMESTAMP())
GROUP BY 1,2;
```

DIRECT_OBJECTS_ACCESSED is a semi-structured array of objects in ACCESS_HISTORY, not a single scalar field. Flatten it first, then read value:objectName.

SRE tasks read v_freshness and page on lag_min > max_lag_min. Product owners review v_usage_7d weekly to see which shared objects are actually being read before cutting over. Because usage is derived from flattened access history, the result reflects real object access rather than a guessed dependency list.

Operational health checks should read from the base operational surface, not from the consumer interface, unless the SLO is specifically about the interface itself. Freshness usually belongs to ingestion timestamps on the source table; schema and adoption signals usually belong to the product views. That separation keeps the product contract clean and the operating signals meaningful.

26.1.5 Entitlements and Least Privilege

Consumers shouldn't need admin roles to adopt products. Grant access at the product boundary and use **Row Access Policies** to scope by region, tenant, or legal entity if the same product serves multiple audiences.

```
GRANT USAGE ON SCHEMA prod_sales.dataset TO ROLE ROLE_BI;
GRANT SELECT ON VIEW prod_sales.dataset.v_orders_v1 TO ROLE ROLE_BI;

CREATE OR REPLACE ROW ACCESS POLICY rap_region AS (region STRING)
RETURNS BOOLEAN ->
  EXISTS (
    SELECT 1
    FROM gov.role_regions m
    WHERE m.role_name = CURRENT_ROLE()
      AND m.region = region
  );
```

For shared products and BI access paths, role-based row scoping is usually safer than user-based scoping because multiple end users may operate through one Snowflake login.

```
ALTER VIEW prod_sales.dataset.v_orders_v1 ADD ROW ACCESS POLICY rap_region
ON (region);
```

26.1.6 Cost Visibility and Internal Chargeback

Tag objects with cost centers and report compute via QUERY_TAG or warehouse tags. A lightweight chargeback is a view over WAREHOUSE_METERING_HISTORY joined to tags.

```
CREATE TAG IF NOT EXISTS cost_center;

ALTER WAREHOUSE wh_bi SET TAG cost_center='sales_bi';

-- Rollup
SELECT w.WAREHOUSE_NAME, t.TAG_VALUE AS cost_center, TO_DATE(w.START_TIME)
AS day, SUM(w.CREDITS_USED) AS credits
FROM SNOWFLAKE.ACCOUNT_USAGE.WAREHOUSE_METERING_HISTORY w
LEFT JOIN SNOWFLAKE.ACCOUNT_USAGE.OBJECT_TAGS t
```

```
  ON t.OBJECT_NAME = w.WAREHOUSE_NAME AND t.TAG_NAME='COST_CENTER'
WHERE w.START_TIME >= DATEADD(day,-30,CURRENT_TIMESTAMP())
GROUP BY 1,2,3;
```

26.1.7 Promotion and Rollback As Metadata

Build in a **candidate clone**, run contract tests, then swap. The interface (`v_orders`) remains stable; internals change atomically.

```
CREATE OR REPLACE DATABASE prod_sales_candidate CLONE prod_sales;
/* apply migrations, compute, tests */
ALTER DATABASE prod_sales_candidate SWAP WITH prod_sales;
```

If something misbehaves, swap back. Because you didn't hand-edit prod, rollback is seconds—not a weekend.

26.1.8 Documentation That Writes Itself

Generate Markdown from `INFORMATION_SCHEMA` + tags, and publish to an internal stage. Consumers click from the catalog to the docs that always match reality.

```
COMMENT ON VIEW prod_sales.dataset.v_orders_v1 IS 'Canonical orders view:
one row per order; hourly refresh; see meta.contract for SLOs.';
```

26.1.9 When Products Span Regions or Org Units

Keep **one product per region** fed by database replication from the owner's primary. Publish the same interface in each region to avoid cross-region reads; ownership and contracts live with the producer.

Define internal datasets as **products** with a stable SQL interface, explicit SLOs, and small, queryable contracts. Add in-place, version when you must, operate from a couple of views (freshness, usage), and promote by swapping—not by editing prod. With that discipline, dozens of teams can share one platform without stepping on each other, and your "data mesh" stays fast, safe, and boring in the best way.

26.2 External Monetization via Marketplace

Selling data is selling **trust on tap**. Buyers want a stable interface, clear freshness, predictable cost, and no surprises when they wire dashboards and models to your product. Snowflake's Marketplace gives you the storefront; your craft is packaging a **thin, durable surface**, operating it like software, and proving value with usage signals and samples.

26.2.1 Product Shape Buyers Can Adopt on Day One

Keep raw tables private and publish **secure views** that read naturally: dates to prune, entity keys to slice, consistent units, and no `SELECT *`. Put everything in a product schema, and attach tags so governance travels.

```
CREATE SCHEMA IF NOT EXISTS provider.prod_weather;
CREATE OR REPLACE SECURE VIEW provider.prod_weather.v_daily_v1 AS
SELECT
  as_of_date, iso_country, region, station_id,
  temp_c, precip_mm, wind_kph
FROM provider.core.weather_daily
WHERE as_of_date >= '2020-01-01';

ALTER VIEW provider.prod_weather.v_daily_v1
  SET TAG owner='data_products', classification='public.commercial';
```

Buyers shouldn't guess freshness. Publish a **heartbeat** alongside the data and keep a tiny contract table in the same schema.

```
CREATE OR REPLACE VIEW provider.prod_weather.v_freshness AS
SELECT MAX(as_of_date) AS newest_dt, CURRENT_TIMESTAMP() AS observed_at
FROM provider.prod_weather.v_daily_v1;

CREATE OR REPLACE TABLE provider.prod_weather.contract AS
SELECT 'V_DAILY_V1' AS object_name, 'daily 02:00 UTC' AS cadence,  'v1.3.0'
AS version, 14 AS max_lag_days;
```

26.2.2 Listings, Pricing, and Access Models

Marketplace supports free, trial, and paid offerings, delivered either as **direct shares** to a consumer's account or via **Reader Accounts** you operate for buyers without Snowflake. Choose one path per listing and optimize the surface for it.

Model	Best for	Compute pays	Notes that matter
Direct share (free/paid)	Existing Snowflake customers	**Consumer**	Lowest friction for power users; publish per-region to avoid cross-region reads
Reader account	Buyers without Snowflake	**Provider**	You meter and cap the BI warehouse; great for trials and simple dashboards
Trial → paid	Evaluation then contract	Trial: Provider (reader), Paid: Consumer	Keep one schema; promote consumers by switching share/entitlements

Price to the *decision*. Buyers prefer predictable tiers (rows/regions included) over opaque usage. For usage-based paid listings, align your product to **prunable queries** so buyers can control their spend without opening tickets.

26.2.3 Readiness Checklist (What You Do Before "Publish")

- **Replicate** your product database to each region you'll list in; point listings to the **regional replica** to eliminate egress and latency.
- Run a **consumer drill**: in a sandbox account, accept the share, query with a least-privilege role, and open dashboards. Fix anything that assumes internal roles, stages, or session settings.
- Add a **sample pack**: 30–90 days of data or a small geographic slice, delivered as a second schema (`prod_weather_sample`) with the same column contracts.
- Write **release notes** (one view) and keep versioned secure views (…`_V1`, later …`_V2`) to avoid breaking changes. Deprecate with dates, not emails.

26.2.4 Operating the Product Like Software

Change happens behind the scenes; the interface remains calm. Add columns in place; ship breaking changes as V2 with an overlap window and a router view for internal use.

```
CREATE OR REPLACE SECURE VIEW provider.prod_weather.v_daily_v2 AS
SELECT as_of_date, iso_country, region, station_id, temp_c, precip_mm,
wind_kph, dewpoint_c
FROM provider.core.weather_daily;

-- Router you use internally and flip later
CREATE OR REPLACE VIEW provider.prod_weather.v_daily AS
SELECT * FROM provider.prod_weather.v_daily_v1;
```

Promote via **clone + SWAP**, so outages are measured in seconds, and roll back just as fast.

```
CREATE OR REPLACE DATABASE provider_prod_candidate CLONE provider_prod;
-- apply migrations, recompute, validate
ALTER DATABASE provider_prod_candidate SWAP WITH provider_prod;
```

26.2.5 Contracts, SLAs, and Incident Hygiene

Buyers trust numbers they can **check with SQL**. Keep two tiny artifacts in the product schema:

- `contract` with cadence, version, support window.
- `release_notes` with dates and changes.

During an incident, write rows to a `service_log` view in the product schema ("freshness exceeded lag due to upstream backlog; mitigation underway"), and fix in a clone. When resolved, your log shows start/stop and impact without a phone call.

```
CREATE OR REPLACE TABLE provider.prod_weather.service_log (
  ts TIMESTAMP_NTZ, severity STRING, summary STRING, status STRING
);
```

26.2.6 Security and Compliance Travel with the Share

Mask PII and scope rows before you publish. For shared data products, apply masking at the base-table column level before the share is created, then project those protected columns through secure views. Consumers inherit the protection from the shared object, but the strongest pattern is to enforce masking on the underlying columns rather than relying only on view-level attachment.

```
CREATE OR REPLACE MASKING POLICY mask_email AS (val STRING) RETURNS
STRING ->
  IFF(IS_ROLE_IN_SESSION('ROLE_PROVIDER_PII'), val, REGEXP_REPLACE(
  val,'(^.).+(@.+$)','\\1***\\2'));

ALTER TABLE provider.core.customers
  MODIFY COLUMN email SET MASKING POLICY mask_email;

CREATE OR REPLACE ROW ACCESS POLICY rap_region AS (region STRING) RETURNS
BOOLEAN ->
  region IN ('EU','US');    -- example: only export allowed regions
ALTER VIEW provider.prod_weather.v_daily_v1 ADD ROW ACCESS POLICY rap_
region ON (region);
```

> Dynamic Data Masking is column-level security in Snowflake and can be applied to table or view columns. For externally shared products, base-table attachment is the more durable default because it keeps the protection anchored to the underlying data surface.

26.2.7 Observability and Monetization Signals

Success is adoption and retained usage. Publish a private provider dashboard fed by flattened ACCESS_HISTORY, organization usage rollups, and listing-specific usage views where available, so you can distinguish broad account activity from actual listing consumption.

- **Reads by listing and region.** Identify hot consumers and cohorts that stall after trials.
- **Query shapes.** Catch broad scans and provide pruning guidance; ship a "best practices" page in the listing.

- **Incident memory.** Join buyer complaints to your `service_log` rows, so you can point to precise minutes and fixes.

```
SELECT f.value:objectName::STRING AS object_name,
       COUNT(*) AS reads
FROM SNOWFLAKE.ACCOUNT_USAGE.ACCESS_HISTORY a,
     LATERAL FLATTEN(input => a.DIRECT_OBJECTS_ACCESSED) f
WHERE a.QUERY_START_TIME >= DATEADD(day,-7,CURRENT_TIMESTAMP())
  AND f.value:objectName::STRING LIKE 'PROVIDER.PROD_WEATHER.%'
GROUP BY 1
ORDER BY reads DESC;
```

For Marketplace listings, also consider `DATA_SHARING_USAGE.LISTING_ACCESS_HISTORY` as the purpose-built source for consumer usage of listing-backed shares. Use flattened `ACCESS_HISTORY` when you want a broader account-level lens and listing usage history when you want listing-specific adoption and object access signals.

For reader accounts, cap spend with a **resource monitor**, and keep warehouses small and elastic.

```
CREATE OR REPLACE RESOURCE MONITOR rm_readers
  WITH CREDIT_QUOTA = 200
  FREQUENCY = MONTHLY
  TRIGGERS ON 80 PERCENT DO NOTIFY ON 100 PERCENT DO SUSPEND;
ALTER WAREHOUSE wh_reader_weather SET RESOURCE_MONITOR = rm_readers;
```

26.2.8 Region Strategy and Data Residency

List **per region** and replicate your product database to match demand. If regulations restrict export, enforce residency with row policies, and only list in compliant regions. For multi-cloud buyers, keep identical schemas and versions across regions, so documentation and release notes don't fork.

26.2.9 Contracts and Legal in the Product, Not Just PDFs

Put the short-form **data contract** right next to the views—columns, semantics, retention, and usage restrictions (e.g., "no model training without license"). Link your full license in the listing UI, but keep the operational promises queryable:

```
CREATE OR REPLACE VIEW provider.prod_weather.v_contract AS
SELECT 'V_DAILY_V1'   AS object_name,
       'daily 02:00 UTC' AS cadence,
       14 AS max_lag_days,
       'redistribution: prohibited; model training: permitted with
       commercial license' AS terms_summary;
```

26.2.10 Trials That Convert

Give buyers a **working sample** and a path to value in ten minutes:

- A sample schema wired to a quick-start notebook and 3–4 SQL examples (freshness check, "your city last seven days," "top stations").
- A short "how to filter cheaply" section so their first dashboard feels fast.
- Clear upgrade steps: same view names appear in the paid share; they just include more regions/history.

A monetizable data product is a **stable SQL interface** backed by visible promises and boring operations. Keep governance on the producer side, replicate per region, publish tiny contract and freshness views, and promote with swaps. Buyers get immediate value and predictable performance; you get adoption you can observe—and revenue you can forecast.

CHAPTER 27

GenAI-Ready Warehousing

Generative AI doesn't replace the warehouse; it **leans on it**. The best answers come from models that see *fresh, permissioned,* and *well-shaped* data. In practice, that means two things inside Snowflake: a **feature layer** that feeds models with point-in-time correct signals and an **embedding layer** that lets you retrieve context (documents, events, product specs) by meaning rather than exact keywords. Both live comfortably next to facts and dims, use the same governance, and ship through the same CI/CD you already run.

The pattern is simple: treat ML signals and embeddings as **first-class datasets** with owners, contracts, and SLOs. Build them from reproducible transforms, refresh them on a named cadence, and publish a small, stable interface for applications (batch scoring, APIs, chat, search). Because it's "just data," your usual playbooks—Time Travel, cloning, contracts, masking—apply unchanged.

27.1 Feature Stores and Embeddings in Snowflake

A feature store answers: *What signals describe an entity at a given time, and how do I serve the same numbers to training and production without drift?*

An embeddings layer answers: *What pieces of content are most relevant to this query, semantically, within this user's permissions and freshness window?* You can keep both inside Snowflake with minimal ceremony.

M. Hander, *Snowflake Data Warehouse Engineering*, https://doi.org/10.1007/979-8-8688-2628-3_27

27.1.1 A Small, Honest Feature Store

Keep three surfaces: **registry** (what exists), **offline features** (batched, point-in-time correct), and **serving views** (stable shapes for inference). Train and serve from the same tables so you avoid skew.

```
-- 1) Registry: declare ownership, grain, refresh promise
CREATE SCHEMA IF NOT EXISTS ml;

CREATE OR REPLACE TABLE ml.feature_registry (
  feature_name STRING,    -- e.g., 'user_7d_txn_cnt'
  entity       STRING,    -- 'user_id'
  owner        STRING,   -- 'growth-ml'
  version      STRING,   -- 'v1'
  cadence      STRING,   -- 'hourly'
  description  STRING
);

-- 2) Offline store (wide feature table; entity + as_of_ts)
CREATE OR REPLACE TABLE ml.user_features_offline (
  user_id STRING,
  as_of_ts TIMESTAMP_NTZ,
  txn_cnt_7d NUMBER,
  amt_sum_30d NUMBER(18,2),
  last_seen_minutes NUMBER,
  plan STRING,
  -- keep the envelope columns first for pruning
  PRIMARY KEY (user_id, as_of_ts)
);
```

Build features as of time (no leakage). Window on event time, not load time; land into the offline table at deterministic "ticks." Generate those ticks with a fixed ROWCOUNT, not TIMELIMIT, so the feature refresh produces a stable and reproducible set of as-of timestamps.

```
-- Idempotent refresh for the last 2 days of ticks
MERGE INTO ml.user_features_offline t
USING (
```

```
WITH ticks AS (
  SELECT DATEADD(
           'hour',
           -ROW_NUMBER() OVER (ORDER BY SEQ4()),
           DATE_TRUNC('hour', CURRENT_TIMESTAMP())
         ) AS as_of_ts
  FROM TABLE(GENERATOR(ROWCOUNT => 48))   -- last 48 hourly ticks
)
SELECT *
FROM ticks
WHERE as_of_ts >= DATEADD(day, -2, CURRENT_TIMESTAMP());

  SELECT
    u.user_id,
    k.as_of_ts,
    COUNT_IF(e.event_ts >= DATEADD(day,-7, k.as_of_ts))      AS txn_cnt_7d,
    SUM(IFF(e.event_ts >= DATEADD(day,-30, k.as_of_ts), e.amount, 0))
    AS amt_sum_30d,
    DATEDIFF('minute', MAX(e.event_ts), k.as_of_ts)   AS last_seen_minutes,
    MAX(u.plan)                                       AS plan
  FROM core.users u
  LEFT JOIN core.events e
    ON e.user_id = u.user_id
  JOIN ticks k
  GROUP BY 1,2
) s
ON t.user_id = s.user_id AND t.as_of_ts = s.as_of_ts
WHEN MATCHED THEN UPDATE SET
  txn_cnt_7d = s.txn_cnt_7d,
  amt_sum_30d = s.amt_sum_30d,
  last_seen_minutes = s.last_seen_minutes,
  plan = s.plan
WHEN NOT MATCHED THEN INSERT VALUES (s.user_id, s.as_of_ts, s.txn_cnt_7d,
s.amt_sum_30d, s.last_seen_minutes, s.plan);
```

Feature ticks should be deterministic. Use a fixed set of hourly boundaries and rebuild a bounded recent horizon, rather than relying on a time-based row generator whose output depends on execution speed. That keeps feature refreshes reproducible and makes backfills line up exactly with training windows.

Serve features through a stable view that selects the **latest tick ≤ request time**. Applications pass a timestamp (or you fix one for batch).

```
CREATE OR REPLACE VIEW ml.v_user_features_serving AS
SELECT f1.*
FROM ml.user_features_offline f1
QUALIFY ROW_NUMBER() OVER (
  PARTITION BY user_id
  ORDER BY as_of_ts DESC
) = 1;  -- "now" serving; for backtests, filter as_of_ts <= :request_ts
```

This keeps training and serving consistent: training samples join to …`_offline AS OF t`, serving reads `v_…_serving` at "now" or at a supplied time.

Contract and freshness stay queryable:

```
CREATE OR REPLACE VIEW ml.v_feature_freshness AS
SELECT 'user_features' AS object_name,
       MAX(as_of_ts) AS newest_tick,
       DATEDIFF('minute', MAX(as_of_ts), CURRENT_TIMESTAMP()) AS lag_min
FROM ml.user_features_offline;
```

27.1.2 Embeddings As First-Class Data

An embedding store is just a **chunk table** plus **metadata**. You chunk content, embed each chunk, store a vector and a small set of filters (tenant/entity, document type, created/validity dates), and query with a similarity search constrained by those filters.

```
CREATE SCHEMA IF NOT EXISTS rag;

-- Core chunk store
CREATE OR REPLACE TABLE rag.chunks (
  doc_id STRING,
  chunk_id STRING,
  tenant_id STRING,
```

```
  created_at TIMESTAMP_NTZ,
  valid_from TIMESTAMP_NTZ,
  valid_to   TIMESTAMP_NTZ,
  source_uri STRING,
  text STRING,
  embedding ARRAY,    -- or VECTOR(FLOAT, 1536)
  PRIMARY KEY (tenant_id, chunk_id)
);
-- Indexable filters live in scalar columns: tenant_id, created_at, doc
type, etc.
```

Store embeddings in Snowflake's native VECTOR type whenever possible and use native similarity functions such as VECTOR_COSINE_SIMILARITY. Fall back to arrays only when you have a hard compatibility reason to do so.

Ingestion pipeline is deterministic: chunk → embed → upsert → mark superseded chunks with valid_to. If you call an external model, wrap it as an External Function or Snowpark job, and cache by hash to avoid re-embedding unchanged text.

```
-- Upsert with idempotence
MERGE INTO rag.chunks t
USING staged.new_chunks s
ON t.tenant_id = s.tenant_id AND t.chunk_id = s.chunk_id
WHEN MATCHED THEN UPDATE SET
  text = s.text, embedding = s.embedding, created_at = s.created_at, valid_
  from = s.valid_from, valid_to = s.valid_to, source_uri = s.source_uri
WHEN NOT MATCHED THEN INSERT (doc_id, chunk_id, tenant_id, created_at,
valid_from, valid_to, source_uri, text, embedding)
VALUES (s.doc_id, s.chunk_id, s.tenant_id, s.created_at, s.valid_from,
s.valid_to, s.source_uri, s.text, s.embedding);
```

Similarity search combines vector distance with governance filters. Prefer Snowflake's native VECTOR type and similarity functions from the start. They are simpler, clearer, and more reliable than custom array math UDFs, and they align with the same retrieval pattern used later in this chapter.

```
-- Example cosine similarity in SQL when embedding stored as array

-- Prefer Snowflake's native VECTOR type and similarity functions
CREATE OR REPLACE TABLE rag.chunks (
  doc_id STRING,
  chunk_id STRING,
  tenant_id STRING,
  created_at TIMESTAMP_NTZ,
  valid_from TIMESTAMP_NTZ,
  valid_to   TIMESTAMP_NTZ,
  source_uri STRING,
  text STRING,
  embedding VECTOR(FLOAT, 1536),
  PRIMARY KEY (tenant_id, chunk_id)
);

-- Top-k with filters and watermark

WITH candidates AS (
  SELECT
    tenant_id,
    chunk_id,
    text,
    source_uri,
    VECTOR_COSINE_SIMILARITY(embedding, :query_embedding) AS sim
  FROM rag.chunks
  WHERE tenant_id = :tenant
    AND CURRENT_TIMESTAMP() BETWEEN valid_from AND COALESCE(valid_to,
    '9999-12-31')
    AND created_at >= DATEADD(day, -90, CURRENT_TIMESTAMP())
)
SELECT *
FROM candidates
ORDER BY sim DESC
LIMIT 10;
```

RAG assembly (retrieve-then-generate) stays in Snowflake too: gather the top chunks, enforce masking/row policies, and hand the model a compact prompt with citations. Store the conversation and retrieved chunk IDs for audit.

```
CREATE OR REPLACE TABLE rag.conversations (
  conv_id STRING, ts TIMESTAMP_NTZ, user_name STRING, tenant_id STRING,
  question STRING, answer STRING, chunks VARIANT
);
```

27.1.3 Serving Patterns That Stay Safe

Keep a **thin API surface**—stable views or procedures that apps call—so you can swap implementations and enforce governance centrally.

- **Features:** `CALL ml.sp_get_user_features(user_id, as_of_ts)` returns a JSON object of features.
- **Embeddings:** `CALL rag.sp_search(tenant_id, query_text, k)` returns chunk IDs and snippets.

This isolates apps from table shapes and lets you add caching, timeouts, and query tagging (`QUERY_TAG='app=chat;tenant=...'`) without touching clients.

27.1.4 Governance, PII, and Row-Level Isolation

Features and embeddings often contain PII. Use the same tag-aware masking and Row Access Policies as the rest of your warehouse.

```
-- Mask emails inside features
CREATE OR REPLACE MASKING POLICY mask_email AS (val STRING) RETURNS
STRING ->
  IFF(IS_ROLE_IN_SESSION('ROLE_PII_READER'), val, REGEXP_REPLACE(
  val,'(^.).+(@.+$)','\\1***\\2'));

ALTER TABLE ml.user_features_offline
  MODIFY COLUMN plan SET MASKING POLICY mask_email;  -- example if plan
  contains PII-like strings
```

```
-- Tenant isolation for embeddings

CREATE OR REPLACE ROW ACCESS POLICY rap_tenant AS (tenant_id STRING)
RETURNS BOOLEAN ->
  EXISTS (
    SELECT 1
    FROM gov.role_tenants m
    WHERE m.role_name = CURRENT_ROLE()
      AND m.tenant_id = tenant_id
  );

ALTER TABLE rag.chunks ADD ROW ACCESS POLICY rap_tenant ON (tenant_id);
```

Because policies evaluate at read, the **same** tables safely power internal notebooks, batch jobs, and multi-tenant applications.

27.1.5 Operating and Testing Like Everything Else

Treat ML tables like products:

- **Contracts**: A tiny table with feature names, types, cadence, owners, and max lag.
- **Freshness**: One view per surface (newest `as_of_ts`, newest chunk `created_at`), with a Task that alerts when lag exceeds the promise.
- **Reproducibility**: Log training datasets (entity list + `as_of_ts` ranges + code hash), and keep them recreatable with Time Travel; store model versions and artifact URIs in a small registry table.
- **CI**: Contract tests that assert "no feature disappeared," "types match," "no PII in unmasked columns," and "embedding count equals chunk count for active docs" before promotion.

A tiny registry sketch:

```
CREATE OR REPLACE TABLE ml.model_registry (
  model_name STRING, version STRING, created_at TIMESTAMP_NTZ,
  training_query STRING, feature_list ARRAY, artifact_stage STRING,
  metrics VARIANT
);
```

27.1.6 Performance and Cost Notes

- **Prune first**: Always filter on time and tenant/entity before any vector math. Keep `as_of_ts`, `tenant_id`, and other envelopes as scalar columns at the front of your tables.
- **Batch embeds**: Embed per chunk in batches; cache by SHA-1 of text to avoid recomputation; only re-embed changed text.
- **Use small elastic lanes**: Short-running, many-small queries benefit from SMALL/XSMALL warehouses with multi-cluster for bursts.
- **Keep chunks compact**: For text, 200–800 tokens per chunk typically balances retrieval accuracy with index size; store the **cleaned** text you actually embed.

Features and embeddings become reliable when you treat them like **warehouse citizens**—with contracts, envelopes for pruning, point-in-time discipline, and the same governance as any mart. Build once for training and serving, search with meaning under your policies, and promote with clones like any other change. Then GenAI is an application concern, not a parallel stack to keep alive.

27.2 Retrieval Patterns and Secure Data Access

Retrieval-augmented generation (RAG) in Snowflake should feel like any other analytic pattern: filter decisively, compute the smallest necessary work, and let governance ride along at read time. The core loop is short. A user asks something; you translate the question into a vector; you fetch the **right** chunks with tight filters; you assemble a prompt with citations; and you let a model answer. The hard parts are not the dot products—they're **isolation, freshness, and determinism**. This section shows how to wire those concerns into the retrieval itself so you don't bolt on security later.

27.2.1 Make Filtering Win Before Similarity

Vectors are expensive if you point them at the whole world. Put envelope columns on every chunk—`tenant_id`, `created_at`, `doc_type`, `valid_from/valid_to`, and often a coarse **topic** or **product** tag. Then run a two-stage plan: eliminate most rows with scalar filters that prune micro-partitions, and do similarity only within that slice. This keeps latency predictable and spend tame.

```
-- Query-time parameters (bound by app): :tenant, :topic, :since_days, :k
WITH base AS (
  SELECT doc_id, chunk_id, text, source_uri, embedding
  FROM rag.chunks
  WHERE tenant_id = :tenant
    AND (:topic IS NULL OR topic = :topic)
    AND created_at >= DATEADD(day, -:since_days, CURRENT_TIMESTAMP())
    AND CURRENT_TIMESTAMP() BETWEEN valid_from AND COALESCE(valid_to,
    '9999-12-31')
),
scored AS (
  SELECT doc_id, chunk_id, text, source_uri,
         VECTOR_COSINE_SIMILARITY(embedding, :query_embedding) AS sim
  FROM base
)
SELECT * FROM scored
ORDER BY sim DESC
LIMIT :k;
```

When the corpus grows, keep the same filters and lean on native vector indexes; the shape of the SQL doesn't change.

27.2.2 Retrieval That Is Policy-Aware by Construction

A secure RAG pipeline doesn't ask the application to remember who can see what. It **pushes the decision into Snowflake** with Row Access Policies (RAP) and masking bound to tags. The retrieval query remains simple; the policies gate the rows and redact the fields.

```
-- Tenant & region isolation

CREATE OR REPLACE ROW ACCESS POLICY rap_tenant_region
AS (tenant_id STRING, region STRING)
RETURNS BOOLEAN ->
```

```
  EXISTS (
    SELECT 1
    FROM gov.role_entitlements e
    WHERE e.role_name = CURRENT_ROLE()
      AND e.tenant_id = tenant_id
      AND e.region = region
  );

ALTER TABLE rag.chunks
  ADD ROW ACCESS POLICY rap_tenant_region ON (tenant_id, region);

-- PII redaction carried by tag
CREATE OR REPLACE MASKING POLICY mask_email AS (v STRING) RETURNS STRING ->
  IFF(IS_ROLE_IN_SESSION('ROLE_PII_READER'), v, REGEXP_REPLACE
  (v,'(^.).+(@.+$)','\\1***\\2'));

-- Any column tagged classification = 'pii.email' will bind to mask_email
via your governance job
```

Because evaluation happens at read, the same tables can back internal notebooks, partner apps, and human-in-the-loop tooling without forks.

27.2.3 Freshness and Watermarks That Keep Answers Consistent

Real-time corpora drift. To avoid "yesterday I said A, today I say B" bugs, choose a **watermark** strategy. For interactive Q&A, cap search to a recent window (e.g., last 90 days) and include the newest timestamp in the answer. For durable decisions (support macros, policy libraries), pin retrieval to a **release tag** stored with the chunks and filter on that tag. Either way, the query declares the cut line; the model inherits determinism.

```
-- Pin to a published release for durable workflows
SELECT * FROM rag.chunks
WHERE tenant_id = :tenant AND release_tag = :release
ORDER BY VECTOR_COSINE_SIMILARITY(embedding, :q) DESC
LIMIT :k;
```

27.2.4 Assembly: Small, Cited, and Cacheable

Once you have chunks, assemble a prompt that is **short and verifiable**. A good approach stores the assembly pieces in SQL, so they can be audited and tuned over time.

```
CREATE OR REPLACE PROCEDURE rag.sp_search_and_cite(q STRING, tenant STRING,
k NUMBER)
RETURNS VARIANT
LANGUAGE SQL
AS
$$
  -- 1) Embed outside or pass in; here we assume :embedding session
  var is set
  WITH topk AS (
    SELECT doc_id, chunk_id, text, source_uri,
           VECTOR_COSINE_SIMILARITY(embedding, :embedding) AS sim
    FROM rag.chunks
    WHERE tenant_id = :tenant
    ORDER BY sim DESC
    LIMIT :k
  ),
  context AS (
    SELECT LISTAGG(CONCAT('[', doc_id, '#', chunk_id, '] ', LEFT(text,
    800)), '\n\n')
           WITHIN GROUP (ORDER BY sim DESC) AS ctx,
           ARRAY_AGG(OBJECT_CONSTRUCT('doc', doc_id, 'chunk', chunk_id,
           'uri', source_uri, 'sim', sim))
           AS citations
    FROM topk
  )
  SELECT OBJECT_CONSTRUCT('context', ctx, 'citations', citations) FROM
  context;
$$;
```

Your application calls the procedure, feeds its context and citations to the model, and stores the resulting `citations` back with the answer for audit. Because the procedure runs with policies, the app never needs to re-implement security.

27.2.5 Re-ranking and Hybrid Search Without a Second System

Many domains benefit from **hybrid search**: combine keyword signals (BM25-like) with vector similarity. You can approximate this in Snowflake by scoring each signal and blending them. The blend remains inside your RAP/masking envelope.

```
WITH cand AS (
  SELECT *,
         VECTOR_COSINE_SIMILARITY(embedding, :q_vec) AS vscore
  FROM rag.chunks
  WHERE tenant_id = :tenant
),
kw AS (
  SELECT chunk_id,
         SEARCH_SCORE() AS kscore
  FROM rag.chunks
  WHERE tenant_id = :tenant
    AND SEARCH(text, :q_text)
)
SELECT c.doc_id, c.chunk_id, c.text, c.source_uri,
       0.6 * c.vscore + 0.4 * COALESCE(k.kscore, 0) AS score
FROM cand c
LEFT JOIN kw k USING (chunk_id)
ORDER BY score DESC
LIMIT :k;
```

If you want keyword signals in Snowflake, use SEARCH() and optionally enable full-text search optimization on the target columns. If that capability is not available in your environment, a lightweight n-gram or trigram gate can still serve as the coarse keyword filter before vector scoring.

27.2.6 Serving Through a Thin, Stable Interface

Expose retrieval through a **stable stored procedure or secure view** and pin session context up front. That gives you one place to assert query tags, timeouts, and safe defaults.

```
-- Session defaults for BI/API lanes
ALTER SESSION SET
  STATEMENT_TIMEOUT_IN_SECONDS = 15,
  QUERY_TAG = 'app=genai;surface=qa';

-- Stable surface (apps call this only)
CREATE OR REPLACE VIEW rag.v_search_latest AS
SELECT doc_id, chunk_id, text, source_uri
FROM rag.chunks
WHERE CURRENT_TIMESTAMP() BETWEEN valid_from AND COALESCE(valid_
to,'9999-12-31');
```

Behind the view, you can switch from brute-force to indexed search, adjust watermarks, or add policy columns without changing clients.

27.2.7 Guardrails Inside the Warehouse

Not every retrieval should reach the model. Add a **cheap refusal path** for empty results and a **prompt budget** limiter, so you don't send megabytes downstream. Both are easy in SQL.

```
-- Refuse on low semantic signal
WITH k AS ( /* top-k select like above */ )
SELECT IFF(COUNT(*)=0 OR MAX(sim) < 0.25, 'REFUSE', 'OK') AS gate
FROM k;
```

If your gate says **REFUSE**, return a templated answer ("I don't have relevant context for that tenant/time range") and log the query for curation.

27.2.8 Caching and Idempotence for Popular Questions

Live systems repeat themselves. Keep a small **answer cache** keyed by (tenant, normalized_question) with an expiry shorter than your freshness SLO. Because retrieval runs inside Snowflake, the cache is a table, not a service.

```
CREATE TABLE IF NOT EXISTS rag.answer_cache (
  tenant_id STRING,
  q_norm STRING,
  answer STRING,
  citations VARIANT,
  expires_at TIMESTAMP_NTZ,
  PRIMARY KEY (tenant_id, q_norm)
);

-- On hit
SELECT answer, citations
FROM rag.answer_cache
WHERE tenant_id=:tenant AND q_norm=:q AND expires_at > CURRENT_TIMESTAMP();
```

Populate the cache on misses after the model responds. Pair this with **usage logging**, so you can curate new chunks where demand is high.

27.2.9 Observability That Explains Behavior

Two tiny SLIs keep you honest: **retrieval hit quality** and **freshness**. Quality can be proxied by top-1 similarity and the fraction of answers that include at least one citation. Freshness is the age of the newest chunk in the filtered slice. Both are single SELECTs you can alert on.

```
-- Freshness under the same filters your app uses
SELECT MIN(DATEDIFF('minute', created_at, CURRENT_TIMESTAMP())) AS min_
age_min,
       MAX(DATEDIFF('minute', created_at, CURRENT_TIMESTAMP())) AS
       max_age_min
FROM rag.chunks
WHERE tenant_id=:tenant AND topic=:topic;
```

Store per-request rows in a small table (`rag.telemetry`) with `query_tag`, top-k sims, and elapsed time. Post-incident, you can reconstruct whether slowness came from pruning failure, vector size, or policy filters shrinking the candidate set too far.

27.2.10 Putting It Together

A secure, fast RAG stack in Snowflake looks ordinary on purpose. The chunk store carries envelope columns so pruning wins; RAP and masking bind to those columns so isolation is automatic; retrieval is a two-stage SELECT; assembly and citations live in a procedure; and freshness is visible as a view. Because every step is SQL, you can test it in a clone, promote with a swap, and answer "why did we say that?" months later by joining conversations to `rag.chunks` and `ACCESS_HISTORY`. GenAI becomes another consumer of your warehouse—not a parallel system to babysit.

Appendixes

1. Appendix A: Snowflake SQL Style Guide

A style guide exists to make SQL **readable, predictable, and safe to change**. Snowflake adds a few twists—Time Travel, zero-copy clones, semi-structured types, Streams/Tasks, masking, and row policies—so conventions should lean into those strengths. The goal here is a calm, consistent voice across DDL, DML, views, and procedures so reviews are about logic, not layout.

1.1. Philosophy

Prefer **clarity over cleverness**. Make shape and intent obvious: prunable envelopes first, stable names, explicit types, defensive joins, idempotent DDL. Every shared query should tell a future reader where to filter, how to aggregate, and how to roll back. Assume your SQL will be skimmed on a laptop at 11 p.m. by someone who didn't write it.

1.2. Names, Schemas, and Object Layout

Keep names short, semantic, and sortable. Reserve each schema for one stage of refinement; publish only through views.

Surface	Purpose	Naming hints
`raw.*`	Landed files/ingest tables	`src_*`, mirror source names, no business logic
`stage.*`	Lightly cleaned, 1:1 with raw	`stg_*`, add envelope columns (`event_dt`, `tenant_id`)
`core.*`	Standardized facts/dims	`fact_*`, `dim_*` (singular columns)
`mart.*`	Consumer views	`v_*` secure views; never expose base tables
`ops.* / gov.*`	Telemetry and governance	`v_*` status, `t_*` logs, `contract`, `release_notes`

M. Hander, *Snowflake Data Warehouse Engineering*, https://doi.org/10.1007/979-8-8688-2628-3

Columns are snake_case. Primary identifiers end with _id. Timestamps carry a suffix: _ts for TIMESTAMP_NTZ, _at for business events (often TIMESTAMP_TZ), and _dt for dates.

1.3. Formatting and Structure

Write SQL so the **shape** is visible:

- One top-level clause per line; align keywords; uppercase keywords, lowercase identifiers.
- Put **filters that prune** up high; push expensive expressions down.
- Use **CTEs** as named paragraphs; keep them small and composable.
- Avoid SELECT * in any shared surface.

Example:

```
WITH orders AS (
  SELECT
    o.order_id,
    o.customer_id,
    o.order_dt,
    o.amount
  FROM core.fact_orders AS o
  WHERE o.order_dt BETWEEN :from_dt AND :to_dt
),

enriched AS (
  SELECT
    o.*,
    c.segment,
    c.territory
  FROM orders AS o
  JOIN core.dim_customer AS c
    ON c.customer_id = o.customer_id
   AND o.order_dt >= c.effective_from
   AND o.order_dt < COALESCE(c.effective_to, '9999-12-31'::DATE)
)
```

```
SELECT
  order_dt,
  segment,
  SUM(amount) AS revenue
FROM enriched
GROUP BY order_dt, segment
ORDER BY order_dt, segment;
```

1.4. CTEs, Comments, and Intent

Write the comment you wish you'd had. Start files with a header that states **grain**, **filters**, and **assumptions**. CTEs are paragraphs—name them as such: deduped, `windowed_7d`, `ranked_latest`.

```
-- PURPOSE: Daily revenue by segment. Grain = (order_dt, segment).
-- ENVELOPES: order_dt prunes; tenant isolation via RAP on schema.
-- ASSUMPTIONS: dim_customer is joined point-in-time using effective_from/
effective_to.
```

```
WITH deduped AS ( ... ),
ranked_latest AS ( ... )
SELECT ...
```

Note In examples involving dimensional history, avoid joining on `is_current = TRUE` unless the query is explicitly current-state only. Snapshot or backdated facts should use point-in-time validity predicates.

1.5. Joins, Nulls, and Determinism

Be explicit about join intent and null semantics.

- Always qualify join conditions; avoid hidden cross joins.
- For slowly changing dimensions, use effective_from/effective_to predicates for point-in-time joins. Use is_current only for current-state reporting, never for historical snapshot joins.

- Use COUNT_IF(...) instead of SUM(CASE WHEN ... THEN 1 END) for intent.
- Decide upfront whether "unknown" equals "other" and cast consistently.

```
SELECT
  COUNT_IF(o.is_won) AS wins,
  COUNT(*)           AS opportunities,
  SAFE_DIVIDE(COUNT_IF(o.is_won), COUNT(*)) AS win_rate  -- custom UDF or
handle div-by-zero
FROM core.f_opportunity_daily AS o;
```

1.6. DDL and Migrations (Idempotent and Safe)

Migrations must be **re-runnable** and small. Guard every DDL; record that it applied.

```
BEGIN;

CREATE TABLE IF NOT EXISTS ops.schema_migrations (id STRING PRIMARY KEY,
applied_at TIMESTAMP_NTZ);

INSERT INTO ops.schema_migrations
SELECT '013_add_order_channel', CURRENT_TIMESTAMP()
WHERE NOT EXISTS (SELECT 1 FROM ops.schema_migrations WHERE id='013_add_
order_channel');

ALTER TABLE IF EXISTS core.fact_orders
  ADD COLUMN IF NOT EXISTS order_channel STRING;

COMMENT ON COLUMN core.fact_orders.order_channel IS 'web|store|partner';

COMMIT;
```

Prefer **SWAP via clone** for risky changes: build in prod_candidate and ALTER DATABASE ... SWAP WITH prod.

1.7. Types, Time, and Time Zones

Be strict with temporal data. Store business events in TIMESTAMP_TZ when provenance matters; transform to DATE for pruning and grouping. Keep one **session policy** for BI/ ELT warehouses (timezone, timestamp type) to avoid invisible drift.

Use	Recommended
Business event	TIMESTAMP_TZ with explicit source TZ handling
Load/process time	TIMESTAMP_NTZ
Partitioning and pruning	DATE envelope column (event_dt, load_dt)
Money	NUMBER(18,2) (or currency-specific rules)

1.8. Semi-structured Data (VARIANT) with Envelopes

Keep a **scalar envelope** for prunes and joins; use VARIANT for the rest. Extract with : paths, cast with TRY_ functions, and avoid schema drift bleeding into marts.

```
CREATE OR REPLACE VIEW stage.events_flat AS
SELECT
  TO_DATE(v:event_ts) AS event_dt,
  v:tenant_id::string AS tenant_id,
  v:user_id::string   AS user_id,
  TRY_TO_NUMBER(v:amount) AS amount,
  v AS payload
FROM raw.events_json;
```

1.9. Views, Dynamic Tables, and Materialization

Views are your API; keep them **secure** and deterministic. For near-real-time aggregates, use **Dynamic Tables** with a clear TARGET_LAG. Avoid volatile functions in shared surfaces.

```
CREATE OR REPLACE SECURE VIEW mart.v_orders AS
SELECT ... FROM core.fact_orders WHERE order_dt >= '2023-01-01';
```

```
CREATE OR REPLACE DYNAMIC TABLE mart.orders_5m
TARGET_LAG = '5 minutes'
WAREHOUSE = wh_marts
AS
SELECT ... FROM core.fact_orders WHERE order_dt >= DATEADD(day,-2,
CURRENT_DATE());
```

1.10. Governance at Read (RAP and Masking)

Policies should ride on tags and envelope columns. Attach once; inherit everywhere.

```
CREATE OR REPLACE MASKING POLICY mask_email AS (v STRING) RETURNS STRING ->
  IFF(IS_ROLE_IN_SESSION('ROLE_PII_READER'), v, REGEXP_REPLACE
(v,'(^.).+(@.+$)','\\1***\\2'));

ALTER TABLE core.dim_customer
  MODIFY COLUMN email SET MASKING POLICY mask_email;

CREATE OR REPLACE ROW ACCESS POLICY rap_tenant AS (tenant_id STRING)
RETURNS BOOLEAN ->
  EXISTS (SELECT 1 FROM gov.user_tenants WHERE user_name=CURRENT_USER() AND
  tenant_id=tenant_id);

ALTER VIEW mart.v_orders ADD ROW ACCESS POLICY rap_tenant ON (tenant_id);
```

1.11. Performance-Aware Habits

Write for **pruning first**, then compute.

- Always filter on envelope columns (`event_dt`, `tenant_id`) **before** touching JSON or UDFs.
- Project only needed columns in shared views; avoid `SELECT *`.
- Cluster **only after** profiling shows broad scans with good predicates; prefer narrow keys (1–2 columns).
- Keep files large and balanced upstream; Parquet ~200 MB with sensible row groups.

A small helper pattern: gate expensive logic.

```
SELECT
  tenant_id,
  event_dt,
  SUM(amount) AS revenue
FROM stage.events_flat
WHERE event_dt BETWEEN :from_dt AND :to_dt
  AND tenant_id = :tenant_id
GROUP BY 1,2;
```

1.12. Query Tags, Telemetry, and Reproducibility

Set QUERY_TAG in every job so cost and behavior are attributable. Record releases and data tests as rows you can join later.

```
ALTER SESSION SET QUERY_TAG='app=elt;pipeline=orders_daily;release=v2025.10.19';
```

```
CREATE TABLE IF NOT EXISTS ops.release_log (
  ts TIMESTAMP_NTZ, release_id STRING, actor STRING, notes STRING
);
```

1.13. Testing and Contracts in SQL

Assert the interface before promotion: column presence, types, invariants, and governance attachments.

```
-- Required columns/types on a public view
WITH expected(col, dtype) AS (
  SELECT 'ORDER_ID','NUMBER' UNION ALL
  SELECT 'ORDER_DT','DATE'  UNION ALL
  SELECT 'AMOUNT','NUMBER'
)
SELECT e.col, e.dtype, c.data_type AS found
FROM expected e
LEFT JOIN prod_candidate.information_schema.columns c
  ON c.table_schema='MART' AND c.table_name='V_ORDERS' AND c.column_
  name=e.col
WHERE c.column_name IS NULL OR UPPER(c.data_type)<>e.dtype;
```

1.14. Error Handling and Transactions

Wrap multi-step changes in BEGIN/COMMIT; prefer **fail fast** in migrations. For ingestion, keep ON_ERROR visible and intentional; capture rejects to a stage/table for audit.

```
COPY INTO stage.orders_rejects
FROM @landing/orders/
FILE_FORMAT=(TYPE=CSV)
ON_ERROR='CONTINUE';
```

1.15. Stored Procedures, UDFs, and Snowpark

Keep UDFs **pure and small**. Pin package versions; avoid heavy imports. For stored procedures that orchestrate SQL, prefer SQL procedures unless you truly need imperative control flow.

```
CREATE OR REPLACE FUNCTION util.safe_divide(n NUMBER, d NUMBER)
RETURNS FLOAT
LANGUAGE SQL
AS 'IFF(d=0, NULL, n/d)';
```

1.16. Review Checklist (Read It Aloud)

- Does the query **prune early** on date/tenant and avoid SELECT *?
- Are **names** consistent with stage (raw/stage/core/mart) and grain clear?
- Is DDL **idempotent** and wrapped in a transaction?
- Do views expose only what consumers need and remain **deterministic**?
- Are policies (masking/RAP) attached via tags/envelopes, not hard-coded?
- Can this run in a **clone** and promote with a **SWAP**?
- Would someone understand it in five minutes next month?

If the answer is yes more often than not, your SQL will age well—and so will the platform it powers.

2. Appendix B: Role and Grant Matrix Templates

This appendix gives you drop-in templates to design **least-privilege, auditable** access in Snowflake. Use the matrices to define who can read, write, operate, and govern—then apply the SQL skeletons to make it real. Keep grants **idempotent**, prefer **future grants**, and separate **platform** from **domain** roles so product teams can move fast without owning the whole account.

2.1. Design Blueprint (At a Glance)

- **Layers of Responsibility**
    - **Account/platform:** Create warehouses, integrations, network policies, and global governance (security).
    - **Domain/product:** Own schemas, tables, views, streams/tasks, and day-to-day grants within a database.
- **Interfaces, Not Tables**
    - Consumers get **secure views**; writers target **stage/core** tables; admins rarely need raw `OWNERSHIP`.
- **Future Grants Everywhere**
    - Rights attach to **all current and future** objects to avoid drift.

2.2. Role Hierarchy Template

Role	Purpose	Inherits	Managed by
ORGADMIN (Snowflake)	Org-wide settings	-	Platform
ACCOUNTADMIN	Account settings, billing, last-resort	SECURITYADMIN, SYSADMIN	Platform
SECURITYADMIN	Create roles/users; grant roles	-	Platform
SYSADMIN	Create databases/schemas/ warehouses	-	Platform
ROLE_PLATFORM_OPS	Ops on platform warehouses, monitors, tasks	SYSADMIN	Platform
ROLE_GOVERNANCE	Tags, masking/row policies (apply)	SECURITYADMIN	DataGov
ROLE_BI_APP	Read marts; run BI warehouse	-	Analytics
ROLE_ANALYST	Ad hoc read in marts; temp tables	ROLE_BI_APP	Analytics
ROLE_ELT_WRITER_<domain>	Write to stage/core; run ELT tasks	-	Domain
ROLE_DATA_READER_<domain>	Read mart views in domain	-	Domain
ROLE_DATA_ADMIN_<domain>	Own domain schemas; manage grants	-	Domain

Create **database roles** for intra-DB delegation when teams own a single database, and you want self-service grants.

2.3. Database-Level Matrix (Templates)

Assume database PROD_SALES with schemas STAGE, CORE, MART, and OPS.

Read/Write/Operate by Schema

Schema	Reader role	Writer role	Admin role
STAGE	(none)	ROLE_ELT_WRITER_SALES	ROLE_DATA_ADMIN_SALES
CORE	(none)	ROLE_ELT_WRITER_SALES	ROLE_DATA_ADMIN_SALES
MART	ROLE_DATA_READER_SALES, ROLE_BI_APP, ROLE_ANALYST	(none)	ROLE_DATA_ADMIN_SALES
OPS	ROLE_PLATFORM_OPS	ROLE_PLATFORM_OPS	ROLE_DATA_ADMIN_SALES

Object Privileges to Apply

Surface	Privileges
Reader on MART	USAGE (DB/Schema), SELECT on **ALL & FUTURE** VIEWS/TABLES
Writer on STAGE,CORE	USAGE, SELECT/INSERT/UPDATE/DELETE on ALL & FUTURE TABLES; USAGE on ALL & FUTURE SEQUENCES; CREATE TASK on schema if needed
Admin	USAGE, OWNERSHIP (selected), MONITOR, CREATE VIEW/TABLE/TASK/STREAM in schemas; grant options as needed

2.4. Warehouse and Execution Matrix

Warehouse	Who uses it	Grants
WH_BI	ROLE_BI_APP, ROLE_ANALYST	USAGE, OPERATE
WH_ELT_SALES	ROLE_ELT_WRITER_SALES	USAGE, OPERATE
WH_OPS	ROLE_PLATFORM_OPS	USAGE, OPERATE

(Keep resource monitors attached to ad hoc and reader lanes.)

2.5. Governance Privileges Matrix

Capability	Where	Privilege
Apply masking/row policies	Account	APPLY MASKING POLICY, APPLY ROW ACCESS POLICY
Manage tags and set values	Account/Objects	APPLY TAG, CREATE TAG (as needed)
See access lineage	ACCOUNT_USAGE.ACCESS_HISTORY	MONITOR, IMPORTED PRIVILEGES on SNOWFLAKE DB

Assumption: governance objects such as tags, masking policies, row access policies, and entitlement lookup tables (e.g., GOV.USER_TENANTS) live in a dedicated governance schema/database managed by ROLE_GOVERNANCE.

2.6. SQL Skeletons (Copy, Adjust, Run)

Create Roles and Basic Inheritance

```
-- Platform & governance
CREATE ROLE IF NOT EXISTS ROLE_PLATFORM_OPS;
CREATE ROLE IF NOT EXISTS ROLE_GOVERNANCE;

GRANT ROLE ROLE_PLATFORM_OPS TO ROLE SYSADMIN;
GRANT ROLE ROLE_GOVERNANCE  TO ROLE SECURITYADMIN;
```

```
-- Domain
CREATE ROLE IF NOT EXISTS ROLE_DATA_ADMIN_SALES;
CREATE ROLE IF NOT EXISTS ROLE_ELT_WRITER_SALES;
CREATE ROLE IF NOT EXISTS ROLE_DATA_READER_SALES;
CREATE ROLE IF NOT EXISTS ROLE_BI_APP;
CREATE ROLE IF NOT EXISTS ROLE_ANALYST;

GRANT ROLE ROLE_BI_APP          TO ROLE ROLE_ANALYST;
```

Database Scaffolding and Admin Rights

```
-- Create database & schemas (by SYSADMIN)
CREATE DATABASE IF NOT EXISTS PROD_SALES;
CREATE SCHEMA   IF NOT EXISTS PROD_SALES.STAGE;
CREATE SCHEMA   IF NOT EXISTS PROD_SALES.CORE;
CREATE SCHEMA   IF NOT EXISTS PROD_SALES.MART;
CREATE SCHEMA   IF NOT EXISTS PROD_SALES.OPS;

-- Domain admin owns schemas
GRANT OWNERSHIP ON SCHEMA PROD_SALES.STAGE TO ROLE ROLE_DATA_ADMIN_SALES
COPY CURRENT GRANTS;
GRANT OWNERSHIP ON SCHEMA PROD_SALES.CORE  TO ROLE ROLE_DATA_ADMIN_SALES
COPY CURRENT GRANTS;
GRANT OWNERSHIP ON SCHEMA PROD_SALES.MART  TO ROLE ROLE_DATA_ADMIN_SALES
COPY CURRENT GRANTS;
GRANT OWNERSHIP ON SCHEMA PROD_SALES.OPS   TO ROLE ROLE_DATA_ADMIN_SALES
COPY CURRENT GRANTS;

-- Base usage
GRANT USAGE ON DATABASE PROD_SALES TO ROLE ROLE_DATA_ADMIN_SALES;
GRANT USAGE ON ALL SCHEMAS IN DATABASE PROD_SALES TO ROLE ROLE_DATA_
ADMIN_SALES;
```

Readers (MART Only)

```
GRANT USAGE ON DATABASE PROD_SALES TO ROLE ROLE_DATA_READER_SALES;
GRANT USAGE ON SCHEMA   PROD_SALES.MART TO ROLE ROLE_DATA_READER_SALES;
```

```
GRANT SELECT ON ALL VIEWS  IN SCHEMA PROD_SALES.MART TO ROLE ROLE_DATA_
READER_SALES;
GRANT SELECT ON FUTURE VIEWS IN SCHEMA PROD_SALES.MART TO ROLE ROLE_DATA_
READER_SALES;

-- If marts include tables (not just views):
GRANT SELECT ON ALL TABLES  IN SCHEMA PROD_SALES.MART TO ROLE ROLE_DATA_
READER_SALES;
GRANT SELECT ON FUTURE TABLES IN SCHEMA PROD_SALES.MART TO ROLE ROLE_DATA_
READER_SALES;

-- Extend to BI/Analyst
GRANT ROLE ROLE_DATA_READER_SALES TO ROLE ROLE_BI_APP;
GRANT ROLE ROLE_BI_APP            TO ROLE ROLE_ANALYST;
```

Writers (STAGE/CORE)

```
GRANT USAGE ON DATABASE PROD_SALES TO ROLE ROLE_ELT_WRITER_SALES;
GRANT USAGE ON SCHEMA   PROD_SALES.STAGE TO ROLE ROLE_ELT_WRITER_SALES;
GRANT USAGE ON SCHEMA   PROD_SALES.CORE  TO ROLE ROLE_ELT_WRITER_SALES;

GRANT SELECT, INSERT, UPDATE, DELETE ON ALL TABLES  IN SCHEMA PROD_SALES.
STAGE TO ROLE ROLE_ELT_WRITER_SALES;
GRANT SELECT, INSERT, UPDATE, DELETE ON FUTURE TABLES IN SCHEMA PROD_SALES.
STAGE TO ROLE ROLE_ELT_WRITER_SALES;

GRANT SELECT, INSERT, UPDATE, DELETE ON ALL TABLES  IN SCHEMA PROD_SALES.
CORE  TO ROLE ROLE_ELT_WRITER_SALES;
GRANT SELECT, INSERT, UPDATE, DELETE ON FUTURE TABLES IN SCHEMA PROD_SALES.
CORE TO ROLE ROLE_ELT_WRITER_SALES;

GRANT USAGE ON ALL SEQUENCES  IN SCHEMA PROD_SALES.CORE  TO ROLE ROLE_ELT_
WRITER_SALES;
GRANT USAGE ON FUTURE SEQUENCES IN SCHEMA PROD_SALES.CORE TO ROLE ROLE_ELT_
WRITER_SALES;
GRANT CREATE TASK ON SCHEMA PROD_SALES.CORE TO ROLE ROLE_ELT_WRITER_SALES;
```

Ops and Tasks

```
-- Allow ops to monitor & run tasks in OPS

GRANT USAGE ON DATABASE PROD_SALES TO ROLE ROLE_PLATFORM_OPS;
GRANT USAGE ON SCHEMA PROD_SALES.OPS TO ROLE ROLE_PLATFORM_OPS;
GRANT MONITOR, OPERATE ON ACCOUNT TO ROLE ROLE_PLATFORM_OPS;

-- Task/pipe execution (schema-level)

GRANT EXECUTE TASK ON ACCOUNT TO ROLE ROLE_PLATFORM_OPS;
```

Note EXECUTE TASK is granted at the account level. Running a task also depends on task ownership or OPERATE on the specific task.

Warehouses

```
-- BI lane
CREATE WAREHOUSE IF NOT EXISTS WH_BI WITH WAREHOUSE_SIZE='SMALL' AUTO_
SUSPEND=120 AUTO_RESUME=TRUE;
GRANT USAGE, OPERATE ON WAREHOUSE WH_BI TO ROLE ROLE_BI_APP;

-- ELT lane
CREATE WAREHOUSE IF NOT EXISTS WH_ELT_SALES WITH WAREHOUSE_SIZE='MEDIUM'
AUTO_SUSPEND=300 AUTO_RESUME=TRUE;
GRANT USAGE, OPERATE ON WAREHOUSE WH_ELT_SALES TO ROLE ROLE_ELT_
WRITER_SALES;
```

Governance Grants

```
-- Apply policies & tags (account-level)
GRANT APPLY MASKING POLICY      ON ACCOUNT TO ROLE ROLE_GOVERNANCE;
GRANT APPLY ROW ACCESS POLICY  ON ACCOUNT TO ROLE ROLE_GOVERNANCE;
GRANT APPLY TAG                ON ACCOUNT TO ROLE ROLE_GOVERNANCE;

-- Visibility into usage (optional)
GRANT IMPORTED PRIVILEGES ON DATABASE SNOWFLAKE TO ROLE ROLE_GOVERNANCE;
```

2.7. Database Roles (Optional Self-Service Pattern)

Database roles let domain admins delegate within their DB without touching account roles.

```
-- Inside PROD_SALES
CREATE DATABASE ROLE IF NOT EXISTS DBR_SALES_READ_MART;
GRANT USAGE ON SCHEMA PROD_SALES.MART TO DATABASE ROLE DBR_SALES_READ_MART;
GRANT SELECT ON ALL VIEWS IN SCHEMA PROD_SALES.MART TO DATABASE ROLE DBR_
SALES_READ_MART;
GRANT SELECT ON FUTURE VIEWS IN SCHEMA PROD_SALES.MART TO DATABASE ROLE
DBR_SALES_READ_MART;

-- Let an account role use the DB role
GRANT DATABASE ROLE DBR_SALES_READ_MART TO ROLE ROLE_DATA_READER_SALES;
```

Note `APPLY TAG / APPLY MASKING POLICY / APPLY ROW ACCESS POLICY` let a governance role attach controls, but the role also needs schema-level `CREATE` privileges to create tags or policy objects used by this pattern.

2.8. Masking and Row Access Attachment Pattern

Use tags to classify sensitive fields, then attach policies consistently through a governance process. The example below shows direct attachment after classification.

```
-- Tag columns
CREATE TAG IF NOT EXISTS classification;
ALTER TABLE PROD_SALES.CORE.DIM_CUSTOMER
  MODIFY COLUMN EMAIL SET TAG classification = 'pii.email';

-- Governance job attaches the right policy where tag == 'pii.*'
-- (RUN BY ROLE_GOVERNANCE)
ALTER TABLE PROD_SALES.CORE.DIM_CUSTOMER
  MODIFY COLUMN EMAIL SET MASKING POLICY MASK_EMAIL;
```

```
-- Tenant/region scoping
CREATE OR REPLACE ROW ACCESS POLICY RAP_TENANT AS (tenant_id STRING)
RETURNS BOOLEAN ->
  EXISTS (SELECT 1 FROM GOV.USER_TENANTS u WHERE u.USER_NAME = CURRENT_
USER() AND u.TENANT_ID = tenant_id);

ALTER VIEW PROD_SALES.MART.V_ORDERS ADD ROW ACCESS POLICY RAP_TENANT ON
(tenant_id);
```

2.9. Review and Drift Control Checklist

- Every consumer role has **USAGE** on DB/Schema and **SELECT** on **ALL & FUTURE** objects it needs—no table-by-table snowflakes.
- Writers have **DML** only where they must; marts remain read-only.
- Warehouses grant only `USAGE/OPERATE`; no platform role leakage.
- Governance rights (`APPLY * POLICY`, `APPLY TAG`) are **centralized**.
- A periodic job re-asserts the matrices (idempotent grants) and reports **drift** (objects without expected grants).

2.10. Idempotent Re-assert (Single Re-runnable Block)

```
-- Re-assert reader rights (safe to run daily)
GRANT USAGE ON DATABASE PROD_SALES TO ROLE ROLE_DATA_READER_SALES;
GRANT USAGE ON SCHEMA   PROD_SALES.MART TO ROLE ROLE_DATA_READER_SALES;
GRANT SELECT ON ALL VIEWS   IN SCHEMA PROD_SALES.MART TO ROLE ROLE_DATA_
READER_SALES;
GRANT SELECT ON FUTURE VIEWS IN SCHEMA PROD_SALES.MART TO ROLE ROLE_DATA_
READER_SALES;
```

How to use this appendix: copy the matrices into your runbook, replace `<domain>` and database names, and keep the SQL blocks in version control. Run them in a **candidate clone**, verify with a short "who can read what" report, and then promote. Your access model will stay stable, auditable, and easy to reason about—even as teams and products grow.

3. Appendix C: Performance Tuning Checklist

Performance tuning in Snowflake is mostly about **doing less work** (pruning and projecting), then **spreading the remaining work** (right warehouse posture), and finally **not repeating work** (caching and reuse). Use this appendix as a calm walk-through you can run during triage and before every release.

3.1. Start Here: A Quick Triage Story

Open **Query Profile** and separate time into **Queued** vs. **Execution**. If queued dominates, you need more lanes (multi-cluster) or staggered starts. If execution dominates, make pruning win: ensure your WHERE clause leads with **date/tenant/entity** scalar filters and that you're not reading columns you don't use. Only after that should you resize the warehouse.

3.2. Pruning First: Make the Table Smaller Before You Touch It

What to Look For and What to Change

Symptom in profile	Likely cause	Smallest fix
Many partitions scanned with good date filter	No scalar envelope column	Add `event_dt/load_dt` column; filter on it *before* JSON
Tenant/entity filters still read broadly	Data interleaves across loads	Land by narrow time slices; add `tenant_id` to path/order; consider clustering `(event_dt, tenant_id)` after proof
JSON-only predicates	VARIANT paths don't prune alone	Keep scalar "envelope" columns (date, tenant, type) and filter on them first
Lots of bytes, little output	`SELECT *` or unused wide columns	Project only needed columns; reduce SELECT list in shared views

A Quick Scan of "Heavy" Queries (Last 24h)

```
SELECT QUERY_ID, WAREHOUSE_NAME, BYTES_SCANNED/POW(1024,3) AS gb_scanned,
       TOTAL_ELAPSED_TIME/1000 AS sec, LEFT(QUERY_TEXT, 200) AS txt
FROM SNOWFLAKE.ACCOUNT_USAGE.QUERY_HISTORY
WHERE START_TIME >= DATEADD(day,-1,CURRENT_TIMESTAMP())
ORDER BY BYTES_SCANNED DESC
LIMIT 20;
```

3.3. Warehouse Posture: Spread Work Without Overspending

Think **small + elastic** over **big + idle**. Keep BI lanes at SMALL/XSMALL with multi-cluster; keep ELT lanes sized to the widest step; prefer raising **cluster count** before jumping size for concurrency spikes.

Situation	Setting to adjust	Good target
Dashboards burst, queue grows	MAX_CLUSTER_COUNT	2–6 clusters, SCALING_POLICY=STANDARD
Nightly ELT saturates but not queued	WAREHOUSE_SIZE	Bump one size for the heavy step only
Warehouses linger warm	AUTO_SUSPEND	60–120s for BI; longer only if tile ripples need it

3.4. Shape Your SQL to Be Cheap

Write predicates that prune **first**, then compute. Avoid volatile functions in shared views so result cache can work.

```
-- Good: prunes by date and tenant up front
SELECT d, tenant_id, SUM(amount) AS revenue
FROM core.fact_orders
WHERE d BETWEEN :from_dt AND :to_dt
  AND tenant_id = :tenant
GROUP BY d, tenant_id;
```

Prefer COUNT_IF, QUALIFY, and explicit column lists. Push expensive expressions and UDFs behind prunable WHERE clauses.

3.5. Caching That Actually Helps

- **Result cache** hits when SQL text is stable. Keep BI parameters as binds; avoid sprinkling CURRENT_TIMESTAMP() into SELECT lists.
- **Warehouse cache** helps repeat access to the same table micro-partitions; grouping related tiles in one lane increases wins.
- **Materialize where churn is tight**: Use **Dynamic Tables** with a small TARGET_LAG (1–5 minutes) for rolling aggregates.

```
CREATE OR REPLACE DYNAMIC TABLE mart.orders_5m
TARGET_LAG='5 minutes' WAREHOUSE=wh_marts AS
SELECT d, tenant_id, SUM(amount) amt
FROM core.fact_orders
WHERE d >= DATEADD(day,-2,CURRENT_DATE())
GROUP BY d, tenant_id;
```

3.6. Micro-partitions, Clustering, and When It's Worth It

Clustering helps **only** when good predicates still scan broadly and the table keeps growing.

Consider clustering when	Verify with	Typical key
Date/tenant filters still read many partitions	SYSTEM$CLUSTERING_INFORMATION and profile scans	(d), (d, tenant_id)
Entity-over-time probes miss	Scan % doesn't drop after envelope fix	(d, user_id) (prove first)

Maintain **after** growth, not hourly. Re-cluster only when average_overlaps and depth drift materially.

3.7. Semi-structured Data: Lift Envelopes, Then Parse

Keep JSON/VARIANT for flexibility but lift prunable scalars (date, tenant, type) into real columns. Use TRY_ casts and avoid deep dot-paths in GROUP BY.

```
CREATE OR REPLACE VIEW stage.events_flat AS
SELECT
  TO_DATE(v:event_ts)          AS event_dt,
  v:tenant_id::string          AS tenant_id,
  v:type::string               AS type,
  TRY_TO_NUMBER(v:amount)      AS amount,
  v                             AS payload
FROM raw.events_json;
```

3.8. Joins and Aggregations: Keep Them Predictable

Join on **clean keys** with clear cardinality; avoid exploding joins on semi-structured arrays by **flatten → aggregate → join**. Aggregate on the smallest necessary grain; stash expensive window logic into a pre-step CTE or Dynamic Table if repeated.

```
WITH sums AS (
  SELECT d, customer_id, SUM(amount) amt
  FROM core.fact_orders
  WHERE d >= :from_dt AND d <= :to_dt
  GROUP BY d, customer_id
)
SELECT * FROM sums WHERE amt > 0;
```

3.9. File Ingest: Make Files Work with You

Small, unbalanced files sabotage parallelism. Compact upstream where possible.

Ingest signal	Better shape
Many tiny files	Consolidate to ~200 MB Parquet (balanced row groups)
Slow external table scans	Row groups ~128 MB; partition paths aligned to prune columns
COPY retries on schema drift	Use `MATCH_BY_COLUMN_NAME=CASE_INSENSITIVE`; stage a SELECT to coerce types

3.10. Fast Checks You Can Keep

Queued vs. Execution (Ten-Minute Window)

```
SELECT
  AVG((QUEUED_OVERLOAD_TIME+QUEUED_PROVISIONING_TIME)/1000) AS queued_s,
  AVG(EXECUTION_TIME/1000) AS exec_s
FROM SNOWFLAKE.ACCOUNT_USAGE.QUERY_HISTORY
WHERE START_TIME >= DATEADD(minute,-10,CURRENT_TIMESTAMP())
  AND WAREHOUSE_NAME = :wh;
```

Top Bytes by Object (2h)

```
SELECT
  f.value:"objectName"::string AS obj,
  SUM(ah.bytes_scanned)/POW(1024,3) AS gb
FROM SNOWFLAKE.ACCOUNT_USAGE.ACCESS_HISTORY AS ah,
LATERAL FLATTEN(input => ah.direct_objects_accessed) AS f
WHERE ah.query_start_time >= DATEADD(hour,-2,CURRENT_TIMESTAMP())
GROUP BY 1
ORDER BY gb DESC
LIMIT 20;
```

Note `ACCESS_HISTORY.direct_objects_accessed` is an array. Flatten it before extracting fields such as objectName or objectDomain.

3.11. Guardrails to Bake into Releases

- **Contract checks** on public views: required columns and types must match; fail deploy if not.
- **Lint** shared SQL for `SELECT *` and missing date/tenant filters.
- **Query tags everywhere** so outliers are attributable (`team=...; surface=...`).
- **Resource monitors** on ad hoc lanes; **statement timeouts** on noisy warehouses.

```
ALTER SESSION SET QUERY_TAG='team=platform;purpose=perf_check';
ALTER WAREHOUSE wh_ad_hoc SET STATEMENT_TIMEOUT_IN_SECONDS=900;
```

3.12. When You Really Must Resize

After pruning and projection are clean, size for **the slowest step's working set**, not for peak concurrency. Upsize to finish in your SLA; downsize once logic is efficient. For BI spikes, raise cluster count before jumping sizes. For ELT, consider a **temporary larger warehouse** just for the heavy stage, then return to normal.

Teach every shared query to prune early and project narrowly; let a small elastic warehouse absorb concurrency; materialize rolling heavy work; and keep one or two SQL probes to spot regressions. Most "tuning" becomes design hygiene you can review and promote like any other change.

4. Appendix D: Cost Governance Checklist

Cost control in Snowflake is mostly about **making spend predictable**: attribute it, cap it, and tune the small habits that cause drift. Treat cost as a product SLO—visible in SQL, guarded by policies, and reviewed on a cadence. This checklist is written as a set of short practices you can drop into your platform without turning everything into process.

4.1. Ground Truth: One FinOps Ledger You Trust

This becomes the single place leadership, SRE, and teams look.

Keep a compact ops mart that turns system views into a daily ledger. Report warehouse consumption by warehouse × day × cost center, and distinguish operational usage from billed account totals: WAREHOUSE_METERING_HISTORY is useful for warehouse-level analysis, while billed credits should be reconciled from METERING_DAILY_HISTORY.

```
CREATE SCHEMA IF NOT EXISTS fin;

-- Credits by warehouse/day
CREATE OR REPLACE VIEW fin.v_credits_daily AS
SELECT
  WAREHOUSE_NAME,
  TO_DATE(START_TIME) AS day,
  SUM(CREDITS_USED)   AS credits
FROM SNOWFLAKE.ACCOUNT_USAGE.WAREHOUSE_METERING_HISTORY
GROUP BY 1,2;

-- Storage composition (latest snapshot per DB)
CREATE OR REPLACE VIEW fin.v_storage_now AS
WITH latest AS (
  SELECT *,
         ROW_NUMBER() OVER (PARTITION BY DATABASE_NAME ORDER BY USAGE_DATE
         DESC) rn
  FROM SNOWFLAKE.ACCOUNT_USAGE.DATABASE_STORAGE_USAGE_HISTORY
)
SELECT DATABASE_NAME,
       ACTIVE_BYTES, TIME_TRAVEL_BYTES, FAILSAFE_BYTES
FROM latest WHERE rn=1;
```

Tag warehouses, databases, and shared views with cost_center, owner, and data_domain, then join tags into the ledger so chargeback and showback are automatic.

4.2. Guardrails That Fail Safe

Set small, explicit limits before you start chasing spikes. Use **resource monitors** to cap lanes, **statement timeouts** on noisy warehouses, and **auto-suspend** tuned to your refresh cadence.

```
-- Cap an ad-hoc lane daily; notify then suspend
CREATE OR REPLACE RESOURCE MONITOR rm_ad_hoc
  WITH CREDIT_QUOTA = 100
  FREQUENCY = DAILY
  TRIGGERS ON 80 PERCENT DO NOTIFY
           ON 100 PERCENT DO SUSPEND;

ALTER WAREHOUSE wh_ad_hoc SET
  RESOURCE_MONITOR = rm_ad_hoc,
  STATEMENT_TIMEOUT_IN_SECONDS = 900,
  STATEMENT_QUEUED_TIMEOUT_IN_SECONDS = 300,
  AUTO_SUSPEND = 120,
  AUTO_RESUME = TRUE;
```

Prefer **small + elastic** warehouses for BI (XSMALL/SMALL with multi-cluster) and **right-sized** single-cluster lanes for ELT steps that are compute-heavy but not concurrent.

4.3. Hygiene That Keeps Work Cheap

Most drift comes from habits. Make these the defaults in shared code and reviews:

- **Prune first.** Tables carry envelope columns (`event_dt`, `tenant_id`) and all public SQL filters on them *before* JSON or UDFs.
- **Project narrowly.** No `SELECT *` in shared views; only columns dashboards actually use.
- **Compact upstream files.** Aim for ~200 MB Parquet with balanced row groups; avoid file explosions that starve parallelism.

- **Manage Time Travel.** Keep retention short on mutable staging (1–2 days) and reasonable on shared prod (7–14 days); use **transient** for intermediates.
- **Normalize schedules.** Stagger top-of-hour bursts; let multi-cluster absorb unavoidable spikes.

A tiny retention pass that won't surprise anyone:

```
ALTER DATABASE prod_staging SET DATA_RETENTION_TIME_IN_DAYS = 1;
ALTER SCHEMA   prod.mart     SET DATA_RETENTION_TIME_IN_DAYS = 7;
```

4.4. Observability: Small Probes and Clear Pages

Alert from SQL, not hunches. Two views usually catch problems early: **credits vs. baseline** and **heavy queries**.

```
-- Today vs 7-day median per warehouse

WITH d AS (
  SELECT
    TO_DATE(start_time) AS day,
    warehouse_name,
    SUM(credits_used) AS credits
  FROM SNOWFLAKE.ACCOUNT_USAGE.WAREHOUSE_METERING_HISTORY
  WHERE start_time >= DATEADD(day, -15, CURRENT_TIMESTAMP())
    AND TO_DATE(start_time) < CURRENT_DATE()
  GROUP BY 1, 2
),
y AS (
  SELECT warehouse_name, credits AS yesterday_credits
  FROM d
  WHERE day = DATEADD(day, -1, CURRENT_DATE())
),
b AS (
  SELECT warehouse_name, MEDIAN(credits) AS med7
  FROM d
```

```
  WHERE day >= DATEADD(day, -8, CURRENT_DATE())
    AND day <= DATEADD(day, -2, CURRENT_DATE())
  GROUP BY 1
)
SELECT
  y.warehouse_name,
  y.yesterday_credits,
  b.med7,
  y.yesterday_credits - b.med7 AS delta
FROM y
JOIN b USING (warehouse_name)
ORDER BY delta DESC;
```

Note Avoid comparing CURRENT_DATE() warehouse credits to completed historical days. ACCOUNT_USAGE.WAREHOUSE_METERING_HISTORY is hourly and can lag by up to three hours (and cloud services by up to six hours), so alerting should usually compare yesterday to a trailing baseline of completed days.

```
-- Top bytes scanned (last 2h)
SELECT QUERY_ID, USER_NAME, WAREHOUSE_NAME,
       BYTES_SCANNED/POWER(1024,3) AS gb_scanned,
       TOTAL_ELAPSED_TIME/1000     AS sec
FROM SNOWFLAKE.ACCOUNT_USAGE.QUERY_HISTORY
WHERE START_TIME >= DATEADD(hour,-2,CURRENT_TIMESTAMP())
ORDER BY BYTES_SCANNED DESC
LIMIT 20;
```

Pair these with a five-minute Task that posts **only new** anomalies to an `ops.alerts` table and sends a single notification.

4.5. Chargeback and Showback That People Accept

Decide how you attribute compute:

- **By warehouse tag** when lanes map cleanly to teams or products.
- **By** QUERY_TAG when multiple teams share a lane (dbt projects, pipelines).

Keep a simple rollup and publish weekly:

```
-- Credits by cost_center via warehouse tag
SELECT
  t.TAG_VALUE AS cost_center,
  TO_DATE(w.START_TIME) AS day,
  SUM(w.CREDITS_USED)   AS credits
FROM SNOWFLAKE.ACCOUNT_USAGE.WAREHOUSE_METERING_HISTORY w
LEFT JOIN SNOWFLAKE.ACCOUNT_USAGE.OBJECT_TAGS t

 ON t.OBJECT_NAME = w.WAREHOUSE_NAME
 AND t.DOMAIN = 'WAREHOUSE'
 AND t.TAG_NAME = 'COST_CENTER'

GROUP BY 1,2;
```

When politics are involved, show **usage AND optimization hints** together ("spent 320 credits; 70% scans lacked date predicate")—it turns invoices into coaching.

Note Warehouse tags work best when a lane maps cleanly to one team or product. For shared lanes, prefer QUERY_TAG-based attribution; Snowflake documents query-tag attribution using QUERY_ATTRIBUTION_HISTORY.

4.6. Known Spend Traps and Simple Fixes

Trap	What's happening	Repair without drama
Dashboards surge then drag	Tiles miss result cache; lanes fan out too late	Normalize parameters; remove `SELECT *`; raise `MAX_CLUSTER_COUNT` a notch; set `AUTO_SUSPEND` to cover the ripple
ELT burns credits with low queue	Broad scans from weak pruning	Add scalar envelopes; project needed columns; cluster narrowly only if proven
Time Travel hoards storage	Retention too long on churny tables	Set staging to 1–2 days; core/marts 7–14; use transient for intermediates
Tiny-file storms	Upstream splits too fine	Compact to ~200 MB Parquet; use manifests or partitioned COPY
Cross-region egress	Reads across regions or chatty external calls	Co-locate compute; replicate only needed schemas; batch external function calls

4.7. Replication, Sharing, and Reader Accounts

Cross-region **database replication** is affordable when cadence is right-sized and only necessary schemas move. Shares to Snowflake customers are compute-free for you; **Reader Accounts** are not—you run and cap their warehouse.

```
-- Cap reader spend
CREATE OR REPLACE RESOURCE MONITOR rm_readers
  WITH CREDIT_QUOTA = 200  FREQUENCY = MONTHLY
  TRIGGERS ON 80 PERCENT DO NOTIFY ON 100 PERCENT DO SUSPEND;

ALTER WAREHOUSE wh_reader_partner SET RESOURCE_MONITOR = rm_readers,
WAREHOUSE_SIZE='XSMALL';
```

Keep per-region products to avoid cross-region reads; buyers get the same schema everywhere.

4.8. External Functions, Snowpipe, and Streaming

Costs here hide in **chattiness**. For external functions, cache results in a lookup table and batch requests. For ingest, prefer **Snowpipe Streaming** when the file tax dominates. When using files, target larger, balanced objects and avoid schema drift that triggers retries.

4.9. Dev/Test/Prod Without Surprise Bills

Give Dev and Test **smaller warehouses**, short autosuspend, and **resource monitors** that suspend instead of notify. Clone Prod into Test for realism; run load tests on capped lanes with QUERY_TAG so they're easy to exclude from chargeback. Promote with **SWAP** to keep rollback cheap.

4.10. A Weekly Rhythm That Keeps You Ahead

Publish one page: **credits by lane vs. budget**, **top ten heavy queries**, **storage delta**, and **open cost alerts**. Before changing warehouse sizes, require a **pruning/projection review** on the top offenders. Small rituals beat big after-the-fact cleanups.

4.11. Final Snippets You'll Reuse

Set economy mode off-hours (BI):

```
-- Example: schedule this nightly
ALTER WAREHOUSE wh_bi SET SCALING_POLICY='ECONOMY';
-- And back to STANDARD at start of business
```

Enforce query tags in jobs:

```
ALTER SESSION SET QUERY_TAG='team=finops;surface=elt;release=v2025.10.19';
```

Close the loop on rejects and errors:

```
COPY INTO stage.rejected FROM @landing/files
FILE_FORMAT=(TYPE=CSV) ON_ERROR='CONTINUE';  -- quantify waste; fix at source
```

Make cost visible as data, put small hard caps where surprise hurts, and tune the everyday patterns—prune first, project narrowly, compact files, and size lanes for behavior. When cost is just SQL you review weekly, governance becomes habit, not heroics.

5. Appendix E: Migration Playbook (From Redshift/ BigQuery/Synapse/Teradata)

A good migration is a **boring release**: data lands the same, queries read faster or equal, and teams barely notice the cutover. In Snowflake, the path is to **lift** data with fidelity, **translate** semantics where needed, and **reshape** only when it materially improves cost/ perf—never as a prerequisite to go-live. This playbook is written to be run: short tables, copy-paste blocks, and habits that make the move predictable.

5.1. Landing Strategy (Keep Shape, Move Bytes Fast)

Pick the native export for your source, land to the matching cloud storage, then `COPY INTO` with strict formats. Keep file names partitioned by business envelopes (date/ tenant) so you gain pruning on day one.

Source	Best export	Lands in	Loader into Snowflake	Notes that matter
Redshift	`UNLOAD` to Parquet/CSV	**S3**	External stage → `COPY INTO`	Prefer Parquet + gzip; include `MANIFEST`; carry sort/keys as tags only (not used)
BigQuery	`EXPORT DATA` to Parquet	**GCS**	External stage (GCS) → `COPY INTO`	Use `snappy` Parquet; map `STRUCT/ARRAY` to `VARIANT`; preserve `partition/time` in path
Synapse	`COPY INTO`/ PolyBase to Parquet	**ADLS Gen2**	External stage (Azure) → `COPY INTO`	Ensure service principal perms; avoid tiny files from CTAS
Teradata	FastExport/TPT to Parquet/CSV	**S3/ADLS/ GCS**	External stage → `COPY INTO`	Export with consistent charset and null markers; wide DECIMALs map to `NUMBER(p,s)`

Skeletons

```
-- Create the external stage (AWS example)
CREATE STORAGE INTEGRATION si_s3
  TYPE = EXTERNAL_STAGE
  STORAGE_PROVIDER = S3
  ENABLED = TRUE
  STORAGE_AWS_ROLE_ARN = '<arn:aws:iam::...:role/snowflake-stage-role>'
  STORAGE_ALLOWED_LOCATIONS = ('s3://company-migration/');

CREATE STAGE stg_mig
  STORAGE_INTEGRATION = si_s3
  URL = 's3://company-migration/source=db/table=fact_orders/';

CREATE FILE FORMAT ff_parquet TYPE = PARQUET;

-- Load to Snowflake landing (one table)

CREATE OR REPLACE TABLE raw.fact_orders_like_source LIKE SOURCE_DB.PUBLIC.
FACT_ORDERS;
```

Note Use LIKE `<source_table>` when copying an existing table definition. Do not use `TEMPLATE_DB` as syntax; `USING TEMPLATE` is a separate Snowflake feature for deriving columns from staged files.

```
COPY INTO raw.fact_orders_like_source
  FROM @stg_mig
  FILE_FORMAT = (FORMAT_NAME = ff_parquet)
  MATCH_BY_COLUMN_NAME = CASE_INSENSITIVE
  ON_ERROR = 'ABORT_STATEMENT';
```

5.2. Semantics Map (Don't Translate Blindly)

Carry the **business** shape first; translate **engine** features only when needed for correctness. Many source constructs have no Snowflake equivalent—and that's fine.

Concept	Redshift	BigQuery	Synapse	Teradata	Snowflake stance
Distribution/ sort keys	DISTKEY, SORTKEY	Partition/ cluster	Hash/round-robin, indexes	PI, partitions, secondary idx	**Ignore** at load; rely on micro-partitions; consider clustering *after* proof
Vacuum/ analyze	Needed	Managed	Statistics	Collect stats	**Not needed**; services layer manages
Indexes	Limited	N/A	Yes	Yes	**None**; design for pruning + projection
Identity	IDENTITY	GENERATE_ UUID, sequences	Identity	Identity	Use SEQUENCE or IDENTITY (auto-increment)
Constraints	Enforced	Optional	Enforced	Enforced	Not enforced; informational only—validate in CI
UDFs	SQL/ PLpgSQL	SQL/JS	T-SQL	SQL	Snowflake JS/ Snowpark UDF/Proc; keep pure, small
Semi-structured	SUPER (Redshift Spectrum)	STRUCT/ ARRAY/JSON	JSON OPENJSON	JSON	VARIANT, OBJECT, ARRAY; schema-on-read
Materialized views	Basic	Native	Yes	Yes	Snowflake supports; design narrowly, prunable

Function and Syntax Nits

Pattern	Source	Snowflake
Null coalesce	COALESCE, NVL	COALESCE, NVL, IFNULL
Safe casts	SAFE_CAST (BQ)	TRY_TO_*(...) e.g., TRY_TO_NUMBER
Qualify	BQ has QUALIFY	Supported (QUALIFY ROW_NUMBER()...=1)
Date trunc	DATE_TRUNC, TIMESTAMP_TRUNC	DATE_TRUNC, TIMESTAMP_TRUNC
Regex	Engine-specific flags	REGEXP_* (POSIX)
Arrays	UNNEST	LATERAL FLATTEN

5.3. Data Type Mapping (Practical Defaults)

Source type	Snowflake	Notes
BOOLEAN	BOOLEAN	Same
INT2/4/8, INTEGER, BIGINT	NUMBER(38,0)	Keep integers precise
DECIMAL(p,s)	NUMBER(p,s)	Up to 38 digits
FLOAT/REAL/DOUBLE	FLOAT	Accept IEEE rounding
VARCHAR(n)	VARCHAR(n)	Consider generous n at land; tighten later
DATE	DATE	Same
TIMESTAMP	TIMESTAMP_NTZ/TZ	Prefer TZ for business events
JSON/STRUCT/ARRAY	VARIANT/OBJECT/ARRAY	Use envelopes (date/tenant) as scalars

5.4. Workload Migration: Get to Green, Then Improve

Migrate **ELT first**, **consumer views next**, **procedures last**. Publish compatible marts quickly; refactors (Snowpark, Dynamic Tables) can follow without delaying adoption.

ELT Pipelines

- Recreate staging → core transforms as **idempotent SQL** (MERGE/CTAS).
- Lift **envelope columns** at land (date/tenant/type) to unlock pruning.
- Replace procedural loops with set-based SQL; where imperative control is needed, use **Tasks/Streams** or a **SQL stored procedure**.

```
MERGE INTO core.dim_customer t
USING stage.customer_delta s
ON t.customer_id = s.customer_id
WHEN MATCHED THEN UPDATE SET name=s.name, segment=s.segment, updated_
at=CURRENT_TIMESTAMP()
WHEN NOT MATCHED THEN INSERT (customer_id,name,segment,created_at)
VALUES (s.customer_id,s.name,s.segment,CURRENT_TIMESTAMP());
```

Stored Procs and Macros

- **Teradata macros/BTEQ** → SQL + Tasks; conditional blocks → SQL stored procedure (or JS when needed).
- **Redshift PL/pgSQL procs** → JS stored procedures or orchestration outside Snowflake.
- **BigQuery scripting** → SQL procedures; JS UDFs port to Snowflake JS UDFs or Snowpark.

Schedulers

Map to **Streams & Tasks** (intraday) or keep external orchestrators (Airflow/Prefect/dbt). Use QUERY_TAG so lineage and cost attribution remain intact.

5.5. Validation (Prove Shape, Row Counts, and Sums)

Automate **shape** and **content** checks. Treat mismatches as bugs you fix before cutover.

Shape Parity

```
-- Columns/types expected in mart view
WITH expected(col, dtype) AS (
  SELECT 'ORDER_ID','NUMBER' UNION ALL
  SELECT 'ORDER_DT','DATE'   UNION ALL
  SELECT 'AMOUNT','NUMBER(18,2)'
)
SELECT e.col, e.dtype, c.data_type AS found
FROM expected e
LEFT JOIN prod_candidate.information_schema.columns c
  ON c.table_schema='MART' AND c.table_name='V_ORDERS' AND c.column_
  name=e.col
WHERE c.column_name IS NULL OR UPPER(c.data_type) <> e.dtype;
```

Content Parity (Day Slices)

```
-- Count and sum deltas by day (source exported vs. Snowflake landed)
SELECT d, src_cnt, tgt_cnt, src_amt, tgt_amt,
       (tgt_cnt-src_cnt) AS cnt_delta, (tgt_amt-src_amt) AS amt_delta
FROM (
  SELECT d, COUNT(*) src_cnt, SUM(amount) src_amt FROM src.orders
  GROUP BY d
) s
JOIN (
  SELECT d, COUNT(*) tgt_cnt, SUM(amount) tgt_amt FROM raw.fact_orders_
  like_source GROUP BY d
) t USING (d)
ORDER BY d DESC;
```

Row-Set Fingerprint Checks

```
-- Deterministic hash on canonical column list

-- Deterministic row-set fingerprint on a canonical column list
SELECT HASH_AGG(
         HASH(
           order_id,
           order_dt,
           amount
         )
       ) AS sig
FROM raw.fact_orders_like_source
WHERE d BETWEEN '2025-10-01' AND '2025-10-07';
```

Note Use a canonical column list for both source and target, with matching casts where needed. Avoid ARRAY_AGG-based signatures for large validation windows because aggregate arrays have size limits.

5.6. Cutover (Dual-Run → Freeze → Flip → Backfill)

- **Dual-run**: Keep source pipelines feeding users; run Snowflake ELT in parallel; compare marts and dashboards side by side.
- **Freeze**: Pick a checkpoint (e.g., "T-0 at 01:00 UTC"), **clone** Snowflake prod candidate, compute deltas from source after checkpoint.
- **Flip**: Repoint BI to Snowflake by changing connection/warehouse; keep the old system read-only for a short window.
- **Backfill**: Replay the gap window into Snowflake using idempotent merges.

Atomic Promotion

```
CREATE OR REPLACE DATABASE prod_clone CLONE prod_candidate;
-- sanity checks...
ALTER DATABASE prod_clone SWAP WITH prod;   -- instant cutover
```

Rollback is the same SWAP in reverse.

5.7. Post-migration Tuning (After Users Are On)

Tune only after you see real workloads.

- Add **narrow clustering** if good predicates still scan broadly (e.g., (order_dt, tenant_id)); measure first.
- Replace hot rolling queries with **Dynamic Tables** (small TARGET_LAG).
- Normalize BI parameters so result cache hits; ensure stable SQL text.
- Right-size warehouses: **SMALL** + **multi-cluster** for dashboards; **step-up just for heavy ELT**.

5.8. Security and Governance Parity

Map users/roles to a **clean RBAC** that matches responsibilities, not legacy grants.

Legacy notion	Snowflake mapping
"DBA for sales"	ROLE_DATA_ADMIN_SALES owning PROD_SALES.* schemas
"ETL service account"	ROLE_ELT_WRITER_<domain> + warehouse WH_ELT_<domain>
"BI consumers"	ROLE_BI_APP + ROLE_ANALYST with **view-only** on MART

Carry classifications as tags; re-attach masking and row policies once, so protections travel with objects.

```
CREATE TAG classification;
ALTER TABLE core.dim_customer MODIFY COLUMN email SET TAG
classification='pii.email';
-- governance job binds mask_pii to all columns tagged 'pii.*'
```

5.9. Source-Specific Notes

Redshift → Snowflake

- UNLOAD to Parquet with GZIP and manifest; include ADDQUOTES only for CSV.

- Drop DISTKEY/SORTKEY at DDL; do not emulate; consider clustering only **after** profiling.
- Replace Spectrum external tables with **Snowflake External Tables** or load to internal tables.

Export

```
UNLOAD ('SELECT * FROM public.fact_orders')
TO 's3://company-migration/source=db/table=fact_orders/part_'
IAM_ROLE 'arn:aws:iam::...:role/redshift-unload'
FORMAT AS PARQUET;
```

BigQuery → Snowflake

- EXPORT DATA to GCS; keep partition/cluster path parts so Snowflake prunes by date/tenant without refactor.
- STRUCT/ARRAY map cleanly to VARIANT; preserve scalar envelopes separately.

Export

```
EXPORT DATA WITH PARTITION OPTIONS(partition_expiration_days=30)
OPTIONS(
  uri='gs://company-migration/fact_orders/part-*.parquet',
  format='PARQUET',
  compression='SNAPPY') AS
SELECT * FROM `project.dataset.fact_orders`;
```

Synapse → Snowflake

- Use PolyBase/COPY INTO to ADLS with Parquet; avoid heaps of tiny files.
- T-SQL procedural logic translates to Snowflake SQL procedures or Tasks.

Export

```
COPY INTO 'https://account.dfs.core.windows.net/migration/fact_orders/'
FROM fact_orders
WITH (FILE_TYPE = 'PARQUET', MAXERRORS = 0);
```

Teradata → Snowflake

- Favor **TPT FastExport** to Parquet; ensure consistent DATEFORMAT and null tokens.
- BTEQ macros become **parameterized views** or SQL procedures; secondary indexes are dropped.

5.10. Documentation and Sign-Off Inside the Warehouse

Keep a small **migration ledger** so audit and memory live with the data.

```
CREATE SCHEMA IF NOT EXISTS mig;

CREATE TABLE IF NOT EXISTS mig.run_log (
  run_id STRING, source STRING, table_name STRING,
  started_at TIMESTAMP_NTZ, finished_at TIMESTAMP_NTZ,
  files NUMBER, rows_loaded NUMBER, checksum STRING, status STRING,
  notes STRING
);

CREATE TABLE IF NOT EXISTS mig.cutovers (
  system STRING, at TIMESTAMP_NTZ, actor STRING, scope STRING, rollback_
  plan STRING
);
```

A nightly page showing **rows by day**, **bytes by table**, and **top query deltas** during dual-run gives stakeholders confidence and catches drifts early.

5.11. What "Done" Looks Like

- Users point BI and SQL clients to Snowflake; the old system is read-only or decommissioned.
- Row counts, sums, and spot hashes match for the defined window; exceptions documented.
- ELT is idempotent, scheduled, and tagged; freshness is visible via a view.
- Security matches intent (least privilege), with masking/row policies active.
- Cost is predictable: BI on small multi-cluster, ELT sized to jobs, resource monitors on ad hoc lanes.

Move bytes first, prove sameness, flip with a clone, and only then take advantage of Snowflake-native features. When each step is just SQL you can test and promote, a "migration" becomes a sequence of small, reversible releases—not a leap.

6. Appendix F: Sample dbt Project Structure

A dbt project should read like a small application: clear layers, explicit contracts, and minimal surprises. The structure below is tuned for Snowflake—sources land in external or internal stages, staging models lift envelopes (date/tenant), intermediate models shape facts/dims, and marts publish secure views. Tests and macros live beside the code they guard. CI runs the same commands locally and in your pipeline.

6.1. Repository Layout (One Glance, One Meaning)

```
.
├── dbt_project.yml
├── packages.yml
├── models/
│   ├── staging/          # 1:1 with sources; light coercion & envelopes
│   │   ├── sources.yml
│   │   └── stg_orders.sql
```

```
│   ├─ intermediate/       # joins, dedup, SCD helpers
│   │  └─ int_orders_dedup.sql
│   ├─ marts/              # published interfaces (facts/dims/views)
│   │  ├─ sales/
│   │  │  ├─ dim_customer.sql
│   │  │  ├─ fact_orders.sql
│   │  │  └─ marts.yml
│   │  └─ ops/
│   │     └─ freshness_heartbeats.sql
│   └─ _models.yml         # defaults (materializations, tags)
├─ macros/
│   ├─ surrogate_key.sql
│   ├─ incremental_filters.sql
│   └─ contracts.sql
├─ seeds/
│   └─ currency_rates.csv
├─ snapshots/
│   └─ customer_scd.yml
├─ tests/
│   ├─ constraints.yml     # generic tests (unique, not_null,
                             accepted_values)
│   └─ contracts.yml       # column-level contracts for public views
└─ .github/workflows/
   └─ dbt.yml              # CI: deps → seed → seed tests → run → test
```

A small table to keep intent obvious:

Area	Why it exists
`staging`	Normalize source quirks; add `event_dt`, `tenant_id`, type; no business logic.
`intermediate`	Deterministic shaping (dedup, SCD joins, windows) shared by many marts.
`marts`	The contract. Facts/dims/views teams read; policies attach here.
`snapshots`	Slowly changing dimensions when source lacks change flags.

(*continued*)

Area	Why it exists
macros	Reusable keys, incremental filters, contract assertions.
seeds	Small reference tables under version control.
tests	Generic and custom data tests separate from model SQL.

6.2. dbt_project.yml Tuned for Snowflake

```
name: warehouse
version: 1.0.0
config-version: 2

profile: snowflake

model-paths: ["models"]
seed-paths:  ["seeds"]
macro-paths: ["macros"]
snapshot-paths: ["snapshots"]
test-paths: ["tests"]

models:
  +persist_docs: {relation: true, columns: true}
  +query_tag: "app=dbt;project=warehouse;env={{ target.name }}"
  +on_schema_change: append_new_columns
  warehouse:
    staging:
      +materialized: view
      +tags: ["stage"]
    intermediate:
      +materialized: ephemeral
      +tags: ["int"]
    marts:
      +materialized: table     # or view if you publish secure views via
      SQL wrappers
      +tags: ["mart"]
```

```
      sales:
        +schema: sales_mart
        +grants:
          select: ["ROLE_BI_APP","ROLE_ANALYST"]
```

Profiles (example `profiles.yml`) should map environments to **distinct roles/warehouses/schemas**:

```
snowflake:
  target: dev
  outputs:
    dev:
      type: snowflake
      account: "{{ env_var('SNOWFLAKE_ACCOUNT') }}"
      user:     "{{ env_var('SNOWFLAKE_USER') }}"
      password:"{{ env_var('SNOWFLAKE_PASSWORD') }}"
      role:     "ROLE_ELT_WRITER_SALES"
      warehouse: "WH_ELT_SALES"
      database: "DEV"
      schema:    "DEV_APP"
      threads: 8
      client_session_keep_alive: false
    prod:
      type: snowflake
      account: "{{ env_var('SNOWFLAKE_ACCOUNT') }}"
      authenticator: oauth
      token: "{{ env_var('SNOWFLAKE_OAUTH_TOKEN') }}"
      role:     "ROLE_ELT_WRITER_SALES"
      warehouse: "WH_ELT_SALES"
      database: "PROD"
      schema:    "PUBLIC"
      threads: 16
```

6.3. Sources and Staging That Prune

models/staging/sources.yml

```
version: 2
sources:
  - name: raw
    database: PROD
    schema: RAW
    tables:
      - name: orders_json
      - name: customers
```

models/staging/stg_orders.sql

```
-- Grain: one row per event
with src as (
  select
    to_timestamp_ntz(value:ts)          as event_ts,
    to_date(value:ts)                   as event_dt,      -- envelope
    value:tenant_id::string            as tenant_id,     -- envelope
    value:order_id::string             as order_id,
    try_to_number(value:amount)        as amount,
    value:type::string                 as type,
    value                              as payload

 from {{ source('raw','orders_json') }} as r,
   lateral flatten(input => r.payload)
)
select * from src
where event_dt >= dateadd(day,-30,current_date());
```

Staging lifts **scalar envelopes** so downstream queries prune by date/tenant before touching VARIANT.

6.4. Intermediate Shaping That's Deterministic

models/intermediate/int_orders_dedup.sql

```
-- Keep latest per (tenant_id, order_id)
select *
from (
  select
    tenant_id, order_id, event_dt, event_ts, amount, type, payload,
    row_number() over (partition by tenant_id, order_id order by event_ts
    desc) as rn
  from {{ ref('stg_orders') }}
)
where rn = 1
```

Use **ephemeral** materialization to inline this logic where used; switch to table if many marts need it hot.

6.5. Marts: Facts, Dims, and Contracts

models/marts/sales/fact_orders.sql

```
{{ config(
    materialized='incremental',
    unique_key=['tenant_id','order_id'],
    incremental_strategy='merge'
) }}

with base as (
  select tenant_id, order_id, event_dt as d, amount
  from {{ ref('int_orders_dedup') }}
  {% if is_incremental() %}
    where d >= dateadd(day,-7,current_date())    -- small, idempotent
    backfill
  {% endif %}
)
select * from base;
```

models/marts/sales/marts.yml

```
version: 2
models:
  - name: fact_orders
    description: "Canonical orders fact; one row per order."
    columns:
      - name: tenant_id
        tests: [not_null]
      - name: order_id
        tests: [not_null, unique]
      - name: d
        tests: [not_null]
```

Prefer attaching masking and row access policies outside dbt in admin-managed migrations. If you use dbt post-hooks, make the statement match the actual relation type (ALTER VIEW for views, ALTER TABLE for tables/incremental models).

```
models:
  marts:
    sales:
      +post-hook:
        - "{% if model.config.materialized in ['view'] %}ALTER VIEW {{ this
        }} ADD ROW ACCESS POLICY RAP_TENANT ON (tenant_id){% elif model.
        config.materialized in ['table', 'incremental'] %}ALTER TABLE {{
        this }} ADD ROW ACCESS POLICY RAP_TENANT ON (tenant_id){% endif %}"
```

6.6. Macros You'll Actually Use

macros/surrogate_key.sql

```
{% macro skey(cols) -%}
md5({{ cols | map('string') | join(" || '|' || ") }})
{%- endmacro %}
```

macros/incremental_filters.sql

```
{% macro recent_days(column, days=7) -%}
{% if is_incremental() -%}
  where {{ column }} >= dateadd(day,-{{ days }}, current_date())
{%- endif %}
{%- endmacro %}
```

Use them to keep models terse and consistent.

6.7. Snapshots for SCD When Needed

snapshots/customer_scd.yml

```
snapshots:
  - name: snap_customer_scd
    target_database: PROD
    target_schema: SNAPSHOTS
    strategy: timestamp
    unique_key: customer_id
    updated_at: updated_at
    source:
      name: raw
      table: customers
```

Run snapshots on a cadence that matches source volatility.

6.8. Seeds As Controlled Reference Data

seeds/currency_rates.csv lives under version control; define types in dbt_project.yml:

```
seeds:
  +column_types:
    rate: numeric
    iso:  string
```

6.9. Tests That Keep Surfaces Honest

```
tests/fact_orders_no_future_dates.sql
{{ config(severity='warn') }}

select 1
from {{ ref('fact_orders') }}
where d > current_date()
```

`tests/contracts.yml` can assert schemas for public views:

```
version: 2
models:
  - name: v_orders_public
    columns:
      - name: order_id
        tests: [not_null]
      - name: d
        tests: [not_null]
      - name: amount
        tests: [not_null]
```

6.10. Packages and Exposures

```
packages.yml

packages:
  - package: dbt-labs/dbt_utils
    version: "1.1.1"
```

Expose critical marts to document downstream dependencies:

```
version: 2
exposures:
  - name: sales_executive_dashboard
    type: dashboard
    maturity: high
```

```
    owner:
      name: Analytics
      email: analytics@example.com
    depends_on:
      - ref('fact_orders')
```

6.11. CI That Mirrors Local

A minimal pipeline: install deps → seed (and test) → run models → run tests. Example GitHub Action:

```
name: dbt
on: [push, pull_request]
jobs:
  run:
    runs-on: ubuntu-latest
    steps:
      - uses: actions/checkout@v4
      - uses: actions/setup-python@v5
        with: { python-version: '3.11' }
      - run: pip install dbt-snowflake
      - run: dbt deps
      - run: dbt seed --target dev --full-refresh
      - run: dbt run  --target dev --fail-fast
      - run: dbt test --target dev
```

Tag heavy models and exclude them on PRs if needed; full runs happen on `main`.

6.12. Snowflake Specifics That Pay Off

- Set `+query_tag` globally so cost and lineage are attributable.
- Keep **schemas per layer** (`STAGE`, `CORE`, `MART`) and **grant select only** on marts to BI roles.
- Use `on_schema_change: append_new_columns` to make additive evolution boring; rely on contract tests to catch breaks.

- For near-real-time aggregates, prefer dbt models that materialize to **Dynamic Tables** (via run-ops or admin scripts) when the cadence is sub-5 minutes.

Put shape and intent where people can see them—folders map to refinement stages, staging models lift envelopes, intermediate logic stays deterministic, marts are the contract, and tests/macros make it safe to move quickly. With that spine, your dbt project feels like code you can change on Friday and trust on Monday.

7. Appendix G: Common Error Codes and Resolutions

Snowflake's errors are usually **plain-English messages** with numeric codes. You don't need to memorize codes; learn the **message patterns** and the **one-minute checks** that fix them. This appendix groups the most frequent issues by what you were trying to do, offers quick triage, and gives copy-paste probes you can run immediately.

7.1. Connection and Authentication

Message snippet	What it really means	Try this first
`Incorrect username or password/Authentication failed`	Bad creds or wrong auth method (OAuth/ SSO vs. password)	Confirm the **authenticator** in your client profile; for SSO, test with `snowsql -a <acct> -u <user> --authenticator externalbrowser`
`Unable to connect to server`	Account locator/region mismatch, network egress blocked	Verify account URL and region; for private connectivity, confirm **Network Policy** or **PrivateLink** endpoint
`JWT token is invalid or expired`	OAuth token expired or audience mismatch	Refresh token; ensure the token's audience matches your Snowflake account URL

Useful Quick Checks

```
-- Where am I connecting?
SELECT CURRENT_ACCOUNT(), CURRENT_REGION(), CURRENT_ROLE(), CURRENT_
WAREHOUSE();
```

7.2. Privileges and RBAC

Message snippet	What it really means	Quick fix
`Insufficient privileges to operate on <object>`	Role lacks **USAGE** on db/schema and/or object privilege	Grant **USAGE** on database and schema *and* the object privilege (`SELECT`, `INSERT…`)
`Not authorized or does not exist`	Either missing privilege **or** wrong name	Try a fully qualified name: `DB.SCHEMA.OBJECT`; list objects from `INFORMATION_SCHEMA`

Grant Skeletons (Idempotent)

```
-- Reader
GRANT USAGE ON DATABASE DB TO ROLE R;
GRANT USAGE ON SCHEMA DB.SCH TO ROLE R;
GRANT SELECT ON ALL VIEWS IN SCHEMA DB.SCH TO ROLE R;
GRANT SELECT ON FUTURE VIEWS IN SCHEMA DB.SCH TO ROLE R;

-- Writer
GRANT INSERT, UPDATE, DELETE, SELECT ON ALL TABLES IN SCHEMA DB.SCH
TO ROLE R;
GRANT USAGE ON ALL SEQUENCES IN SCHEMA DB.SCH TO ROLE R;
```

See What You Can't See

```
-- If this returns rows, the name is right but your role can't read it
SELECT * FROM DB.INFORMATION_SCHEMA.TABLES WHERE TABLE_NAME = 'FACT_
ORDERS';
```

7.3. SQL Compilation and Object State

Message snippet	Likely cause	Resolution pattern
SQL compilation error: Object does not exist or not authorized	Typo, wrong schema, or reader role	Use DB.SCH.OBJ; SHOW OBJECTS LIKE '...'; switch role if needed
Object already exists	CREATE without guard	Use CREATE OR REPLACE (caution: replaces!) or IF NOT EXISTS
Cannot perform <DDL>: dependent objects exist	Views/streams depend on it	SHOW DEPENDENTS ON ...; drop or recreate dependents, or prefer **SWAP via CLONE**

Safe DDL Habits

```
CREATE OR REPLACE TABLE core.dim_product LIKE core.dim_product_template;
ALTER TABLE core.fact_orders ADD COLUMN IF NOT EXISTS order_channel STRING;
```

7.4. Warehouse, Concurrency, and Timeouts

Message snippet	What happened	Quick move
No active warehouse selected	Session lacks a warehouse	USE WAREHOUSE WH_BI; or set in connection profile
Query has reached its statement timeout	Long execution; statement cap hit	Raise warehouse size (single step), add pruning filters, or increase STATEMENT_TIMEOUT_IN_SECONDS temporarily
Insufficient resources to execute the query/long Queued time	Concurrency spike	Enable **multi-cluster** or raise MAX_CLUSTER_COUNT; stagger schedules

Profile at a Glance

```
SELECT AVG((QUEUED_OVERLOAD_TIME+QUEUED_PROVISIONING_TIME)/1000) queued_s,
       AVG(EXECUTION_TIME/1000) exec_s
FROM SNOWFLAKE.ACCOUNT_USAGE.QUERY_HISTORY
WHERE START_TIME >= DATEADD(minute,-10,CURRENT_TIMESTAMP())
  AND WAREHOUSE_NAME='WH_BI';
```

7.5. COPY/Ingestion

Message snippet	Likely cause	Fix fast
`File not found`/`No files found to copy`	Stage URL/prefix wrong; permissions	`LIST @stage/prefix`; check **STORAGE INTEGRATION** and path
`Numeric value '...' is not recognized`/type parse errors	Bad cast in file vs. column type	Add a staging table with relaxed types; coerce with `TRY_TO_*` in `SELECT`-style `COPY`
`JSON parsing error`	Non-uniform JSON/ arrays	`STRIP_OUTER_ARRAY=TRUE` if needed; load to `VARIANT` then normalize
`Permission denied for storage location`	Cloud role/service principal missing rights	Re-grant bucket/container access to the **integration's** principal/role

Diagnostic and Pattern

```
LIST @mystage/path/;

CREATE OR REPLACE FILE FORMAT ff_json TYPE=JSON STRIP_OUTER_ARRAY=TRUE;

COPY INTO stage.raw_events
FROM @mystage/path
FILE_FORMAT=(FORMAT_NAME=ff_json)
ON_ERROR='CONTINUE';  -- capture rejects for inspection
```

7.6. Semi-structured (VARIANT)

Message snippet	What it means	Resolution
`Invalid JSON`	Source row malformed	Load with `ON_ERROR='CONTINUE'` and review reject files; fix upstream
`Cannot cast VARIANT to <type>`	Mixed shapes	Use `TRY_TO_*` and defaulting; avoid deep dot-paths in GROUP BY

Safer Extraction

```
SELECT
  v:tenant::string              AS tenant_id,
  TO_DATE(v:event_ts)           AS event_dt,
  TRY_TO_NUMBER(v:amount)       AS amount
FROM stage.events;
```

7.7. Streams, Tasks, and Dynamic Tables

Message snippet	Meaning	What to do
`Task <name> has failed to run`	Privilege, warehouse, or SQL error	`SHOW TASK HISTORY LIKE '<name>';` check `ERROR_MESSAGE`, set `WAREHOUSE`, grant `EXECUTE TASK`
`Stream is stale and needs to be recreated`	Source table recreated; offset invalid	Recreate stream; consider using **MERGE** for backfills instead of consuming the stream
`Dynamic table refresh error/lag` exceeds target	Upstream failure or heavy dependency chain	Check `DYNAMIC_TABLE_REFRESH_HISTORY`; reduce windowed range or fix upstream object

Quick Peeks

```
SELECT *
FROM TABLE(<db_name>.INFORMATION_SCHEMA.TASK_HISTORY())
ORDER BY SCHEDULED_TIME DESC
LIMIT 20;
```

Note TASK_HISTORY is an Information Schema table function. If INFORMATION_SCHEMA is not the active schema, use a fully qualified database name.

```
SELECT NAME, LAST_REFRESH_TIME, ERROR_MESSAGE
FROM SNOWFLAKE.ACCOUNT_USAGE.DYNAMIC_TABLE_REFRESH_HISTORY
ORDER BY LAST_REFRESH_TIME DESC LIMIT 20;
```

7.8. Governance (Masking/Row Access)

If the role still cannot apply a policy after receiving the account-level APPLY privilege, verify it also has access to the specific policy object and the target table/view.

Message snippet	Likely cause	Fix
`Operation cannot be performed on a masked column`	DDL conflicts with masking policy	Detach policy, perform DDL, reattach; or apply DDL compatible with policy
`Insufficient privileges to apply masking/row access policy`	Role lacks account-level APPLY privilege	Grant `APPLY MASKING POLICY`/`APPLY ROW ACCESS POLICY` to the governance role
Data appears redacted unexpectedly	Masking policy active for your role	Use a role with proper entitlements; verify `IS_ROLE_IN_SESSION(...)` logic

Attach Once, Inherit Everywhere

```
GRANT APPLY MASKING POLICY, APPLY ROW ACCESS POLICY, APPLY TAG ON ACCOUNT
TO ROLE ROLE_GOVERNANCE;

ALTER TABLE core.dim_customer
  MODIFY COLUMN email SET MASKING POLICY mask_email;

ALTER VIEW mart.v_orders ADD ROW ACCESS POLICY rap_tenant ON (tenant_id);
```

7.9. Shares, Reader Accounts, and Replication

Message snippet	What it means	Fix
`Object referenced by a share cannot be modified`	You're altering an object that is part of an **outbound share**	Perform changes in a **clone** and **SWAP**, or pause the share during migration
`Replication failed`	Privileges, region mismatch, or large backlog	Check replication lag; confirm `IMPORTED PRIVILEGES`/role grants; reduce frequency or scope
Reader queries fail intermittently	Reader warehouse too small or suspended	Bump reader warehouse size; tune `AUTO_SUSPEND`; attach a **resource monitor**

7.10. Query Results and Data Quality

Message snippet	Clue	Move
`Division by zero`	Data edge case	Use safe divide UDF or `NULLIF(d,0)`
`Invalid value [..] for cast`	Dirty source	Wrap with `TRY_TO_*` and audit rejects
"Fast, then slow" dashboards	Cache miss, differing SQL text	Parameterize queries; remove `SELECT *`; align session settings

Handy Helpers

```
CREATE OR REPLACE FUNCTION util.safe_divide(n NUMBER, d NUMBER)
RETURNS FLOAT LANGUAGE SQL AS 'IFF(d=0, NULL, n/d)';

-- Find heavy queries by bytes (last 2h)
SELECT QUERY_ID, BYTES_SCANNED/1e9 AS gb, LEFT(QUERY_TEXT,200)
FROM SNOWFLAKE.ACCOUNT_USAGE.QUERY_HISTORY
WHERE START_TIME >= DATEADD(hour,-2,CURRENT_TIMESTAMP())
ORDER BY gb DESC LIMIT 20;
```

7.11. Snowpark, UDFs, and External Functions

Message snippet	Meaning	Fix
`ModuleNotFoundError/` package import errors	Missing/unsupported package	Add to `PACKAGES=('...')`; pin versions; for heavy libs, use **Anaconda** availability
`Python runtime error`	Handler threw an exception	Add input guards; test UDF locally; log with `SYSTEM$LOG_INFO`
`External function call failed`	Network or auth to your service	Verify integration, secret, and outbound allow-list; batch calls to reduce cost

UDF Skeleton

```
CREATE OR REPLACE FUNCTION util.parse_json(s STRING)
RETURNS VARIANT
LANGUAGE JAVASCRIPT
AS $$
  try { return JSON.parse(s); } catch (e) { return null; }
$$;
```

7.12. Storage, Time Travel, and Fail-Safe

Message snippet	Symptom	Action
Storage rose unexpectedly	Long retention on mutable tables	Set DATA_RETENTION_TIME_IN_DAYS sensibly (1–2 staging; 7–14 prod); use **TRANSIENT** for intermediates
Need old state restored	Data changed/ dropped recently	Use **Time Travel** (AT/BEFORE) or **CLONE AS OF**; beyond window → **Fail-safe** via support

Recover Patterns

> **Note** In Time Travel queries and clones, the TIMESTAMP value should be explicitly cast to a TIMESTAMP type.

```
-- Point-in-time query

SELECT * FROM core.fact_orders AT (TIMESTAMP => '2025-10-19
01:00:00'::TIMESTAMP_NTZ);

-- Instant environment for hotfix

CREATE DATABASE prod_fix CLONE prod AT (TIMESTAMP => '2025-10-19
01:00:00'::TIMESTAMP_NTZ);
```

7.13. One-Minute Triage Loop (Keep This Handy)

1. **Read the message literally.** Most tell you the missing privilege or object.
2. **Qualify fully.** Try DB.SCHEMA.OBJECT and SHOW OBJECTS LIKE '...';
3. **Check your role/warehouse.** SELECT CURRENT_ROLE(), CURRENT_WAREHOUSE();
4. **If it's a grant:** Add **USAGE** on db/schema + object privilege.

5. **If it's ingest:** `LIST @stage;` confirm file format, `ON_ERROR`, and path.
6. **If it's slow:** Open Query Profile; if **Queued** → lanes; if **Execution** → pruning and projection.
7. **If it's governance:** Verify **APPLY** privileges; review masking/row policies on the object.

Most Snowflake errors collapse to **naming, privilege, or pruning**. Use fully qualified names, grant from the **boundary in** (db → schema → object), keep envelope filters up front, and lean on clones/Time Travel when you need a safe place to fix things.

8. Appendix H: Literature (with Where to Find Each Source)

Below is a practical, annotated shelf for day-to-day use. For books, I include publisher, year, and ISBN; for websites, I point to the official home page or canonical docs.

8.1. Primary References (Snowflake)

- **Snowflake Documentation and SQL Reference**: The canonical source for features, limits, and system views.

  Find it: `https://docs.snowflake.com`

- **Snowflake Release Notes/What's New**: Track behavior changes and newly shipped capabilities.

  Find it: `https://docs.snowflake.com/release-notes`

- **Snowflake Best Practices and Architecture Guides**: Design patterns for storage/compute separation, pruning, sharing.

  Find it: `https://docs.snowflake.com/en/user-guide/best-practices`

- **Snowflake Community and Cookbook**: Short recipes and discussion from practitioners.

  Find it: `https://community.snowflake.com`

8.2. Architecture and Data Warehousing

- **Ralph Kimball and Margy Ross:** *The Data Warehouse Toolkit* **(3rd ed.)**

  Wiley, 2013. ISBN-13: 978-1118530801. Dimensional modeling patterns.

  Find it: `https://www.wiley.com/en-us/The+Data+Warehouse+Toolkit%2C+3rd+Edition-p-9781118530801`

- **W. H. Inmon:** *Building the Data Warehouse* **(4th ed.)**

  Wiley, 2005. ISBN-13: 978-0764599446. Enterprise info factory perspective.

  Find it: `https://www.wiley.com/en-us/Building+the+Data+Warehouse%2C+4th+Edition-p-9780764599446`

- **Martin Kleppmann:** *Designing Data-Intensive Applications*

  O'Reilly, 2017. ISBN-13: 978-1449373320. Immutability, streams vs. batch, correctness.

  Find it: `https://www.oreilly.com/library/view/designing-data-intensive-applications/9781491903063/`

- **Zhamak Dehghani:** *Data Mesh*

  O'Reilly, 2022. ISBN-13: 978-1492092391. Product thinking for data.

  Find it: `https://www.oreilly.com/library/view/data-mesh/9781492092384/`

8.3. ELT, Orchestration, and Analytics Engineering

- **dbt Labs: dbt Documentation and Discourse**

  Project structure, tests, contracts, macros.

  Find it: `https://docs.getdbt.com` and `https://discourse.getdbt.com`

- **Joe Reis and Matt Housley:** *Fundamentals of Data Engineering*

  O'Reilly, 2022. ISBN-13: 978-1098108304. End-to-end platform and pipeline thinking.

  Find it: `https://www.oreilly.com/library/view/fundamentals-of-data/9781098108298/`

- **Maxime Beauchemin: "The Rise of the Data Engineer" (Essay)**

  Context for roles and tooling.

  Find it: `https://maximebeauchemin.medium.com/the-rise-of-the-data-engineer-91be18f1e603`

8.4. Streaming, CDC, and Near-Real-Time

- **Tyler Akidau, Slava Chernyak, Reuven Lax:** *Streaming Systems*

  O'Reilly, 2018. ISBN-13: 978-1491983874. Watermarks, windowing, late data.

  Find it: `https://www.oreilly.com/library/view/streaming-systems/9781491983874/`

- **Debezium Documentation**: Open-source CDC patterns and connectors.

  Find it: `https://debezium.io/documentation/`

- **Confluent Kafka Connect and Ecosystem Docs**: Managed connectors and design guides.

  Find it: `https://docs.confluent.io`

8.5. Governance, Security, and Compliance

- **NIST SP 800-53 (Security and Privacy Controls)**

  Baseline control families.

  Find it: `https://csrc.nist.gov/publications/detail/sp/800-53/rev-5/final`

- **NIST SP 800-207 (Zero Trust Architecture)**

  Architectural posture for access enforcement.

  Find it: `https://csrc.nist.gov/publications/detail/sp/800-207/final`

- **Evans and Wiggins:** *Data Governance: The Definitive Guide*

  O'Reilly, 2023. ISBN-13: 978-1098138400. Practical governance programs.

  Find it: `https://www.oreilly.com/library/view/data-governance-the/9781098138417/`

8.6. Performance and FinOps

- **J. R. Storment and Mike Fuller:** *Cloud FinOps* **(2nd ed.)**

  O'Reilly, 2023. ISBN-13: 978-1492054627. Cost visibility and guardrails.

  Find it: `https://www.oreilly.com/library/view/cloud-finops-2nd/9781492054627/`

- **Gajendran:** *Cost Management for Cloud Data Platforms*

  Apress, 2022. ISBN-13: 978-1484289302. Tactics for warehouse sizing and scheduling.

  Find it: `https://link.springer.com/book/10.1007/978-1-4842-8931-9`

8.7. ML, Snowpark, and GenAI

- **Snowpark Developer Guides (Python/Scala/Java)**

  Packaging, dependencies, secure execution.

  Find it: `https://docs.snowflake.com/en/developer-guide/snowpark`

- **Ben Wilson, Luca Massaron, Micheleen Harris:** *Machine Learning Engineering in Action*

  Manning, 2022. ISBN-13: 978-1617298719. Reproducibility, feature pipelines, deployment.

  Find it: `https://www.manning.com/books/machine-learning-engineering-in-action`

- **Vector Search/RAG Primers (Vendor-Agnostic Overviews)**

  Background for embeddings and hybrid search.

  Find it: (intro) `https://www.pinecone.io/learn/` (general learning hub), or academic overview: `https://arxiv.org/abs/2004.06197`

8.8. Observability and SRE

- **Beyer, Jones, Petoff, Murphy:** *Site Reliability Engineering*

  O'Reilly, 2016. ISBN-13: 978-1491929124. SLOs, error budgets, incident response.

  Find it: `https://sre.google/sre-book/table-of-contents/`

- **Beyer et al.:** *The SRE Workbook*

  O'Reilly, 2018. ISBN-13: 978-1492029502. Hands-on practices and case studies.

  Find it: `https://sre.google/workbook/table-of-contents/`

- **Charity Majors: Essays on Observability**

  Practical guidance for metrics/traces/logs.

  Find it: `https://charity.wtf` and `https://www.honeycomb.io/blog/`

8.9. SQL Craft and Style

- **Joe Celko:** *SQL Programming Style*

  Morgan Kaufmann, 2005. ISBN-13: 978-0120887972. Conventions for clarity.

  Find it: https://www.elsevier.com/books/sql-programming-style/celko/978-0-12-088797-2

- **Refactoring SQL (Assorted Authors and Blogs)**

  Patterns to turn procedural code into set-based SQL.

  Find it: https://docs.getdbt.com/best-practices and curated posts at https://blog.getdbt.com

8.10. Migration Aides (Engine-Specific)

- **Amazon Redshift UNLOAD and Migration Guides**

  Find it: https://docs.aws.amazon.com/redshift/latest/dg/r_UNLOAD.html and https://docs.aws.amazon.com/schema-conversion-tool/latest/userguide/Welcome.html

- **BigQuery EXPORT and Storage Docs**

  Find it: https://cloud.google.com/bigquery/docs/exporting-data and https://cloud.google.com/bigquery/docs

- **Azure Synapse PolyBase/COPY**

  Find it: https://learn.microsoft.com/azure/synapse-analytics/sql-data-warehouse/sql-data-warehouse-load-from-azure-blob-storage-with-polybase and https://learn.microsoft.com/azure/synapse-analytics/sql/develop-tables-cetas

- **Teradata TPT/FastExport**

  Find it: https://docs.teradata.com and search for "Teradata Parallel Transporter" and "FastExport".

8.11. Keeping Current (Suggested Habit)

- **Snowflake What's New (RSS/Email)**: Skim weekly and annotate your backlog with any item that affects cost, security, or SLOs.

  Find it: `https://docs.snowflake.com/release-notes`

- **Engineering Blogs from dbt Labs, Confluent, Google SRE, and vendor-agnostic FinOps groups**: For patterns that translate well to Snowflake.

  Start here: `https://blog.getdbt.com`, `https://www.confluent.io/blog/`, `https://sre.google`, `https://www.finops.org`

Index

M. Hander, *Snowflake Data Warehouse Engineering*, https://doi.org/10.1007/979-8-8688-2628-3

D

G

H

I

N

O

P

Q

R

S

T

Z

GPSR Compliance
The European Union's (EU) General Product Safety Regulation (GPSR) is a set of rules that requires consumer products to be safe and our obligations to ensure this.

If you have any concerns about our products, you can contact us on

ProductSafety@springernature.com

In case Publisher is established outside the EU, the EU authorized representative is:

Springer Nature Customer Service Center GmbH
Europaplatz 3
69115 Heidelberg, Germany

www.ingramcontent.com/pod-product-compliance
Lightning Source LLC
Chambersburg PA
CBHW080916180826
50016CB00035B/595
* 9 7 9 8 8 6 8 8 2 6 2 7 6 *